Membranes, Molecules, Toxins, and Cells

Edited by

Konrad Bloch, PhD
Liana Bolis, PhD
Daniel C. Tosteson, MD

John Wright • PSG Inc.
Boston • Bristol • London

Library of Congress Cataloging in Publication Data

Main entry under title:

Membranes, molecules, toxins, and cells.

Bibliography: p.
Includes index.
1. Membranes (Biology) 2. Cell interaction.
3. Biological transport. I. Bloch, Konrad Emil,
1912– II. Bolis, Liana. III. Tosteson, D. C.,
1925– [DNLM: 1. Cell membrane. QH601 M53294]
QH601.M483 574.87'5 80-16595
ISBN 0-88416-309-1

Medicine is an ever-changing science. As new research and clinical experience broaden our knowledge, changes in treatment and drug therapy are required. The editors and the publisher of this work have made every effort to ensure that the treatment and drug dosage schedules herein are accurate and in accord with the standards accepted at the time of publication. Readers are advised, however, to check the product information sheet included in the package of each drug they plan to administer to be certain that changes have not been made in the recommended dose or in the indications and contraindications for administration. This recommendation is of particular importance in regard to new or infrequently used drugs.

Printed in the United States of America.

International Standard Book Number: 0-88416-309-1

Library of Congress Catalog Card Number: 80-16595

EDITORS

Konrad Bloch, PhD
Department of Chemistry
Harvard University
Boston, Mass

Liana Bolis, PhD
Istituto di Fisiologia Generale
Universitá di Messina
Messina
Italy

Daniel C. Tosteson, MD
Dean
Harvard Medical School
Boston, Mass

Con tanto affetto
[illegible]
Boston 20.11.81
Liana

CONTRIBUTORS

Bernard M. Babior, MD
Hematology Service
New England Medical Center Hospital, and Department of Medicine
Tufts University School of Medicine
Boston, Massachusetts

John A. Badwey, PhD
Departments of Biological Chemistry and Medicine
Harvard Medical School
Boston, Massachusetts

Etienne-Emile Baulieu, MD
Laboratoire Hormones
Bicêtre
France

Patrice Boquet, MD, PhD
Institut Pasteur
Paris
France

Mark S. Bretscher, PhD
MRC Laboratory of Molecular Biology
University Medical School
Cambridge
England

Michael S. Brown, MD
Department of Molecular Genetics and Internal Medicine
The University of Texas Health Science Center
Dallas, Texas

Max M. Burger, MD, PhD
Department of Biochemistry
Biocenter, University of Basel
Basel
Switzerland

Werner Burkart, PhD
Department of Biochemistry
Biocenter, University of Basel
Basel
Switzerland

Paulo Canciglia, PhD
Istituto di Fisiologia Generale
Facoltà di Scienze
Messina
Italy

Giorgio Carmignoto, PhD
FIDIA Research Laboratories
Abano Terme, Padova
Italy

Richard A. Cooper, MD
Hematology-Oncology Section
Department of Medicine
University of Pennsylvania School of Medicine
Philadelphia, Pennsylvania

Milton J. Cormier, PhD
Department of Biochemistry
University of Georgia
Athens, Georgia

Angelo D'Acunto, PhD
International Institute of Genetics and Biophysics
Naples
Italy

Rui C. De Sousa, MD
Departments of Physiology and Medicine
School of Medicine
Geneva
Switzerland

Mario Finesso, MS
FIDIA Research Laboratories
Abano Terme, Padova
Italy

Richard A. Finkelstein, PhD
Department of Microbiology
School of Medicine
University of Missouri-Columbia
Columbia, Missouri

Irwin Fridovich, PhD
Department of Biochemistry
Duke University Medical Center
Durham, North Carolina

Francesco S. Gaeta, PhD
International Institute of Genetics and Biophysics
Naples
Italy

François Godeau, MD
Laboratoire Hormones
Bicêtre
France

Ira M. Goldstein, MD
Division of Rheumatology
Department of Medicine
New York University Medical Center
New York, New York

Joseph L. Goldstein, MD
Department of Molecular Genetics and Internal Medicine
The University of Texas Health Science Center
Dallas, Texas

Alfredo Gorio, PhD
FIDIA Research Laboratories
Abano Terme, Padova
Italy

Alda Grosso, DSc
Departments of Physiology and Medicine
School of Medicine
Geneva
Switzerland

H. Moustafa Hassan, PhD
Department of Biochemistry
Duke University Medical Center
Durham, North Carolina

Ari Helenius, PhD
Europäisches Laboratorium für Molekularbiologie
Heidelberg
Germany

Sylvia T. Hoffstein, PhD
Division of Rheumatology
Department of Medicine
New York University Medical Center
New York, New York

Emma R. Jakoi, PhD
Department of Anatomy
Duke University
Durham, North Carolina

Harold P. Jones, PhD
Department of Biochemistry
University of Georgia
Athens, Georgia

Manfred L. Karnovsky, PhD
Departments of Biological Chemistry and Medicine
Harvard Medical School
Boston, Massachusetts

Eugene P. Kennedy, PhD
Department of Biological Chemistry
Harvard Medical School
Boston, Massachusetts

Helen M. Korchak, PhD
Division of Rheumatology
Department of Medicine
New York University Medical Center
New York, New York

G. Richard Lee, MD
Howard Hughes Medical Institute, and Hematology/Oncology Division
Department of Medicine
University of Utah Medical Center
Salt Lake City, Utah

Robert E. Lynch, MD
Howard Hughes Medical Institute, and Hematology/Oncology Division
Department of Medicine
University of Utah Medical Center
Salt Lake City, Utah

Richard B. Marchase, PhD
Department of Anatomy
Duke University Medical Center
Durham, North Carolina

Joe M. McCord, PhD
Department of Biochemistry
University of South Alabama
Mobile, Alabama

Damiano G. Mita, PhD
International Institute of Genetics
and Biophysics
Naples
Italy

Joel Moss, MD, PhD
Laboratory of Cellular Metabolism
National Heart, Lung, and Blood
Institute
National Institutes of Health
Bethesda, Maryland

Robert Nielsen, PhD
Institute of Biological Chemistry A
August Krogh Institute
University of Copenhagen
Copenhagen
Denmark

Barbara M.F. Pearse, PhD
MRC Laboratory of Molecular Biology
University Medical School
Cambridge
England

H. Daniel Perez, MD
Division of Rheumatology
Department of Medicine
New York University Medical Center
New York, New York

William F. Petrone, PhD
Department of Biochemistry
University of South Alabama
Mobile, Alabama

J. David Robertson, MD, PhD
Department of Anatomy
Duke University Medical Center
Durham, North Carolina

Michel Schorderet, PhD
Département de Pharmacologie
Reproduction
Ecole de Médecine
Geneva
Switzerland

Sabine Schòrderet-Slatkine, PhD
Département d'Obstétrique
et de Gynécologie
Hôpital Cantonal
Geneva
Switzerland

Alan Sher, PhD
Departments of Pathology
and Medicine
Harvard Medical School
Boston, Massachusetts

Donald M. Small, MD
Biophysics Institute
Departments of Medicine
and Biochemistry
Boston University School of Medicine
Boston, Massachusetts

James E. Smolen, PhD
Division of Rheumatology
Department of Medicine
New York University Medical Center
New York, New York

Alfred I. Tauber, MD
Departments of Biological Chemistry
and Medicine
Harvard Medical School
Boston, Massachusetts

L. Tredese, MS
FIDIA Research Laboratories
Abano Terme, Padova
Italy

Hans H. Ussing, PhD
Institute of Biological Chemistry A
August Krogh Institute
University of Copenhagen
Copenhagen
Denmark

Martha Vaughan, MD
Laboratory of Cellular Metabolism
National Heart, Lung, and Blood
Institute
National Institutes of Health
Bethesda, Maryland

Juan A. Vergara, MD
Department of Anatomy
Duke University Medical Center
Durham, North Carolina

Anne E. Warner, PhD
Department of Anatomy and Embryology
University College London
London
England

Gerald Weissmann, MD
Division of Rheumatology
Department of Medicine
New York University Medical Center
New York, New York

CONTENTS

Section III Membranes and Superoxide

Section IV Membranes in Cell-Cell Interactions

Section V Membrane Transport

PREFACE

This volume brings together a series of papers describing recent investigations of the interactions between biologic membranes and various chemicals. It is mainly through such interactions that living cells detect and respond to their environment. One theme of the papers is the attempt to identify the particular ligands on membrane molecules that react with specific substances to which cells are exposed. Thus, the book is an early step toward understanding the molecular and membrane basis of environmental biology and medicine.

The first group of papers concentrates on the lipids and lipoproteins present in biologic membranes and on the transport of lipoproteins into cells. A second group of articles deals with the interactions between membranes and toxins, viruses, and parasites. The third part of the volume comprises a series of chapters on membranes and superoxide. This is followed by four papers on membrane-mediated, cell-cell interactions. The book concludes with a section on transport across biologic membranes.

This volume is a progress report and an assessment of the state of knowledge about important aspects of membrane biology. The investigations described in the several chapters give hope for improved understanding of the subtle relations between living cells and their environment.

Daniel C. Tosteson, MD
September 1980

ACKNOWLEDGMENTS

We wish to acknowledge and thank the following institutions for their help and support in making the International Conference on Biological Membranes possible through their financial assistance: Burroughs Wellcome Co; E.I. DuPont De Nemours & Co; Fidia Laboratories; Hoffman-La Roche, Inc; Lirca Laboratories; Merck Sharp & Dohme; Packard Instrument Company; Schering-Plough Corp; Smith Kline & French Laboratories; and The Upjohn Company.

Also for their hospitality, we thank the Hotel du Golf in Crans-sur-Sierre and the Commune of Chermignon in Valais, Switzerland.

SECTION I
Membranes, Lipids, and Lipoproteins

1 Biosynthesis of Membrane Lipids
Topologic and Enzymic Studies

Eugene P. Kennedy, PhD

The asymmetric distribution of phospholipids appears to be a general feature of biologic membranes. Asymmetry of this type has been studied in greatest detail in the erythrocyte membrane in which the choline-containing phospholipids are principally localized in the outer half of the bilayer, and phosphatidylethanolamine and phosphatidylserine in the inner layer.[1]

Our own work in this field has been directed to the problem of the biogenesis of lipid asymmetry in membranes, and to the mechanism of the rapid transbilayer movement of phospholipids that occurs in certain types of biologic membranes, but not in lipid model membranes. We have found that *Bacillus megaterium* is in an attractive organism for the study

Supported by the National Institutes of Health, Division of General Medical Sciences, Grants 19,822 and 22,057.

of such problems. The gram-positive *B megaterium* with its single plasma membrane has a much simpler cell envelope than *Escherichia coli* and other gram-negative organisms. Because of its rapid rate of growth *B megaterium* is much more attractive for the study of problems of biosynthesis than the mammalian erythrocyte. Indeed, the mammalian erythrocyte has been described as essentially a fossil by Bretscher,[2] who was careful to add, however, that much can be learned from the study of fossils.

Asymmetric Distribution of Phospholipids in Membranes of *B megaterium*

Rothman and Kennedy[3] investigated the transbilayer distribution of phosphatidylethanolamine (PE) in *B megaterium* with the aid of the impermeant reagent 2,4,6-trinitrobenzenesulfonic acid (TNBS) and isethionyl acetimidate. It was found that the reaction with TNBS at 15 C is distinctly biphasic (Figure 1-1). About 34% of the total phosphatidylethanolamine of the membrane reacts rapidly with TNBS with a half-life of 2 minutes. It was concluded that this represents the PE in the external layer of the membrane. The slow-reacting fraction of PE, with a half-life of 41 minutes, is assigned to the inner half of the bilayer.

It is recognized that this assignment of the pools of phosphatidylethanolamine to the external and internal halves of the bilayer, respectively, is an operational one based on the differential reactivity with TNBS. There is strong evidence that the fast-reacting pool is indeed external, since under the conditions of the experiment, the labeling of cytoplasmic components is minimal. The conclusion that the slow-reacting pool of PE is in the inner half of the bilayer is less direct. Bishop et al[4] have suggested that the lack of reactivity of PE in experiments of this kind may be the result of the accumulation of net negative charge, and the accumulation of the bulky trinitrophenyl residues that may limit the reactivity of the remainder of the PE. However, essentially similar results were obtained by Rothman and Kennedy[3] using isethionyl acetimidate. The amidinated derivative of PE produced by this reagent is very different from the trinitrophenyl derivative.

The most direct evidence, that the distinction between the internal and external pools is described on the basis of their reactivity with TNBS, comes from the pulse-label experiments of Rothman and Kennedy[5] and Langley and Kennedy,[6] to be described briefly below. These experiments have revealed that the initial specific activity of the internal PE is many times higher than that of the external pool. This result offers strong evidence that these are in fact two biologically distinct pools of phosphatidylethanolamine molecules in the membrane that are distinguished on the basis of differential reactivity with TNBS.

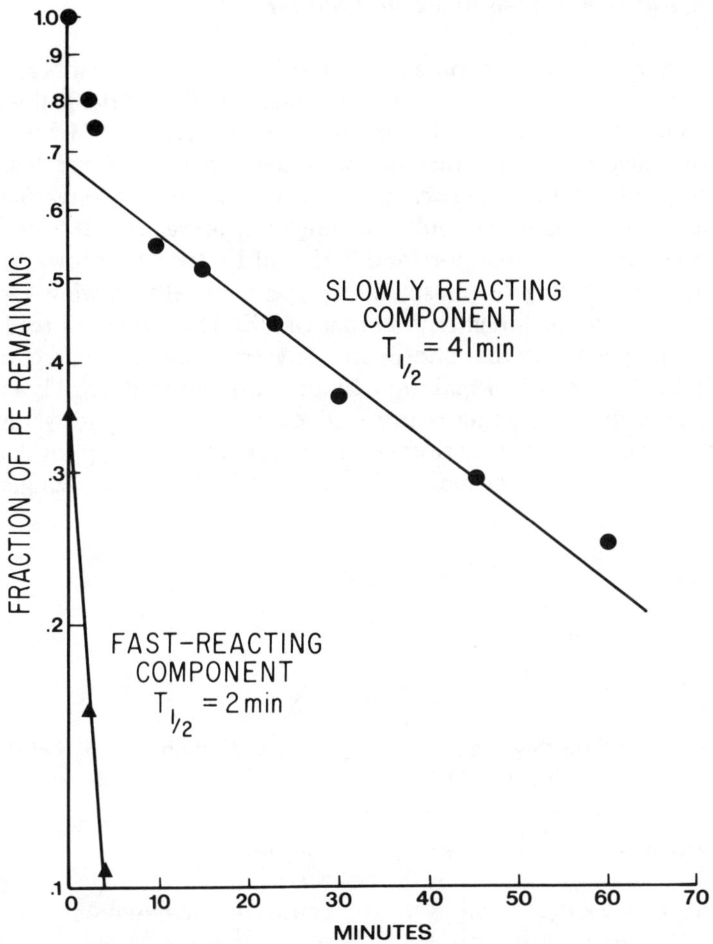

Figure 1-1 Time course for the trinitrophenylation of PE at 15 C. The fraction of PE converted to trinitrophenyl-phosphatidylethanolamine (TNP-PE) was determined from the relative amounts of ^{32}P in the corresponding spots of a thin-layer chromatogram. The fraction of the initial PE remaining is plotted on a logarithmic scale versus time of incubation. Two components are observed. The slope and intercept of the slow-reacting component were obtained by a least-squares fit to all time points after and including 10 minutes. The best fit for the intercept was 0.64, and the 95% confidence interval for the intercept was 0.59 to 0.69. The fast-reacting component (▲-▲, $t_{1/2}$ = 2 minutes) was obtained by subtracting the slow-reacting component (• - •, $t_{1/2}$ = 41 minutes) from the total, assuming the total was the result of only two exponential terms. The linearity of the calculated fast-reacting component confirms this assumption. Note that the fast-reacting component has been through five half-lives by the 10-minute time point, validating the choice of data points for fitting the slow-reacting component. Reproduced with permission from Rothman JE, Kennedy EP: Asymmetrical distribution of phospholipids in the membrane of *Bacillus megaterium*. *J Mol Biol* 110:603–618, 1977. Copyright by Academic Press Inc. (London) Ltd.

Pathways of Biosynthesis of PE in *B megaterium*

Interpretation of experiments on the biogenesis of membrane lipid asymmetry in *B megaterium* requires an understanding of the pathways of phospholipid biosynthesis in this organism. Recent studies of Langley and Kennedy[7] have established that the pathways shown in Figure 1-2, originally worked out for *Escherichia coli,* also occur in *B megaterium.* Investigations of the pattern of pulse-labeling of membrane lipids from ^{32}P in *B megaterium* in our laboratory and in that of Fulco[8] have shown that the initial rate of labeling of phosphatidylglycerol in cells growing logarithmically was very much higher than that of PE. This finding forced us to consider the possibility that phosphatidylglycerol may be the precursor of phosphatidylserine in *B megaterium.* Since phosphatidylglycerol is a highly reactive phospholipid, acting as an effective donor of phosphatidyl units, as well as of glycero-phosphate residues, it appeared likely that phosphatidylglycerol might similarly be a precursor of phosphatidylserine. Langley and

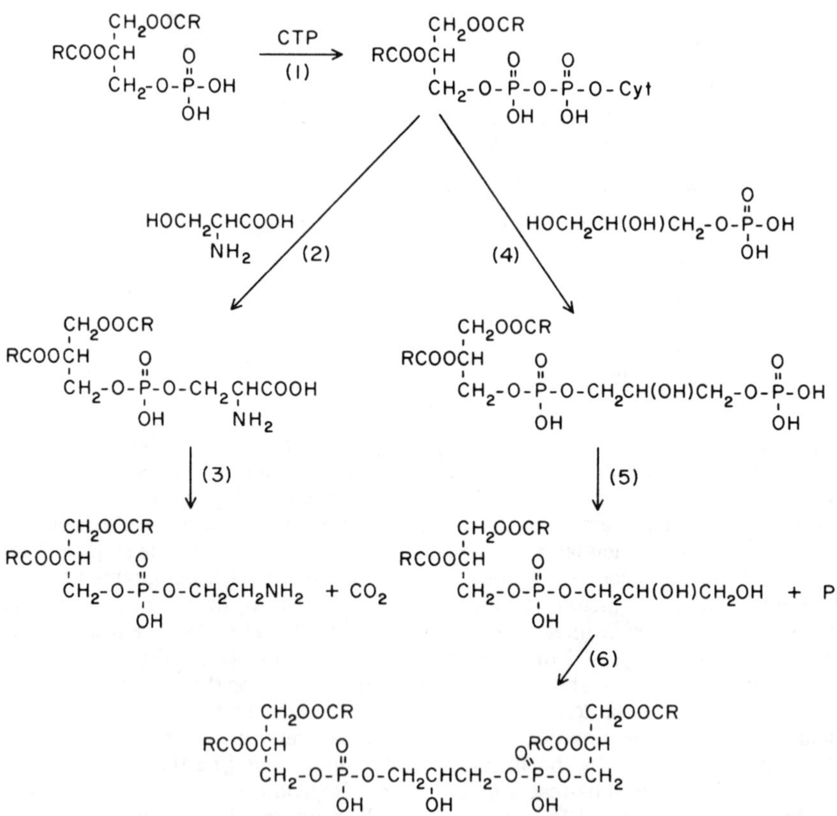

Figure 1-2 Pathways for the biosynthesis of membrane phospholipids in *B megaterium* and *E coli.*

Kennedy,[7] however, found no evidence that phosphatidylglycerol could be the precursor of phosphatidylserine. Instead, the pattern of labeling was found to be the result of an accumulation of phosphatidylserine in certain strains of *B megaterium.* The presence of a large pool (5% to 10% of the total phospholipid) acts by isotope dilution to slow the initial incorporation of ^{32}P into PE. In *E coli,* and in other strains of *B megaterium,* in contrast, levels of phosphatidylserine are low.

Although the activities of CDP-diglyceride:L-serine phosphatidyltransferase in *B megaterium* appeared to be quite low, definite evidence for the occurrence of this enzyme in cell-free extracts of the organism was obtained. This result, together with the data obtained by detailed studies of the pattern of incorporation of isotope in vivo supports the conclusion that the biosynthesis of membrane lipids in *B megaterium* proceeds as indicated in Figure 1-2.

Compositional Asymmetry and Biosynthetic Asymmetry

In considering the rationale of the pulse-label experiments to be described, it is useful to distinguish clearly between compositional asymmetry and biosynthetic asymmetry. Compositional asymmetry refers to the unequal amounts of PE found in the inner and outer halves of the bilayer. Biosynthetic asymmetry refers to the preferential incorporation of newly synthesized PE molecules into the inner half of the bilayer, in membranes of *B megaterium.* In pulse-label experiments this causes the specific activity of the PE in the inner half to be much higher initially than the specific activity of the outer half. However, because of the rapid transmembrane movement of PE, during a period of chase this biosynthetically generated isotopic asymmetry rapidly decays. In contrast, in cells growing in the steady state, the compositional asymmetry remains unaltered.

Rapid Transbilayer Movement of Phosphatidylethanolamine

Rothman and Kennedy[3] showed that the decay of the biosynthetically generated isotopic asymmetry described above could be used to measure the transmembrane movement of PE from the inner face of the membrane to the outer layer.

In experiments such as that of Figure 1-3, cells of *B megaterium* were uniformly labeled by growth on $^{33}P_i$. The cells were then pulsed with $^{32}P_i$. The ratio $^{32}P/^{33}P$ thus represents the specific activity of newly synthesized phospholipids.

At various time intervals during a period of exposure to $^{32}P_i$, samples of the cells were harvested and treated with TNBS. The phospholipids were then extracted and the TNP-PE was separated from unmodified PE by thin-layer chromatography. Conditions for the TNBS treatment were chosen so that the specific activity of the trinitrophenyl derivative could be

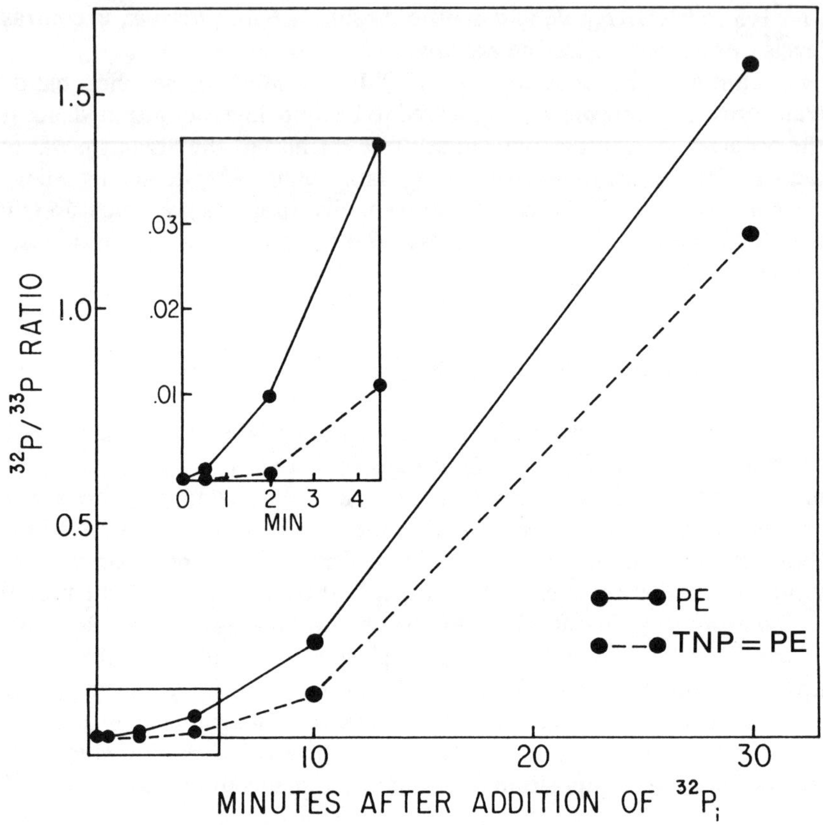

Figure 1-3 Preferential incorporation of newly synthesized PE into inner half of the bilayer in membranes of *B megaterium*. Cells of *B megaterium*, uniformly labeled with $^{33}P_i$, were pulse-labeled at 37 C by adding $^{32}P_i$ (0.17 mCi/ml) to the growth medium with $^{33}P_i$ still present. At the indicated times, samples (0.9 ml) were removed and filtered and washed on Millipore filters (0.45 μ pore) with 0.1 M KCl containing 50 mM potassium phosphate, pH 7.9. The filter was then transferred to a counting vial containing 0.9 ml of 90 mM KCl, 64 mM $NaHCO_3$, 12 mM TNBS, 45 mM potassium phosphate; pH 8.1, at 15 C. These manipulations took about 30 seconds. The vial was vortex-mixed, to remove cells from the filter, and incubated for 4 minutes at 15 C. The cells were precipitated with 5% (wt/vol) trichloroacetic acid after the addition of bovine serum albumin (0.25 mg/ml). The lipids were extracted and separated by thin-layer chromatography. The PE and TNP-PE spots were located by autoradiography and their radioactivity was counted. About 25% of the total PE was converted to the trinitrophenyl derivative by this procedure, representing primarily external PE. Inset shows the early time points in more detail. Reproduced from Rothman and Kennedy.[5]

used to calculate that of the external half of the bilayer, while the specific activity of the unmodified PE represented that of the internal pool.

The initial ratio of $^{32}P/^{33}P$ for the internal PE, recovered as the unmodified phospholipid, was much higher than that of the external phosphatidylethanolamine (see inset of Figure 1-3). However, as growth in the presence of $^{32}P_i$ continued, the specificity of the external PE began to approach that of the internal pool, revealing a rapid transmembrane movement of the newly synthesized molecules. In experiments such as that of Figure 1-3, complete isotopic equilibration is not achieved because of the continuing synthesis of new ^{32}P-labeled molecules, which are preferentially inserted into the inner half of the bilayer.

Possible Mechanisms of the Rapid Transbilayer Movement of Phospholipids

The transbilayer movement of phospholipids ("flip-flop") in lipid model membranes is extremely slow. Thus, Roseman et al[9] showed that the transbilayer movement of PE in vesicles has a half-life on the order of 80 days. In contrast, the transbilayer movement of PE in *B megaterium* is several orders of magnitude faster.

Since there is strong evidence that the lipid bilayer represents a fundamental mode of organization of biologic membranes, it is important to consider the types of mechanisms that may be involved in the very rapid transbilayer movement of phospholipids observed in *B megaterium,* and in certain other biologic membranes.[1] It appears that the mere presence of large amounts of proteins inserted into the lipid bilayer in biologic membranes cannot account for the observed differences. In erythrocyte membranes, which contain about 52% protein by weight, the translocation of phosphatidylcholine across the erythrocyte membrane has a half-life of a few hours.[1] Although this is much more rapid than the process observed in lipid model membranes, it is much slower than that observed in *B megaterium.* It is clear that the simple insertion of proteins into the membrane cannot account for the high rates observed in the latter organism.

We must consider the possibilities that the rapid transmembrane movement of phospholipid in *B megaterium* is driven by metabolic energy. The linkage to metabolic energy could take the form of a system equivalent to active transport that might translocate the phospholipid molecules from the inner to the outer layers. Alternatively, the outward movement of PE could be dependent on the continuing synthesis of new lipid and protein molecules followed by their insertion into the inner half of the bilayer. These possibilities have been tested experimentally in recent work in our laboratory.[6]

The Rapid Transbilayer Movement of PE Is Independent of Metabolic Energy

Cells of *B megaterium* uniformly labeled with ^{33}P were pulsed with ^{32}P, essentially as described above. After two minutes, the cells were treated with a combination of inhibitors and uncouplers consisting of cyanide, fluoride, and carbonyl cyanide trifluoromethylphenylhydrazone. The efficiency of the metabolic poisons was determined in a separate experiment in which it was shown that the synthesis of ATP was blocked and cellular levels of ATP fell sharply, as revealed by the luciferase ATP assay.

The decay of the biosynthetically induced isotopic asymmetry, resulting from the rapid equilibration of the inner and outer pools of PE, was in fact more rapid in the energy-poisoned cells than in the controls. This result is explained by the fact that in the energy-poisoned cells, the synthesis of new lipid is completely blocked. In the control cells, however, new lipid continues to be synthesized and preferentially inserted into the inner half of the bilayer, thus delaying the process of equilibration.

It was shown that the synthesis of lipid and of protein in energy-poisoned cells was completely blocked under the experimental conditions chosen. We conclude that the mechanism of the rapid transmembrane movement of PE is completely independent of the synthesis of new lipid, of protein, and more generally independent of sources of metabolic energy.

Kinetics of the Rapid Transmembrane Movement of PE

Langley and Kennedy[7] have carried out a more detailed analysis of kinetics of the rapid transmembrane movement of PE, as revealed by the decay of biosynthetic asymmetry. The analysis was greatly aided by the discovery, described above, that the process is not affected by energy poisons. After two minutes' exposure of the cells to $^{32}P_i$, the metabolic inhibitors were added, and the equilibration was measured under conditions in which the synthesis of new lipid was completely blocked.

In a treatment essentially similar to that of de Kruijff and van Zoelen,[10] the movement of PE from the inner to the outer half of the bilayer can be described by a rate constant k, defined as that fraction of the internal pool that is exchanged with the external pool per minute. It is assumed that the process consists of a one-for-one exchange under conditions in which the composition of the two pools does not change. If it is further assumed that the size of the inner pool is four times that of the outer pool, as is the case with energy-poisoned cells (see below), then it can be shown that the logarithm of the difference in specific activities *(SA)* of the two pools is a linear function of time *(t)* according to the equation (with C an integration constant):

$$\ln (SA_{in} - SA_{ex}) = -5kt + C$$

From the slopes of plots of ln $(SA_{in} - SA_{ex})$ versus time, in several such experiments, Langley and Kennedy[7] found that k has a value of about 0.1 per minute. That is, about 10% of the inner pool exchanges with the outer pool per minute. Since the outer pool is only one fourth of the inner pool under the particular conditions of the experiments, it follows that 40% of the outer pool must be exchanged per minute if the process consists of one-for-one exchange, as indeed appears very likely. The ratio SA_{in}/SA_{ex} approached 1.0 asymptotically, strongly indicating that the movement of PE represents an equilibration of the entire inner pool with the PE of the outer layer.

Compositional Asymmetry of PE Is Independent of Metabolic Energy

The asymmetric distribution of phospholipids across a membrane may be viewed as a kind of concentration gradient across the membrane. The asymmetry of a given phospholipid species may represent the most stable arrangement of the membrane; that is, an energy minimum. Alternatively, the possibility must be considered that the compositional asymmetry is maintained by a continuous expenditure of metabolic energy, as is required, for example, to maintain a gradient of Na^+ and K^+ across the cytoplasmic membrane of mammalian cells.

The discovery that the rapid transbilayer movement of PE persists in energy-poisoned cells made it possible to determine whether the observed asymmetry of PE is maintained by the continuous expenditure of metabolic energy. Cells of *B megaterium* were treated with energy poisons for a period of ten minutes at 30 C, a period of time which should allow the full equilibration of internal and external pools. If the maintenance of the compositional asymetry is dependent on metabolic energy, the amounts of PE in the inner and outer halves of the bilayer should be equalized at the end of this period because of the rapid transbilayer equilibration. It was found, however, that treatment with energy poisons under these conditions did not destroy the asymmetry of distribution of PE. Indeed, the distribution of this species of phospholipid was significantly more asymmetric in the energy-poisoned cells. In such cells about 20% of the total PE was found in the external pool, and 80% in the internal part of the bilayer. The ratio of internal to external PE was thus 4.0. In the control cells, in confirmation of the previous findings of Rothman and Kennedy,[3] the ratio of internal to external PE was about 2.0.

Clearly, the compositional asymmetry reflects an equilibrium process independent of metabolic energy. Presumably the asymmetry is the result of the differential binding of PE (and of phosphatidylglycerol, the other principal component of the membrane) by proteins and other ligands, principally counterions. If this is the case, the explanation for the increased asymmetry of distribution of phosphatidylethanolamine found in the energy-poisoned cells may lie in the altered ionic environment, and perhaps also in the altered membrane potential, caused by the action of the inhibitors and uncouplers.

REFERENCES

1. Op den Kamp JAF: Lipid asymmetry in membranes. *Annu Rev Biochem* 48:47–72, 1979.

2. Bretscher M: Membrane structure: some general principles. *Science* 181:622–629, 1973.

3. Rothman JE, Kennedy EP: Asymmetrical distribution of phospholipids in the membrane of *Bacillus megaterium. J Mol Biol* 110:603–618, 1977.

4. Bishop DG, Bevers EM, van Meer G, et al: A monolayer study of the reaction of trinitrobenzene sulphonic acid with amino phospholipids. *Biochim Biophys Acta* 551:122–128, 1979.

5. Rothman JE, Kennedy EP: Rapid transmembrane movement of newly synthesized phospholipids during membrane assembly. *Proc Natl Acad Sci USA* 74: 1821–1825, 1977.

6. Langley KE, Kennedy EP: Energetics of rapid transmembrane movement and of compositional asymmetry of phosphatidylethanolamine in membranes of *Bacillus megaterium. Proc Natl Acad Sci USA* 76:6245–6249, 1979.

7. Langley KE, Kennedy EP: Biosynthesis of phospholipids in *Bacillus megaterium. J Bacteriol* 140:996–1007, 1979.

8. Lombardi FJ, Chen SL, Fulco AJ: The conversion of phosphatidyl glycerol to phosphatidylethanolamine in *Bacillus megaterium. Fed Proc* 37:1494, 1978.

9. Roseman M, Litman BJ, Thompson TE: Transbilayer exchange of phosphatidylethanolamine for phosphatidylcholine and N-acetimidoylphosphatidylethanolamine in single-walled bilayer vesicles. *Biochemistry* 14:4826–4830, 1975.

10. de Kruijff B, van Zoelen EJJ: Effect of the phase transition on the transbilayer movement of dimyristoyl phosphatidylcholine in unilamellar vesicles. *Biochim Biophys Acta* 511:105–115, 1978.

2 Membrane and Plasma Lipoproteins—Bilayer-to-Emulsion and Emulsion-to-Bilayer Transitions

Donald M. Small, MD

Biologic membranes in the simplest sense consist of an asymmetric lipid bilayer as a continuous phase in which is anchored or floats a large variety of asymmetrically positioned proteins and glycoproteins. Most lipoproteins[1] isolated from the blood of normal warm-blooded animals are emulsion or microemulsion particles (Figure 2-1).

Fasting human plasma lipoproteins are generally classified into four major types of particles. The very large chylomicrons (CM) are true emulsion particles secreted from the intestine during the absorption of fat, whereas the very low density lipoproteins (VLDL),[2] low density lipoproteins (LDL),[3,4] and high density lipoproteins (HDL) are microemulsion particles. The CM consist of a core containing mainly triglyceride, small amounts of cholesterol ester, and a surface phase of emulsifiers: phospholipid, free cholesterol, and very small amounts of apoproteins A and B. VLDL are similar but smaller particles secreted from the liver. They contain relatively more cholesterol in the core and have small amounts of

Supported by National Health Service Grants HL 18673, HL 07291, and HL 22682.

apoproteins B and C on the surface. LDL are metabolic byproducts of VLDL metabolism. The core consists primarily of cholesterol ester and the surface phase of phospholipid, free cholesterol, and apoprotein B.[5,6] HDL are protein-rich particles (thus giving them the name "high density") with a small core of cholesterol ester and a surface emulsifying layer consisting primarily of phospholipids and A peptides. Some C peptides and cholesterol are also present.

In this chapter I will show that there are two widespread types of biochemical reactions (Table 2-1). One reaction can convert membrane bilayer structures into emulsion particles through the synthesis of relatively apolar lipids such as triglycerides or cholesterol esters; the converse reaction occurs on the surface of emulsion particles to catabolize the triglyceride or cholesterol ester and results in the reformation of bilayers.

An emulsion is a relatively stable suspension of one liquid in another.[7] Two imiscible liquids such as triglyceride and water separate rapidly when shaken together. However, in the presence of an emulsifier a stable suspension of oil in water can be obtained by shaking. Thus, the emulsifier is the molecule that maintains the stability of the emulsion. The emulsifier is usually an amphiphilic molecule containing both hydrophilic and hydrophobic parts, which forms an interface between the two liquids. Emulsifiers

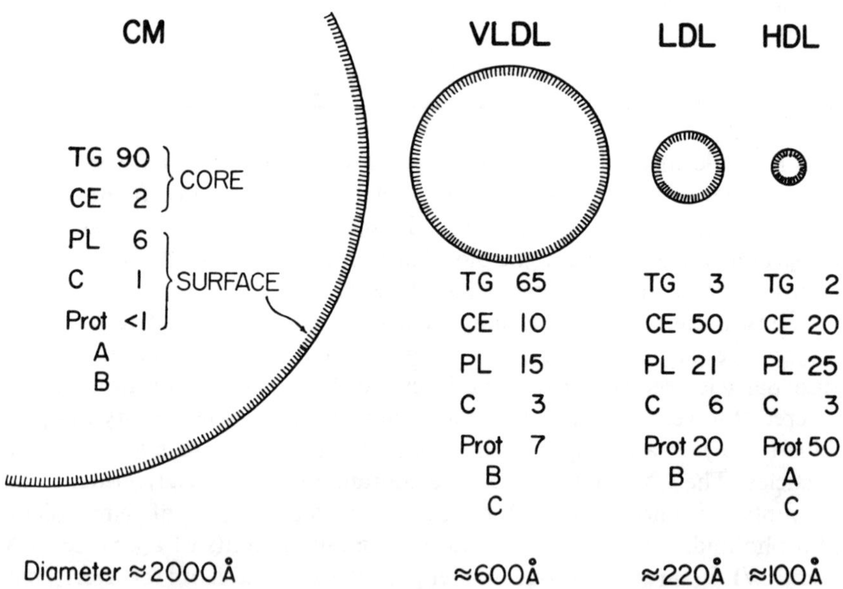

Figure 2-1 Approximate composition[1] and size of normal human plasma lipoproteins. CM = chylomicron; VLDL = very low density lipoprotein; LDL = low density lipoprotein; HDL = high density lipoprotein; TG = triacylglycerol; CE = cholesterol ester; PL = phospholipids; C = cholesterol; Prot = apoproteins classes A, B, and C. The thickness of the surface layer (~ 21 Å) is in the same scale as the core diameter. Numbers are given in weight percent.

Table 2-1
Membrane Biochemical Reactions

Reaction	Examples
1) Bilayer —(Synthesis of "apolar molecules"*)→ Emulsion particles	a) Formation of CM, VLDL. b) Formation of storage droplets in cells of TG, CE, wax esters. c) Lecithin: cholesterol acyltransferase (LCAT) reaction in plasma.
2) Emulsion particles —(Catabolism of "apolar molecules"*)→ Bilayer	a) Lipoprotein lipase reaction. Formation of surface remnants (bilayers) from CM and VLDL catabolism. b) ? Cellular hydrolysis of stored droplets of TG, CE, or wax esters.

*Triglyceride, cholesterol esters, or wax esters.

may be lipids such as soaps, detergents, sterols, monoglycerides, glycolipids, phospholipids, proteins, or complex polysaccharides. I will restrict my remarks to the emulsion particles in the plasma, ie, lipoproteins, in which an oil of triglyceride or cholesterol ester, completely insoluble in water, is emulsified largely by phospholipid, free cholesterol, and some apolipoproteins. Since an emulsion particle consists of a core of relatively apolar lipids surrounded by a monolayer of emulsifying molecules, the particle itself consists of two separate phases, a core phase and a surface phase.

The Physical State of Membrane and Lipoprotein Lipids

The major lipids in lipoproteins are phospholipids, cholesterol and cholesterol esters, and triglycerides. The interactions of equilibrated mixtures of three of these molecules, egg phosphatidylcholine, cholesteryl linoleate, and cholesterol, have been studied in detail in aqueous systems[8-19] and can be represented on triangular coordinates[19,20] in which there are four major zones (Figure 2-2). (Other individual phospholipids such as sphingomyelin, cholesterol esters having different acyl chains, and other lecithins such as dimyristoyl phosphatidylcholine have also been studied.[5,21-23] The exact phase diagrams are slightly different but the general form is similar provided temperature is taken into account.)

Zone I, a lamellar liquid crystalline phase, contains bimolecular leaflets of phospholipid, into which free cholesterol and small amounts of cholesterol ester can be incorporated. The maximum amount of free cholesterol incorporated at equilibrium is about 33% of the amount of phospholipid by weight, or approximately 1 mole of free cholesterol per mole of phospholipid.[8,14,15,25] The maximum amount of cholesterol ester incorporated is about 2%, or approximately 1 mole per 40 of phospholipid.[11,19] In this phase, which is analagous to the lipid regions of many cell membranes and cell-organelle membranes,[26] the lamellar liquid crystals may be more or less fluid, depending on the type of phospholipid, the fatty acyl chains within it, and the other membrane components present, such as cholesterol, glycosphingolipids, or proteins.

Zone II contains a single phase of cholesterol ester, which incorporates small amounts of free cholesterol—about 3% to 5% at 37 C.[13,27] Depending on the fatty acid that is esterified to cholesterol, this phase may be in one of four states: a liquid state, in which the molecules are randomly associated; a cholesteric liquid-crystal state, in which the molecules are nearly oriented along their long axis; a smectic liquid-crystal state, in which the molecules form a layered structure perpendicular to the long axis (Figure 2-2); or a crystalline solid state. Equivalent amounts of cholesterol (3% to 5% by weight) are incorporated by the liquid, cholesteric, and smectic states. Crystalline ester excludes free cholesterol from its lattice.

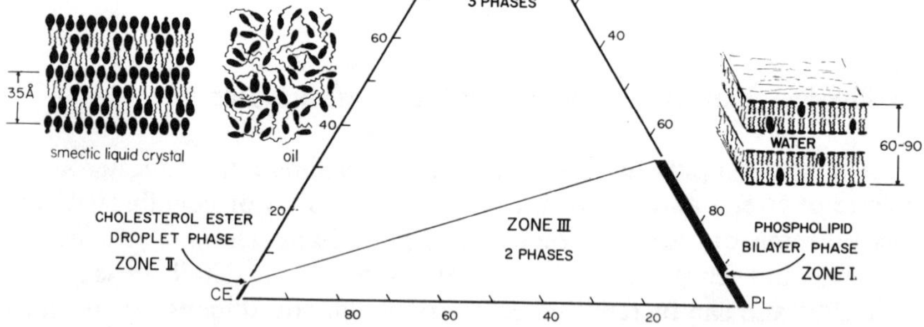

Figure 2-2 Phase diagram of a three-component system showing interactions of cholesterol (C), phospholipid (PL) and cholesterol ester (CE) in excess water at 37 C and 1 at m pressure (adapted from Katz et al[24]). Schematic molecular representations of the phases are shown near each point of the triangle. The irregular lines denote the phospholipid molecule, the solid symbols the cholesterol molecule, and the solid, tailed symbols cholesterol ester.

In Zone III, mixtures of these lipids separate into the cholesterol-ester phase (Zone II), which floats in water, and the lamellar liquid-crystalline phase (Zone I), which sinks. In Zone IV, three phases with fixed composition coexist: the lamellar phospholipid liquid-crystalline phase saturated with cholesterol and cholesterol ester; an oily cholesterol-ester phase saturated with free cholesterol; and a cholesterol-monohydrate-crystal phase.[19]

Each of these phases has physical characteristics that can be identified with the polarizing or electron microscope, by calorimetry, or by x-ray diffraction. The liquid-crystalline phase in Zone I has a characteristic appearance under polarizing and electron microscopes, and its x-ray diffraction is typical of a one-dimensional lamellar lattice. The first-order lamellar spacing reflects the thickness of the bilayer and its accompanying layer of water. This thickness, determined by the lipids and the amount of water present, may vary from about 6 to 9 nm (60 to 90 Å), but it is characteristic for a given lipid at a given temperature. For example, at 25 C, the total thickness of a typical egg-lecithin bilayer in water is 6.4 nm (64 Å),[16] whereas the total thickness of a sphingomyelin bilayer is 7.6 nm (76 Å).[28]

Zone II, the cholesterol-ester phase, may be liquid, liquid-crystalline, or crystalline. If liquid, it is isotropic (nonbirefringent between crossed polars) and produces x-ray scattering with very diffuse maxima at approximately 0.5 nm (5 Å) and 3 nm (30 Å). The cholesteric and smectic phases have different birefringent textures under the polarizing microscope.[17,29] Furthermore, the cholesteric phase scatters x-rays like the liquid, but the smectic phase produces a sharp maximum, from which the organization of ester molecules in layers has been deduced (see Figure 2-2).[30] Each ester has a characteristic transition temperature that can be quantitated by microscopy and calorimetry,[17] and mixtures of esters undergo transitions in a predictable manner; for example, the more double bonds in the acyl chains of the mixture, the lower the transition temperature. Thus, the transition temperature provides evidence of the chemical composition of the esters in Zone II.

The cholesterol-monohydrate-crystal phase in Zone IV has been characterized extensively.[12,31] The crystals melt at 85 C, have a characteristic crystal appearance by polarizing microscope, and show characteristic x-ray spacings.[12,24] In addition their crystal structure has been determined.[31]

In the above system triglyceride, a major component of some lipoproteins, was not considered. The measured solubility of triglyceride in egg-lecithin bilayers in excess water is about 10% by weight at 37 C. Furthermore, the solubility of free cholesterol in either triglyceride or cholesterol ester is virtually the same in excess water.[13,27] Finally, the solubility of cholesterol in mixtures of triglyceride and cholesterol ester at 37 C in excess water is the same as in pure triglyceride or cholesterol ester (4% by weight).

Therefore, I have estimated the phase relationships of mixed triglyceride and cholesterol ester, phospholipid, and free cholesterol in excess water (Figure 2-3). The exact solubility of triglyceride and cholesterol ester in the phospholipid (right side, Figure 2-3) is not known, but the total solubility of the two relatively apolar lipids must be small (< 10% [TG + CE] in phospholipid).

The composition of a number of lipoproteins (CM, VLDL, LDL, HDL, nascent HDL from perfused rat livers,[32] and HDL from LCAT-deficient patients[33,34]) is plotted. The HDL from LCAT-deficient patients and nascent HDL from perfused rat livers fall very close to the single bilayered phase region. By electron microscopy they appear largely as bilayered structures.[35] The other plasma lipoproteins fall in a zone where two separate phases would be expected within the particle.

Thus, with the possible exceptions of the nascent HDL isolated from liver[32] and intestinal perfusates,[36] and of HDL from patients with LCAT deficiency,[35] each of which has a bilayered structure by electron microscopic examination and very little triglyceride or cholesterol ester, all of the other major lipoprotein classes are emulsion particles, consisting of both core and surface phases. Low density lipoproteins (LDL) and high density lipoproteins (HDL) may be considered microemulsions, since they are too small to scatter visible light, whereas chylomicrons are true emulsions. All of the major lipoprotein classes are quasi-spherical particles whose size may be predicted from their chemical composition.[37-40]

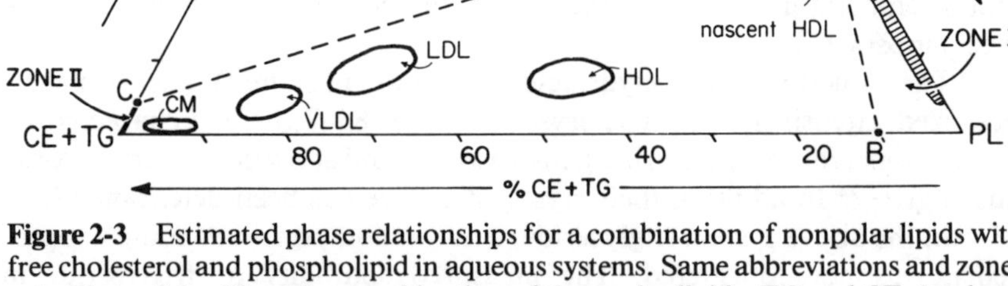

Figure 2-3 Estimated phase relationships for a combination of nonpolar lipids with free cholesterol and phospholipid in aqueous systems. Same abbreviations and zones as in Figure 2-2, although a combination of the apolar lipids, TG and CE, has been substituted in Zone II in the left-hand corner. The three points A, B, and C represent: the measured solubility of free cholesterol in egg lecithin, A[8,14,25]; the solubility of triolein in egg lecithin, B; and the solubility of free cholesterol in mixtures of triglyceride (TG) and cholesterol ester all done in an excess of water, C.[13,27] The lines A-B and A-C are estimates of the phase boundaries. The composition of the major lipoproteins is also plotted (see text).

A good empirical relationship between size and composition was found by assuming that all of the phospholipid, free cholesterol, and protein were at the surface, and all of the cholesterol ester and triglyceride were in the core. Using these assumptions, Sata et al[41] calculated the thickness of the surface of the lipoprotein particle to be about 21.5 Å. This corresponds reasonably to the length of one phospholipid molecule.

That all of the free cholesterol occurs in the surface is probably an incorrect assumption, as some fraction must be dissolved in the core. Also incorrect is the assumption that all triglyceride is in the core, since small amounts of triglyceride (up to 10% by weight, as shown in Figure 2-3) may be present in the interface. Even some cholesterol ester (~ 2%) must be in the surface phase if equilibrium lipid phase diagrams (see Figures 2-2 and 2-3) apply. Nevertheless, the Sata[41] approximation is reasonable and correlates with independent size measurements. Thus, the diameter (*d*, in Å) of any spherical lipoprotein may be roughly estimated from its composition by the equation:

$$d = \frac{21.5 \times 2}{[1 - (1 - F)^{1/3}]}$$

The volume fraction *(F)* of any component of the particle is calculated using its weight % and the estimated partial specific volumes below:

TG	= 1.093 gm/ml
CE	= 1.044 gm/ml
C	= 0.968 gm/ml
PL	= 0.970 gm/ml
Protein	= 1705 gm/ml

The derived relationship between the volume fraction of the lipoprotein constituents and the size of the lipoprotein is given in Figure 2-4.

Thus, as the fraction of total volume of the particle occupied by surface constituents decreases,[41] the size of the particle increases. The relative sizes of chylomicrons, very low density lipoproteins, low density lipoproteins, and high density lipoproteins are indicated on the graph. Chylomicrons have the smallest proportion of surface lipids, and high density lipoproteins the largest.

Formation of Emulsion Particles from Membranes in Cells

Two different kinds of emulsion particles may be produced in cells. One enters the cytoplasm and becomes a storage droplet of lipid (triglyceride, cholesterol ester, wax ester, etc). The other enters the cisternal side of the endoplasmic reticulum and is ultimately destined for secretion as a lipoprotein (VLDL from the liver and CM from the intestinal epithelial cells). Consider now the physical mechanisms for the formation of such emulsion particles.

Plasma membranes of cells contain mainly phosphatidylcholine and sphingomyelin on the exterior surface, and phosphatidylethanolamine and phosphatidylserine on the cytoplasmic surface.[10,42] Similar findings have been suggested for the endoplasmic reticulum membranes.[43] Thus, the surface facing the cytoplasm would be rich in phosphatidylserine and phosphatidylethanolamine, whereas the surface facing the cisterna would be rich in phosphatidylcholine and sphingomyelin. Lipid composition of newly secreted lymph CM[44] (Redgrave and Small, unpublished data, 1978) and nascent VLDL contain mainly phosphatidylcholine and sphingomyelin. This is consistent with the concept that the phospholipids of these particles came from the side of the membrane that contained the high proportion of phosphatidylcholine and sphingomyelin (ie, the cisternal surface of the endoplasmic reticulum).

In the liver cells the first appearance of fat globules (ie, the emulsion particle) during the formation of VLDL are seen in the smooth endoplasmic reticulum or in the smooth ends of the Golgi apparatus.[45] These

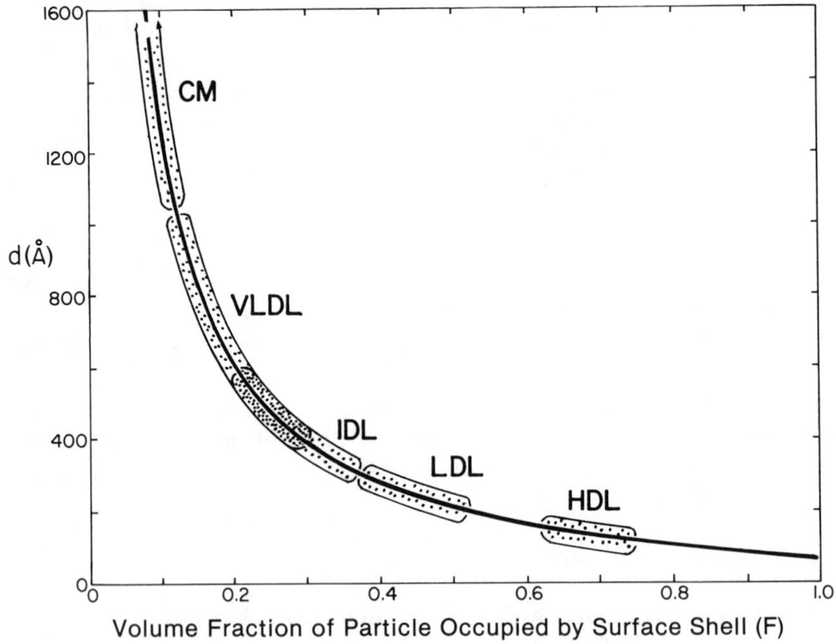

Figure 2-4 The calculated diameter (*d*, in Å) of spherical lipoproteins having different compositions, plotted against the volume fraction of the surface emulsifying layer *(F)* of the particle. The size ranges of CM, VLDL, IDL (intermediate density lipoproteins), LDL, and HDL are indicated in the figure.

particles are often quite similar in size to plasma VLDL and VLDL within the Golgi apparatus. Particles are often seen touching each other. The fact that particles can touch and not fuse into large globs of fat indicates that they are covered by an emulsifier which prevents coalescence. Alexander et al[45] have shown that protein components of the VLDL appear to be synthesized in the rough endoplasmic reticulum and are added to the particle at the confluence of the rough endoplasmic reticulum with the Golgi. Thus, the original VLDL in the smooth endoplasmic reticulum particle probably does not contain the protein as the emulsifier. Rather, the emulsifier is most likely the phospholipid. Once the particle is in the Golgi it is transferred into a secretory vesicle, and ultimately released into the space of Disse where it enters the bloodstream.

A number of electron micrographic studies on the time course of fat absorption by intestinal epithelial cells appear to show a similar pathway.[46-49] Furthermore, using biochemical and electron micrographic studies during fat absorption, we found that the very first triglyceride droplets were formed in the smooth endoplasmic reticulum at the base of microvillus membranes in the intestinal epithelial cells.[50] Some of the CM protein (B protein) appears to be synthesized throughout the cell except in the nucleus and microvilli.[51] Thus, it is possible that some protein is added during initial CM particle formation. These emulsion particles, the precursors to CM, are later transported to the Golgi apparatus and are finally released, apparently by exocytosis, into the intestinal lacteal spaces where they pass into the lymphatics. Both newly secreted CM and VLDL are covered mainly with phospholipid, but some proteins and free cholesterol are also present.

I would now like to speculate on mechanisms for the formation of such particles (Figure 2-5). The enzyme systems necessary for triglyceride synthesis appear to be on the cytoplasmic surface of the endoplasmic reticulum.[52] The soluble precursors of triglyceride synthesis, including glycerol 6 phosphate and long chain fatty acyl coenzyme A, may diffuse or be transported through the cytoplasm to the cytoplasmic surface of the smooth endoplasmic reticulum membrane. The enzyme system then converts the precursors into triglycerides, and when a small amount of triglyceride is formed it can enter either or both sides of the bilayer of the membrane, consistent with its limited solubility shown in the phase diagram (see Figure 2-3). If synthesis were relatively slow, then triglyceride could diffuse down a concentration gradient from the site of synthesis and diffuse through the membranes of the cell. Therefore there is probably some barrier to the lateral diffusion of triglycerides and perhaps phospholipids, which is suggested by the bar on the left of Figure 2-5A. As synthesis of triglyceride continues, it saturates the bilayer system, and because it is insoluble in both cytoplasmic and cisternal aqueous compartments, it must partition into the only hydrophobic area possible, the center part of the membrane (Figure 2-5B). A few molecules here will split the bilayer and begin to form a small lens of nonpolar lipids in the center of the membrane. As synthesis continues the lens will increase in size and cause a bulge in the membrane.

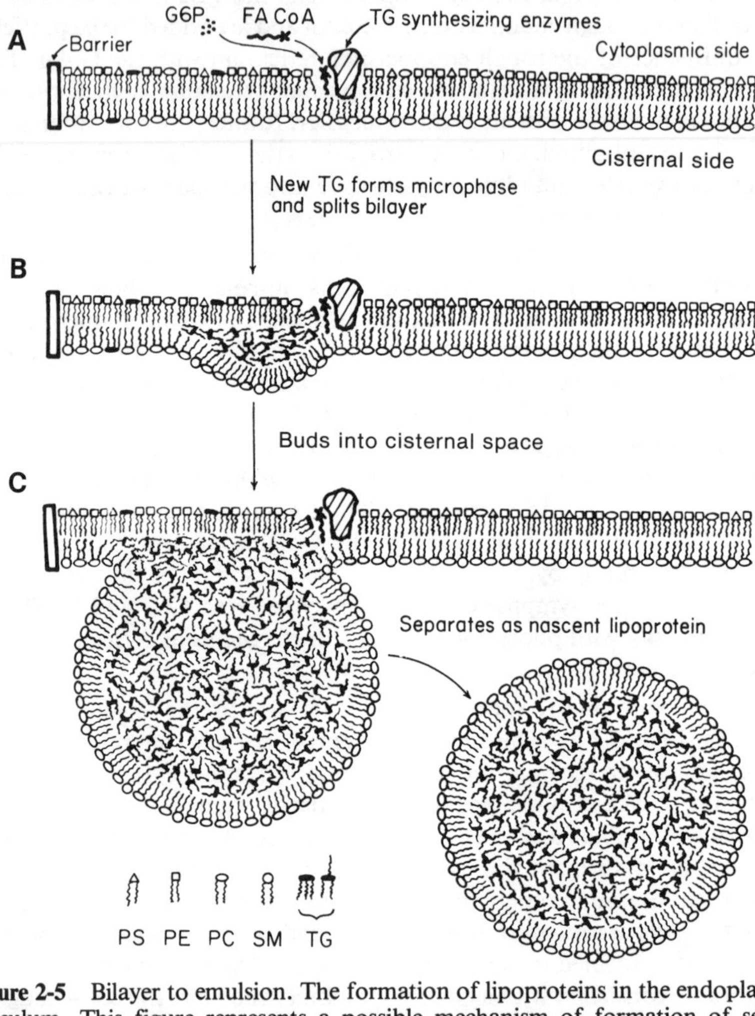

Figure 2-5 Bilayer to emulsion. The formation of lipoproteins in the endoplasmic reticulum. This figure represents a possible mechanism of formation of serum lipoproteins from the smooth endoplasmic reticulum. The mechanism would apply to the formation of VLDL in the liver or CM in the intestinal epithelial cells. **A** The enzyme complex carries out a series of reactions resulting in the formation of TG. **B** As the reaction proceeds the apolar product, TG, separates as an individual phase between the two halves of the bilayer. **C** As the reaction proceeds further the bud gets larger and finally pinches off to form nascent lipoprotein. G6P = glycerol 6 phosphate; FACoA = fatty acyl coenzyme A; TG = triglycerides; PS = phosphatidylserine; PE = phosphatidylethanolamine; PC= phosphatidylcholine; SM = sphingomyelin.

Such bulges can apparently involve either the cytoplasmic space or the cisternal space. For instance, fat droplets formed in adipocytes, muscle cell, or in tissue culture cells, or cholesterol ester droplets stored in the cytoplasm as the result of the acyl-coA:cholesterol acyltransferase (ACAT) reaction may bud into the cytoplasmic space. On the other hand, VLDL and CM bud in the other direction, into the cisternal space, and are ultimately processed for secretion into the extracellular space. What determines the direction of budding—cytoplasmic or cisternal—is not known, but it is presumably related to the relative ability of the two halves of the bilayer to expand, and the role of membrane proteins, divalent cations (Mg^{++}, Ca^{++}), and asymmetric fluidity of the membrane could be important factors. When the bud has reached a certain size it is pinched off to form the nascent lipoprotein (or cytoplasmic) particle (Figure 2-5C).

The mechanism in its simplest form involves the synthesis of a relatively apolar hydrophobic product, which cannot diffuse rapidly enough within the bilayer system, is not soluble in the aqueous space, and has very limited solubility in the membrane bilayer. Therefore, it splits the bilayer, ie, it undergoes phase separation within the bilayer, to produce an emulsion particle. The surface of the particle is covered largely by the phospholipid composition of the side of the bilayer from which it buds. In the case of CM and VLDL the lipid composition of these newly secreted particles reflects that of the cisternal surface of the endoplasmic reticulum membrane. The apoproteins could be added at the budding stage if they were present on the cisternal surface of the endoplasmic reticulum, or could be added later in the Golgi by adsorption to the nascent particle.

The Formation of Bilayers from Emulsion Particles— The Effect of Lipoprotein Lipase on CM and VLDL

When the CM is secreted from the intestinal cell, it enters the lymph and then the plasma. During this short period of time, marked changes occur in the surface but not the core of the chylomicron. First, the composition of the surface phase changes rapidly.[44,53,54] Simultaneously proteins and free cholesterol are absorbed to the surface, and an approximately equal mass of phospholipid is lost.[53] Some of the apoproteins which are adsorbed to the surface are those apoproteins which act as cofactors for the lipoprotein lipase reaction.[55] The CM then circulate to capillaries where they stick to the endothelium. The triglyceride-splitting enzyme, lipoprotein lipase, is secreted by the endothelial cell and begins to catabolize the triglyceride (presumably that small quantity of triglyceride that partitions into the emulsifying phase). The triglyceride is rapidly hydrolyzed to fatty acids and partial glycerides, which in part enter the capillary wall or are carried away by serum albumin.

Therefore, as triglyceride is lost the volume of the particle decreases. We have found that chylomicrons circulating for one-half hour in a rat in which the liver has been diverted from the circulation will reduce its triglyceride content by about 95% and its volume by about 90%.[44] It thus produces an emulsion particle which is much smaller, the so-called CM remnant.[56,57] During this process both protein and phospholipid are lost from the surface, such that chylomicrons lose about 88% of their phospholipid and about 40% of their protein.[44] These proteins and phospholipids can be found in HDL fractions.[44]

Figure 2-6 is a schematic diagram showing the general features of the mechanism.[58] In this system, as the core is removed the surface has very little ability to change its surface area, ie, it is not readily compressible.

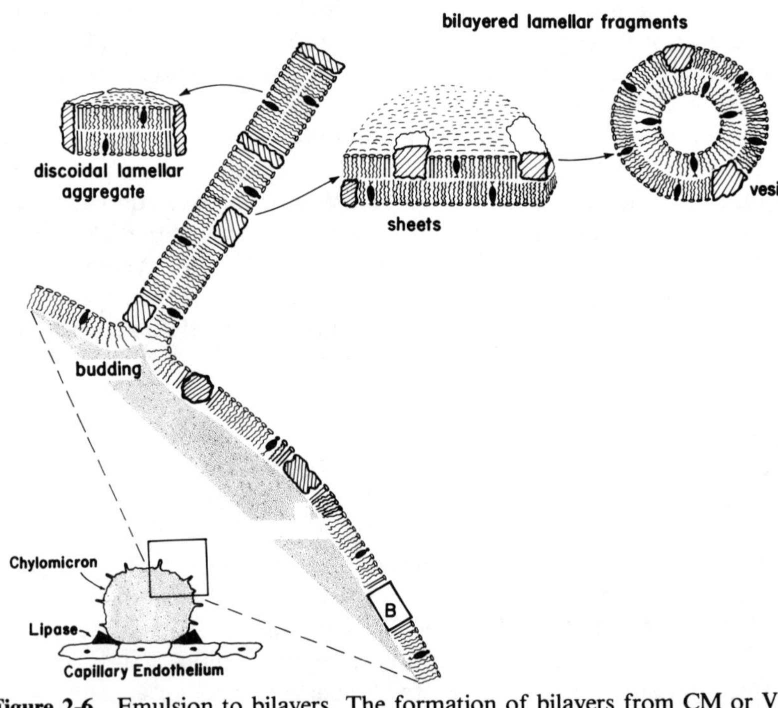

Figure 2-6 Emulsion to bilayers. The formation of bilayers from CM or VLDL during lipolysis of TG. The chylomicron lying on the capillary endothelium comes into contact with lipoprotein lipase. The lipase hydrolyzes triglyceride and the products are dissipated into the vessel wall and into serum albumin. Thus, the chylomicron shrinks, leaving excess surface. The surface of the chylomicron contains phospholipids, free cholesterol, soluble A and C peptides (shown as the hatched bodies), and apo B protein (labeled B). As triglyceride is removed during lipolysis the core shrinks, and the redundant surface constituents form lipid bilayer folds projecting from the CM. These bilayered excrescences may then be discharged into the plasma as bilayered sheets, vesicles, or even as discoidal aggregates surrounded by soluble apoproteins.

Therefore when the core volume is decreased the surface buckles and forms wrinkles of bilayer on the surface, which can then break off into the plasma as bilayer sheets, vesicles, or even discs.[58] Thus, bilayers have been regenerated from the emulsion particle by the catabolic action of the enzyme, lipoprotein lipase. We will discuss the fate of these regenerated bilayers in the next section.

The CM remnant, ie, what remains of the core after lipolysis, is rapidly taken up by the liver and delivers back to the liver what is left in the core. The mechanism for its uptake is not known. Very low density lipoprotein undergoes a very similar triglyceride lipase catabolic reaction to produce bilayers from the surface[59] and another familiar microemulsion particle, the so-called LDL particle.[60-62] LDL is probably the end core remnant of VLDL formed after hydrolysis of most of the triglyceride from VLDL.

The Generation of Microemulsions from Bilayers in the Plasma

The smallest microemulsion particles in plasma are the HDL. As isolated from the plasma, HDL are small heterogeneous particles containing roughly 50% protein and 50% lipid. Considerations of the physiochemical properties of the lipids (see Figures 2-2 and 2-3) and proteins[63] involved, as well as x-ray scattering[64,65] and electron microscopic studies,[35,66] indicate that plasma HDL is a quasi-spherical particle in which the phospholipid, protein, and some of the free cholesterol form the surface, and the less polar lipids (cholesterol ester and triglyceride) form the core. The composition of HDL as found in the plasma has been plotted in Figure 2-3, which illustrates that the HDL falls in a zone where one would expect two phases to exist: a core phase made up mainly of cholesterol ester and trigylceride; and a surface phase made up of phospholipids. Free cholesterol would be expected to partition between these phases, while protein would be found in the surface phase.

Since HDL is about 50% protein we need to consider the role of the protein in the structure of HDL. Important concepts of the structure of the proteins came from the sequences of several different apoproteins. Of the major apoproteins so far sequenced (A peptides and C peptides),[67-71] Segrest et al[63,72] pointed out that most contain sequences that can form amphiphilic helices, ie, α-helical segments, with one side hydrophobic, and one side hydrophilic. This is illustrated in Figure 2-7 by the sequence[73] of residues 33 to 55 of apo C-I; one side is clearly hydrophilic, the other hydrophobic. The major apoprotein on HDL is apo A-I, an apoprotein of MW $\approx$ 28,000 with high α-helicity in solution, which contains 245 amino acids (Figure 2-8) including eight specific sequences of 10 to 18 amino acids which should be amphipathic α-helices (Figure 2-9).[67,70,74] In simplified form, the structure is rather like a series of amphiphilic rods connected with structure-breaking or β-turn-forming amino acids. Although the definite tertiary structure of A-I, or any of the other soluble apoproteins for that matter, is not known, it is clear that they have rather unique structures among proteins in general. In dilute solutions, these amphiphilic helices presumably attempt to hide the hydrophobic parts of the helices from the water by intrahelical hydrophobic binding. The monomer appears to form a rod-shaped particle in aqueous solution with an axial ratio of about 1:5 (Figure 2-10).[74]

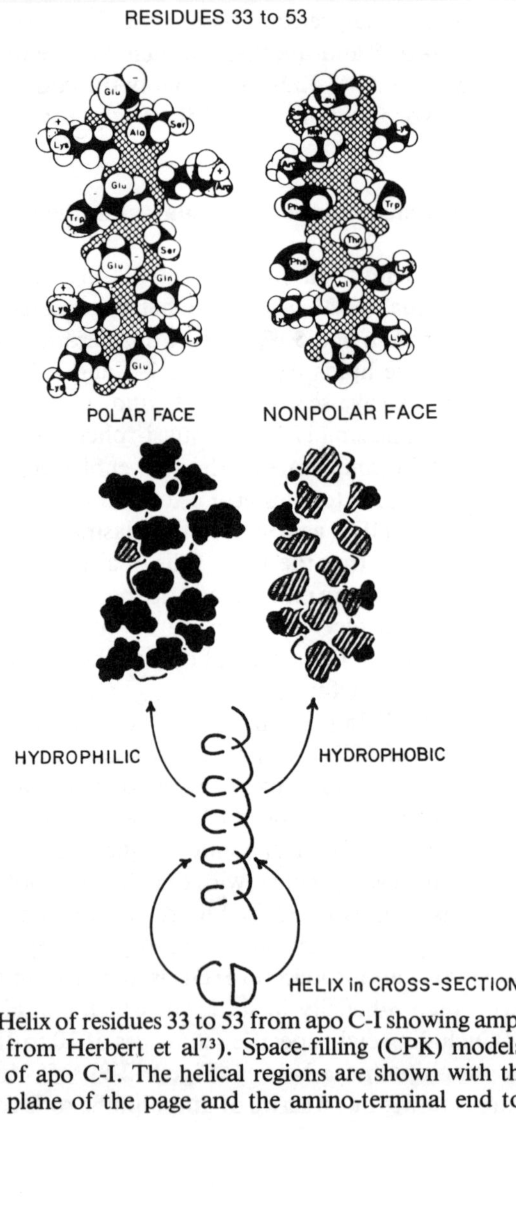

Figure 2-7 α-Helix of residues 33 to 53 from apo C-I showing amphiphilic nature of helix (adapted from Herbert et al[73]). Space-filling (CPK) models of amphipathic helical regions of apo C-I. The helical regions are shown with their axes oriented parallel to the plane of the page and the amino-terminal end toward the top of the page.

APO A-I

Asp-Glu-Pro-Pro-Gln-Ser-Pro-Trp-Asp-Arg-Val-Lys-Asp-Leu-Ala-Thr-Val-Tyr-Val-Asp-
20
Val-Leu-Lys-Asp-Ser-Gly-Arg-Asp-Tyr-Val-Ser-Gln-Phe-Gln-Gly-Ser-Ala-Leu-Gly-Lys-
40
Gln-Leu-Asn-Leu-Lys-Leu-Leu-Trp-Asp-Asp-Val-Thr-Ser-Thr-Phe-Ser-Lys-Leu-Arg-Gln-
60
Glu-Leu-Gly-Pro-Val-Thr-Glu-Glu-Trp-Phe-Asn-Asp-Leu-Gln-Glu-Lys-Leu-Asn-Leu-Glu-
80
Lys-Glu-Thr-Gly-Glu-Leu-Arg-Gln-Glu-Met-Ser-Lys-Asp-Leu-Glu-Glu-Val-Lys-Ala-Lys-
100
Val-Gln-Pro-Tyr-Leu-Asp-Asp-Phe-Gln-Lys-Lys-Trp-Gln-Glu-Met-Glu-Leu-Tyr-Arg•Gln-
120
Lys-Val-Glu-Pro-Leu-Arg-Ala-Glu-Leu-Gln-Glu-Gly-Ala-Arg-Gln-Lys-Leu-His-Glu-Leu-
140
Gln-Glu-Lys-Leu-Ser-Pro-Leu-Gly-Glu-Glu-Met-Arg-Asp-Arg-Ala-Arg-Ala-His-Val-Asp-
160
Ala-Leu-Arg-Thr-His-Leu-Ala-Pro-Tyr-Ser-Asp-Glu-Leu-Arg-Gln-Arg-Leu-Ala-Ala-Arg-
180
Leu-Glu-Ala-Leu-Lys-Glu-Asn-Gly-Ala-Gly-Arg-Leu-Ala-Glu-Tyr-His-Ala-Lys-Ala-Thr-
200
Glu-His-Leu-Ser-Thr-Leu-Ser-Glu-Lys-Ala-Lys-Pro-Ala-Leu-Glu-Asp-Leu-Arg-Gln-Gly-
220
Leu-Leu-Pro-Val-Leu-Glu-Ser-Phe-Lys-Val-Ser-Phe-Leu-Ser-Ala-Leu-Glu-Glu-Tyr-Thr-
240
Lys-Leu-Asn-Thr-Gln
245

Figure 2-8 The sequence of lipoprotein apo A-I.[67,70] Solid underlining indicates those sequences that form α-helices; dashed underlining shows those sequences that form β-pleated sheets.

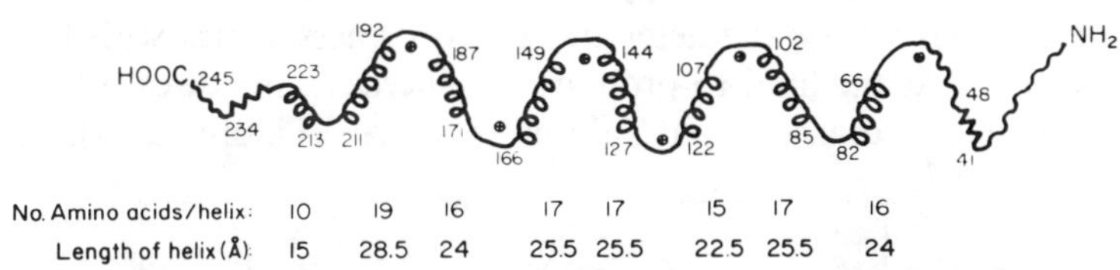

Figure 2-9 The possible tertiary structure of apo A-I, according to Edelstein et al.[38] There are eight putative amphiphilic α-helices varying in length from 15 to 28.5 Å. At the ends of the molecule there are two β-pleated sheet regions (amino acids 41 to 48 and 223 to 234). Dots at the turns indicate putative β-turns.

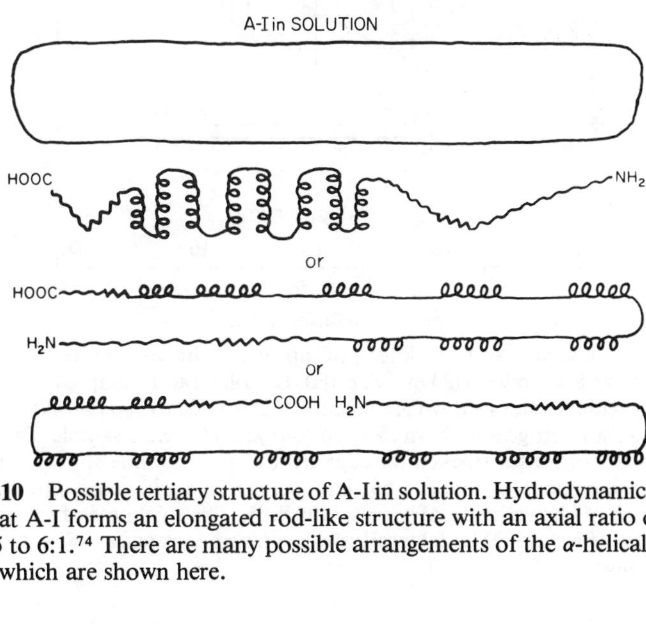

Figure 2-10 Possible tertiary structure of A-I in solution. Hydrodynamic studies indicate that A-I forms an elongated rod-like structure with an axial ratio of approximately 5 to 6:1.[74] There are many possible arrangements of the α-helical structure, three of which are shown here.

Calorimetric studies allowed us to assess the stability of apo A-I in different systems.[75-77] Compared to many other globular proteins,[70,78] whose free energy at 37 C in solution (ΔG°_{37}) is greater than 10 kcal/mole, the stability of apo A-I in aqueous systems is very low (ΔG°_{37} = 2.4 kcal/mole).[76] Furthermore, one might expect that these proteins would self-associate in water to form aggregates that are somewhat analogous to micelles. Recently, it has been shown that the A-I peptides form aggregates of up to five to eight monomers, depending on the concentration.[6,79-81]

As we have seen, phospholipids such as lecithin form myelin figures that, on a molecular basis, are bimolecular leaflets of lecithin interspaced with layers of water[16] (see Figure 2-2). If apo A-I or a mixture of HDL apoproteins (apo HDL) are added to a cloudy suspension of myelin figures of phospholipid, such as dipalmitoyl or dimyristoyl lecithin, within a few minutes or hours the suspension becomes optically clear.[75,77] This is analogous to the interaction of detergents such as bile salts with lecithin.[82,83] When these A-I-phospholipid suspensions are observed by negative-stain electron microscopy, they are found to be a series of small disc-like particles whose diameter but not thickness varies with lipid-to-protein ratio.[77] As the lipid-to-protein ratio increases the discs enlarge, up to a maximum diameter of about 160 Å (Figure 2-11). From

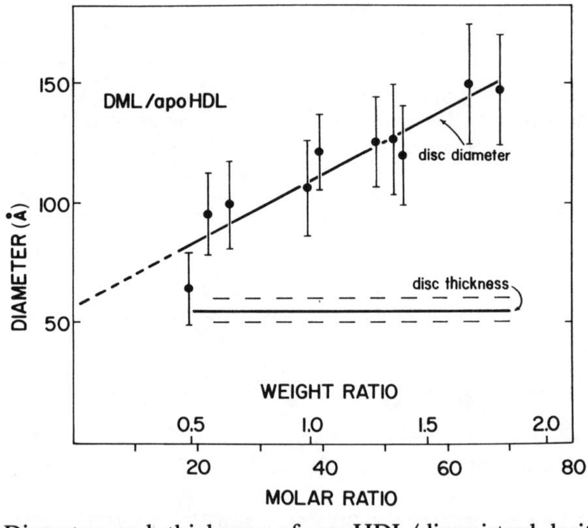

Figure 2-11 Diameter and thickness of apoHDL/dimyristoyl lecithin (DML) recombinants as a function of DML/apoHDL molar ratio (adapted from Tall et al[77]). The individual values shown are mean ± SD for 200 to 300 particles measured from six to eight micrographs from three to four grids for each sample. Dimensions were obtained from discs stacked on edge in rouleux. The line shown is the least square regression line, $y = 1.37x + 56$. The correlation coefficient was 0.94 and the standard error of the estimate of y on x was 8.05. The thickness of the discs was similar to that of a pure DML bilayer[21] and remained constant at 55 ± 5 Å despite diameter changes.

calorimetric,[75,77] x-ray scattering,[64] and neutron scattering studies,[84] the structure of the particle appears to be a bimolecular disc of phospholipid, encircled at its perimeter by the apoprotein. The most reasonable structure would have the hydrophobic sides of the amphiphilic α-helices of the A-I and other apopeptides in contact with the hydrophobic parts of the lecithin. The hydrophilic half of the α-helix could contact the aqueous system (Figure 2-12). There is a striking analogy between these discs and the disc-like bile salt-lecithin micelles formed at high lecithin-bile salt ratios.[83]

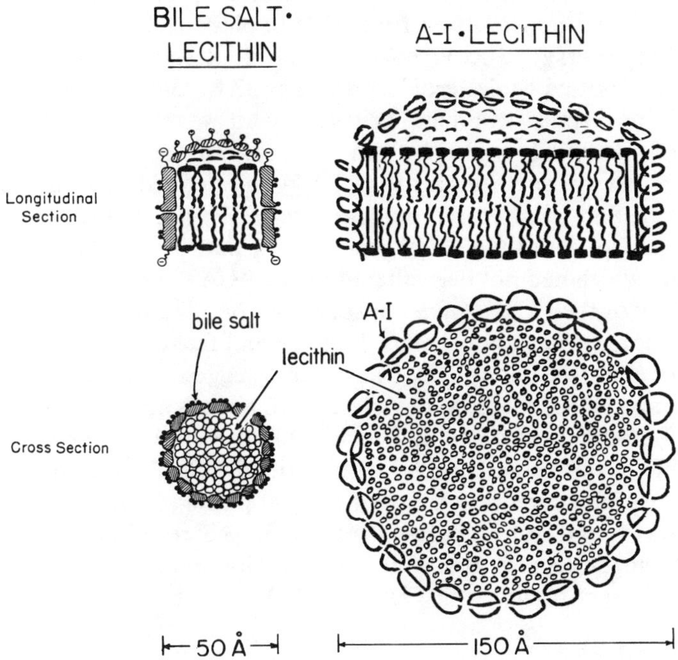

Figure 2-12 Discoidal bilayered aggregates. The lecithin-bile salt micelle and the apo A-I-lecithin disc. (Adapted from Small.[83] Reproduced with permission.) Bile salts (eg, sodium cholate, sodium deoxycholate) and lecithin in molar ratios from 1 bile salt to 1 lecithin, to 1 bile salt to 2 lecithins, form small discoidal aggregates in which the hydrophobic side of the bile salt shields the hydrophobic aliphatic chains of the lecithin from the water.[83] The hydrophilic parts of the bile salt (hydroxyls and carboxyl groups) extend into the aqueous phase on the perimeter of the disc while the ends of the discs are rendered hydrophilic by the phosphoryl choline groups of the lecithin.

The arrangement of the apo A-I-lecithin disc is similar. The amphipathic helices of the proteins cover the hydrophobic chains on the perimeter of the disc, presumably with the hydrophobic half of the helices facing the aliphatic chain region. The hydrophilic part of the helix then extends into the water, rendering the whole aggregate hydrophilic. In fact, the exact conformation of the protein on the perimeter of the disc is not known. In this picture it is shown with the axis of the helices perpendicular to the plane of the disc. Alternate models with the helices running parallel are equally possible.

Such disc-like apoprotein-phospholipid micelles or aggregates can, like bile salt-lecithin micelles, solubilize a certain amount of cholesterol. The maximum amount of cholesterol solubilized by discs of apo A-I and dimyristoyl appears to be about 1 cholesterol to 2 lecithins.[85,86] Tall and Lange[86] concluded from calorimetric studies that the cholesterol interacted only with the lecithin and was excluded from the protein interface. Phospholipid-cholesterol-apoprotein discs can be formed using a number of apoproteins as well as a variety of different phospholipids. What do these disc-like micellar apoprotein-cholesterol apoprotein-lipid aggregates have to do with microemulsion formation in plasma?

In 1935, Sperry[87] discovered a reaction in plasma that increased the amount of esterified cholesterol in plasma at 37 C. Glomset and Norum[88] described this reaction again and believed that it was important in the formation of esterified cholesterol in plasma. He termed this enzyme lecithin:cholesterol acyltransferase (LCAT) and defined its characteristics more clearly. Shortly after, Norum and others[33-35,89] described the lipoproteins of a patient with LCAT deficiency. These patients had lipoproteins in their serum which had not been altered by the LCAT system. Interestingly, the majority of their HDL were not spherical, but disc-like particles. They contained appropriate apoproteins, lecithin, and free cholesterol, but little cholesterol ester (see Figure 2-3). The authors suggested that these were the newly formed HDL, which were secreted into plasma and in the absence of the LCAT reaction remained as discoidal bilayers. Indeed, the incubation of the patients' discoidal particles with purified LCAT converted them to spherical particles similar to those of normal plasma HDL. Interestingly, apo A-I was found to be a cofactor for the LCAT reaction.

A schematic representation of the formation of microemulsions from discoidal bilayers is shown in Figure 2-13. Lamellar sheets and discs are

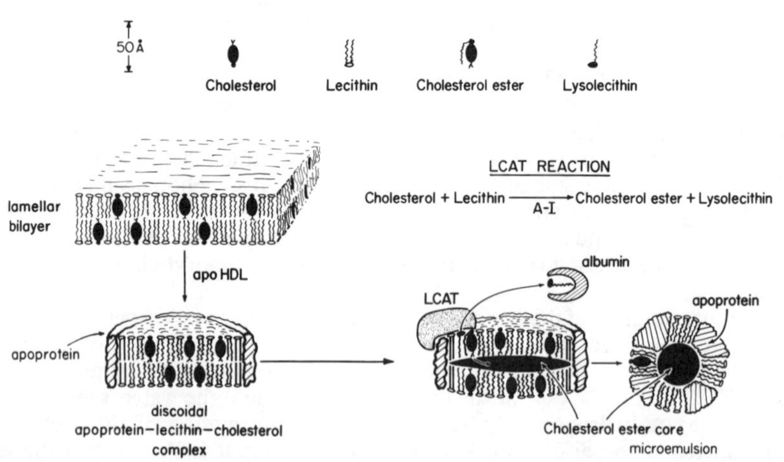

Figure 2-13 The formation of microemulsions from lamellar bilayers in plasma. The LCAT reaction and the formation of spherical HDL.

generated from the surface of chylomicrons and VLDL (as indicated in Figure 2-6) and enter the plasma. Bilayered structures contain both lecithin and free cholesterol, the substrates for the lecithin:cholesterol acyltransferase (LCAT) reaction. When the LCAT enzyme binds to a discoidal particle containing apo A-I, cholesterol and lecithin are converted to cholesterol ester and lysolecithin. In plasma the lysolecithin is removed from the particle surface by albumin, whereas the cholesterol ester is insoluble in the interface and partitions into the oily parts of the particle, ie, between the ends of the hydrocarbon chains of the phospholipid. As the reaction proceeds the bilayered particle is converted into a sphere with a cholesterol ester core. Such spherical microemulsion particles are consistent with the structure and composition of circulating human plasma HDL.

In animals some nascent discoidal HDL similar in composition to HDL of LCAT-deficient patients (see Figure 2-3) appear to be secreted directly by the liver and intestine. The rat liver secretes discoidal HDL with a high concentration of E apoprotein,[32] an apoprotein with similar physical properties to A-I. The intestine also secretes some disc-like particles rich in apo A-I.[36] These particles are good substrates for LCAT.[32]

The major quantity of plasma HDL, however, may arise from surface bilayers generated from CM and VLDL lipolysis (see Figure 2-6), and their subsequent interaction with circulating HDL and LCAT. When we injected large amounts of CM into rats and stimulated their lipoprotein lipase we noted that after one-half hour most of the phospholipid was lost from CM and was found in the HDL fraction.[44] We expected to see discs, vesicles, and other bilayer fragments in this fraction, but very few were seen even after an extensive electron microscopic search. In fact, the HDL particles appeared larger than control HDL and were richer in phospholipid. We concluded that at least some of the phospholipid lost from CM during lipolysis fused with circulating HDL to produce the larger particle. The particle would be an emulsion with excess surface, and thus an excellent substrate for LCAT.

Summary

The formation and catabolism of lipoproteins is intimately associated with membranes. The biochemical reactions that produce triglycerides in the liver and intestine, destined for secretion as VLDL and chylomicrons, are present in the endoplasmic reticulum on the cytoplasmic side. These soluble substrates are converted into insoluble products (triglycerides) that separate as an oil phase in the center of the bilayer. Particular characteristics of the asymmetric endoplasmic reticulum membrane allow these lipids to bulge into the cytoplasmic cisterna where they are pinched off to form nascent VLDL and CM. After association with proteins, passage through the Golgi, and secretion into the extracellular spaces, these lipoproteins pass into the blood circulation. There they undergo some rapid surface modifications and quickly stick to capillary endothelium, where lipoprotein lipase begins their catabolism.

The action of lipoprotein lipase hydrolyzes the triglyceride and decreases the volume of the triglyceride emulsion particles. The excess surface is released into the circulation as bilayer fragments (discs, sheets, vesicles, etc). These fragments pick up free cholesterol from other cell membranes, and interact with lecithin:cholesterol acyltransferase to produce cholesterol ester from cholesterol and lecithin. This reaction converts bilayers back into small spherical microemulsion particles called high density lipoproteins. Other bilayered fragments secreted from the liver and the intestine are probably also transformed into microemulsions through the action of LCAT. Thus, the synthesis of lipoproteins involves a bilayer-to-emulsion transformation due to the synthesis of triglyceride, while the catabolism of triglyceride-rich lipoprotein particles involves an emulsion-to-bilayer transition due to the removal of the triglyceride. Such bilayer → emulsion → bilayer transformations are universal parts of the processes of transport and storage of relatively nonpolar lipids such as triglycerides, cholesterol esters, and wax esters.

REFERENCES

1. Nelson GJ: *Blood Lipids and Lipoproteins: Quantitation, Composition and Metabolism.* New York, John Wiley & Sons Inc, 1972.

2. Deckelbaum RJ, Tall AR, Small DM: The interaction of cholesterol ester and triglyceride in human plasma very low density lipoproteins. *J Lipid Res* 18:164–168, 1977.

3. Deckelbaum RJ, Shipley GG, Small DM: Structure and interaction of lipids in human plasma low density lipoproteins. *J Biol Chem* 252:744–754, 1977.

4. Deckelbaum RJ, Shipley GG, Small DM, et al: Thermal transitions in human plasma low density lipoproteins. *Science* 190:392–394, 1975.

5. Untracht SH, Shipley GG: Molecular interactions between lecithin and sphingomyelin: temperature and composition dependent phase separation. *J Biol Chem* 252:4449–4457, 1977.

6. Vitello LB, Scanu AM: Studies on human serum high density lipoproteins. *J Biol Chem* 251:1131–1136, 1976.

7. Becher P: *Emulsions—Teaching and Practice,* 2nd ed. New York, Reinhold Publishing Co, 1965.

8. Bourges M, Small DM, Dervichigan DG: Biophysics of lipidic associations. II. The ternary systems: cholesterol-lecithin-water. *Biochim Biophys Acta* 137: 157–167, 1967.

9. Brecher P, Chobanian J, Small DM, et al: The use of phospholipid vesicles for in vitro studies on cholesteryl ester hydrolysis. *J Lipid Res* 17:239–247, 1976.

10. Bretscher MS: Phosphatidyl-ethanolamine: differential labelling in intact cells and cell ghosts of human erythrocytes by a membrane-impermeable reagent. *J Mol Biol* 71:523–528, 1972.

11. Janiak MJ, Loomis CR, Shipley GG, et al: The ternary phase diagram of lecithin, cholesteryl lineolenate and water. *J Mol Biol* 86:325–339, 1974.

12. Loomis CR, Shipley GG, Small DM: The phase behavior of hydrated cholesterol. *J Lipid Res* 20:525–535, 1979.

13. North BE, Katz SS, Small DM: The dissolution of cholesterol monohydrate crystals in atherosclerotic plaque lipids. *Atherosclerosis* 30:211–217, 1978.

14. Small DM, Bourges M, Dervichian DG: Ternary and quaternary aqueous systems containing bile salts, lecithin and cholesterol. *Nature* 211:816–818, 1966.

15. Small DM, Bourges M: Lyotropic paracrystalline phases obtained with ternary systems of amphiphilic substances in water. *Molec Cryst* 1:541–561, 1966.

16. Small DM: Observations on lecithin. Phase equilibria and structure of dry and hydrated egg lecithin. *J Lipid Res* 8:551–557, 1967.

17. Small DM: The physical state of lipids of biological importance—cholesterol esters, cholesterol and triglyceride, in Blank M (ed): *Surface Chemistry of Biological Systems.* New York, Plenum Publishing Corporation, 1970, pp 55–84.

18. Small DM, Loomis C, Janiak M, et al: Liquid crystalline behavior of biologically important lipids. Polyunsaturated cholesterol esters and phospholipids, in Porter RS, Johnson JF (eds): *Ordered Fluids and Liquid Crystals.* New York, Plenum Publishing Corporation (Advances in Chemistry Series), 1974, vol 2, pp 11–22.

19. Small DM, Shipley GG: Physical-chemical basis of lipid deposition in atherosclerosis. *Science* 185:222–229, 1974.

20. Masing G: *Ternary Systems. Introduction to the Theory of Three Component Systems.* New York, Dover Publications Inc, 1944.

21. Janiak MJ, Small DM, Shipley GG: The thermal pre-transition of synthetic phospholipids: dimyristoyl and dipalmitoyl lecithin. *Biochemistry* 15:4575–4580, 1976.

22. Janiak MJ, Small DM, Shipley GG: Interactions of cholesterol esters with phospholipids: cholesterol myristate and dimyristoyl lecithin. *J Lipid Res* 20:183–199, 1979.

23. Janiak MJ, Small DM, Shipley GG: Temperature and compositional dependence of the structure of hydrated dimyristoyl lecithin. *J Biol Chem* 254:6068–6078, 1979.

24. Katz SS, Shipley GG, Small DM: Physical chemistry of the lipids of human atherosclerotic lesions. Demonstration of a lesion intermediate between fatty streaks and advance plaques. *J Clin Invest* 58:200–211, 1976.

25. Lecuyer H, Dervichian DG: Structure of aqueous mixtures of lecithin and cholesterol. *J Mol Biol* 45:39–57, 1969.

26. Small DM: Liquid crystals in living and dying systems. *J Colloid Interface Sci* 58:581–602, 1977.

27. Jandacek RJ, Webb MR, Mattson FH: Effect of an aqueous phase on the solubility of cholesterol in an oil phase. *J Lipid Res* 18:203, 1977.

28. Shipley GG, Avecilla L, Small DM: Phase behavior and structure of aqueous dispersions of sphingomyelin. *J Lipid Res* 15:124–131, 1974.

29. Gray GW: The mesomorphic behavior of the fatty esters of cholesterol. *J Chem Soc* 3:3733–3739, 1956.

30. Atkinson D, Deckelbaum RJ, Small DM, et al: The structure of human plasma low density lipoproteins. The molecular organization of the central core. *Proc Natl Acad Sci USA* 74:1042–1046, 1977.

31. Craven BM: Crystal structure of cholesterol monohydrate. *Nature* 260:727–729, 1976.

32. Hamilton RL, Williams MC, Fielding CJ, et al: Discoidal bilayer structure of nascent high density lipoprotein from perfused rat liver. *J Clin Invest* 58:667–680, 1976.

33. Glomset JA, Nichols AV, Norum KR, et al: Plasma lipoproteins in familial lecithin:cholesterol acyltransferase deficiency. *J Clin Invest* 52:1078–1092, 1973.

34. Glomset JA, Norum KR, Nichols AV, et al: Plasma lipoproteins in familial lecithin:cholesterol acyltransferase deficiency: effects of dietary manipulation. *Scand J Clin Lab Invest* 35(suppl 142):3–30, 1975.

35. Forte T, Norum KR, Glomset JA, et al: Plasma lipoproteins in familial lecithin:cholesterol acyltransferase deficiency: structure of low and high density lipoproteins as revealed by electron microscopy. *J Clin Invest* 50:1141–1148, 1971.

36. Green PHR, Tall AR, Glickman RM: Rat intestine secretes discoid high density lipoprotein. *J Clin Invest* 61:528–534, 1978.

37. Dole VP, Hamilin JT III: Particulate fat in lymph and blood. *Physiol Rev* 42:674–701, 1962.

38. Edelstein C, Kezdy FJ, Scanu AM, et al: Apolipoproteins and the structural organization of plasma lipoproteins: human plasma high density lipoproteins-3. *J Lipid Res* (review) 20:143–153, 1979.

39. Fraser R: Size and lipid composition of chylomicrons of different Svedberg units of flotation. *J Lipid Res* 11:60–65, 1970.

40. Fraser R, Cliff WJ, Courtice FC: The effect of dietary fat load on the size and composition of chylomicrons in thoracic duct lymph. *Q J Exp Physiol* 53:390–398, 1968.

41. Sata T, Havel RJ, Jones AL: Characterization of subfractions by gel chromatography from blood serum of normolipemic and hyperlipemic humans. *J Lipid Res* 13:757–768, 1972.

42. Zwaal RFA, Roelofsen B, Colley CM: Localization of red cell membrane constituents. *Biochim Biophys Acta* 399:159–182, 1973.

43. Nilsson OS, Dallner G: Enzyme and phospholipid asymmetry in liver microsomal membranes. *J Cell Biol* 72:568–583, 1977.

44. Redgrave TG, Small DM: Quantitation of the transfer of surface components of chylomicrons to the high density lipoprotein fraction density during the catabolism of chylomicrons in the rat. *J Clin Invest* 64:162–171, 1979.

45. Alexander CA, Hamilton RL, Havel RJ: Subcellular localization of B apoprotein of plasma lipoproteins in rat liver. *J Cell Biol* 69:241–263, 1976.

46. Friedman HI, Cardell RR Jr: Effects of puromycin on the structure of rat intestinal epithelial cells during fat absorption. *J Cell Biol* 52:15–40, 1972.

47. Jersild RA Jr: A time sequence study of fat absorption in the rat jejunum. *Am J Anat* 118:135, 1966.

48. Strauss EW: Electron microscopic study of intestinal fat absorption in vitro from mixed micelles containing linolenic acid, monoolein, and bile salt. *J Lipid Res* 7:307–323, 1966.

49. Strauss EW: Morphological aspects of triglyceride absorption, in Field J (ed): *Handbook of Physiology*. Washington, DC, American Physiology Society, 1968, p 1377.

50. Robins SJ, Small DM, Trier JS, et al: Localization of fatty acid reesterification in the brush border region of intestinal absorptive cells. *Biochim Biophys Acta* 233:550–561, 1971.

51. Glickman RM, Khorana J, Kilgore A: Localization of apolipoprotein B in intestinal epithelial cells. *Science* 193:1254–1255, 1976.

52. Coleman R, Bell RM: Evidence that biosynthesis of phosphatidylethanolamine, phosphatidylcholine, and triacylglycerol occurs on the cytoplasmic side of microsomal vesicles. *J Cell Biol* 76:245–253, 1978.

53. Imaizumi K, Fainaru M, Havel RJ: Composition of proteins of mesenteric lymph chylomicrons in the rat and alterations produced upon exposure of chylomicrons to blood serum and serum proteins. *J Lipid Res* 19:712–722, 1978.

54. Minari O, Zilversmit DB: Behavior of dog lymph chylomicron lipid constituents during incubation with serum. *J Lipid Res* 4:424–436, 1963.

55. Havel RJ, Kane JP, Kashyap ML: Interchange of apolipoproteins between chylomicrons and high density lipoproteins during alimentary lipemia in man. *J Clin Invest* 52:32–38, 1973.

56. Mjos OD, Faergeman O, Hamilton RL, et al: Characterization of remnants produced during the metabolism of triglyceride-rich lipoproteins of blood plasma and intestinal lymph in the rat. *J Clin Invest* 56:603–615, 1975.

57. Redgrave TG: Formation of cholesteryl ester-rich particulate lipid during metabolism of chylomicrons. *J Clin Invest* 49:465–471, 1970.

58. Tall AR, Small DM: Plasma high density lipoproteins. *N Engl J Med* 299:1232–1236, 1978.

59. Chajek T, Eisenberg S: Very low density lipoproteins: metabolism of phospholipids, cholesterol and apolipoprotein C in the isolated perfused rat heart. *J Clin Invest* 62:1654–1665, 1978.

60. Bilheimer DW, Eisenberg S, Levy RI: The metabolism of very low density lipoprotein proteins. I. Preliminary in vitro and in vivo observations. *Biochim Biophys Acta* 260:212–221, 1972.

61. Eisenberg S, Bilheimer DW, Levy RI: The metabolism of very low density lipoprotein proteins. II. Studies on the transfer of apoproteins between plasma lipoproteins. *Biochim Biophys Acta* 280:94–104, 1972.

62. Eisenberg S, Rachmilewitz D: Metabolism of rat plasma very low density lipoprotein. II. Fate in circulation of apoprotein subunits. *Biochim Biophys Acta* 326:391–405, 1973.

63. Segrest JP, Gotto AM, Fredrickson DS: Familial lipoprotein deficiency (abetalipoproteinemia, hypobetalipoproteinemia, and Tangier Disease), in Stanbury JB, Wyngaarden JW, Fredrickson DS (eds): *The Metabolic Basis of Inherited Disease,* 4th ed. McGraw-Hill Inc, 1978, pp 544–588.

64. Atkinson D, Smith HM, Dickson J, et al: Interaction of apoprotein from porcine high density lipoprotein with dimyristoyl lecithin. 1. The structure of the complexes. *Eur J Biochem* 64:541–547, 1976.

65. Shipley GG, Atkinson D, Scanu AM: Small angle x-ray scattering of human serum high density lipoproteins. *J Supramol Struct* 1:98–104, 1972.

66. Ohtskuki M, Edelstein C, Sogard M, et al: Electron microscopy of negatively stained and freeze-etched high density lipoproteins-3 from human serum. *Proc Natl Acad Sci USA* 74:5001, 1977.

67. Baker HN, Gotto AM Jr, Jackson RL: The primary structure of human plasma high density apolipoprotein glutamine I (apo A-I). II. The amino acid sequence and alignment of cyanogen bromide fragments IV, III and I. *J Biol Chem* 250:2725–2738, 1975.

68. Brewer HB Jr, Lux SE, Ronand R, et al: Amino acid sequence of human apoLp-Gln-II (apo A-II), an apolipoprotein isolated from the high-density lipoprotein complex. *Proc Natl Acad Sci USA* 69:1304–1308, 1972.

69. Brewer HB Jr, Schulman R, Herbert P, et al: The complete amino acid sequence of alanine apolipoprotein (apo C-III), an apolipoprotein from human plasma very low density lipoproteins. *J Biol Chem* 249:4975–4984, 1974.

70. Delahunty T, Baker HN, Gotto AM Jr, et al: The primary structure of human plasma high density apolipoprotein glutamine I (apo A-I). I. The amino acid sequence of cyanogen bromide fragment II. *J Biol Chem* 250:2718–5313, 1975.

71. Jackson RL, Sparrow JT, Baker HN, et al: The primary structure of apolipoprotein-serine. *J Biol Chem* 249:5308–5313, 1974.

72. Segrest JP, Feldmann RJ: Amphipathic helixes and plasma lipoproteins: a computer study. *Biopolymers* 16:2053–2065, 1977.

73. Herbert PN, Gotto AM, Fredrickson DS: Familial lipoprotein deficiency (abetalipoproteinemia and Tangier Disease), in Stanbury JB, Wyngaarden JW, Fredrickson DS (eds): *The Metabolic Basis of Inherited Disease,* 4th ed. McGraw-Hill Inc, 1978, pp 544–588.

74. Barbeau DL, Jonas A, Teng T-L, et al: Asymmetry of apolipoprotein A-I in solution as assessed from ultracentrifugal, viscometric, and fluorescence polarization studies. *Biochemistry* 18:362–369, 1979.

75. Tall AR, Small DM, Shipley GG, et al: Apoprotein stability and lipid protein interaction in human plasma high density lipoproteins. *Proc Natl Acad Sci USA* 72:4940–4942, 1975.

76. Tall AR, Shipley GG, Small DM: Conformational and thermodynamic properties of apo A-I from human plasma high density lipoproteins. *J Biol Chem* 251:3749–3755, 1976.

77. Tall AR, Small DM, Deckelbaum RJ, et al: Structure and thermodynamic properties of high density lipoprotein recombinants. *J Biol Chem* 252:4701–4717, 1977.

78. Privalov PL, Khechinashivili NN: A thermodynamic approach to the problem of stabilization of globular protein structure: a calorimetric study. *J Mol Biol* 86:665–684, 1974.

79. Fromisano S, Brewer HB Jr, Osborne JC Jr: Effect of pressure and ionic strength on the self-association of apo A-I from the human high density lipoprotein complex. *J Biol Chem* 253:354–360, 1978.

80. Stone WL, Reynolds JA: The self-association of the apo-Gln-I and apo-Gln-II polypeptides of human high density serum lipoproteins. *J Biol Chem* 250:8045–8048, 1975.

81. Swaney JB, O'Brien K: Cross-linking studies of the self-association properties of apo-A-I and apo-A-II from human high density lipoprotein. *J Biol Chem* 253:269–277, 7069–7077, 1977.

82. Small DM: The physico-chemical properties of cholanic acids, in Nair PP, Kritchevsky D (eds): *The Bile Acids—Chemistry, Physiology and Metabolism.* New York, Plenum Publishing Corporation, 1971, vol I, pp 247–354.

83. Small DM: Bile salts of the blood. High density lipoprotein systems and cholesterol removal, in Bianchi L, Gerok W, Sickinger K (eds): *Liver and Bile.* (Falk Symposium #23), Lancaster, England, Medical & Technical Publishing Co Ltd, 1977, pp 89–100.

84. Atkinson D, Small DM, Shipley GG: Neutron small-angle scattering of reassembled plasma lipoproteins. American Chemical Society Meeting. Boston, Mass, August 1979 (abstract).

85. Jonas A, Krajnovich DJ: Effect of cholesterol on the formation of micellar complexes between A-l apolipoproteins and L-α-dimyristoylphosphatidylcholine. *J Biol Chem* 253:5758–5763, 1978.

86. Tall AR, Lange Y: Incorporation of cholesterol into high density lipoprotein recombinants. *Biochem Biophys Res Commun* 80:206–212, 1978.

87. Sperry W: Cholesterol esterase in blood. *J Biol Chem* 111:467–478, 1935.

88. Glomset JA, Norum KR: The metabolic role of lecithin:cholesterol acyltransferase: perspectives from pathology. *Adv Lipid Res* 11:1–65, 1973.

89. Norum K, Glomset JA, Nichols AV, et al: Plasma lipoproteins in familial lecithin:cholesterol acyltransferase deficiency: effects of incubation with lecithin: cholesterol acyltransferase in vitro. *Scand J Clin Lab Invest* 35(suppl 142):31–55, 1975.

3 Receptor-Mediated Endocytosis in Animal Cells

Michael S. Brown, MD
Joseph L. Goldstein, MD

Receptor-mediated endocytosis is a coupled two-step process by which certain extracellular proteins are first bound to specific cell surface receptors and are then internalized by animal cells. This process is now recognized to have an important role in cellular growth, nutrition, and differentiation.[1,2] Biologically important molecules known to be taken up by this mechanism include plasma transport proteins (such as low density lipoproteins and transcobalamin II), certain polypeptide hormones (such as insulin and epidermal growth factor), asialoglycoproteins, and lysosomal enzymes.

Properties of Receptor-Mediated Endocytosis

Protein transport systems that utilize the mechanism of receptor-mediated endocytosis are characterized by the following four properties:

1. The binding component on the cell surface is a receptor in the strictest sense, ie, it is a molecule whose function is to recognize an endogenous ligand and thereby achieve a physiologic

effect. This criterion distinguishes these receptors from other cell surface components that serve as binding sites for foreign molecules, such as viruses, lectins, and toxins.
2. Internalization of the protein ligand is efficiently coupled to binding. Once the protein is bound to its receptor, the half-life for internalization is less than ten minutes.
3. In all cases for which ultrastructural data are available, the receptor-bound proteins enter cells through specialized regions of the surface membrane, called coated pits, that invaginate rapidly into the cell during endocytosis to form coated vesicles.
4. The internalized proteins are usually delivered to lysosomes, where they are degraded completely to amino acids; occasionally processing of the internalized protein involves delivery to cellular structures other than lysosomes.

To date, at least 13 systems of receptor-mediated endocytosis have been described. These systems, which have been reviewed in detail elsewhere,[1,2] are listed in Table 3-1. The low density lipoprotein (LDL)-receptor pathway, which functions to transport cholesterol from plasma

Table 3-1
Systems for Receptor-Mediated Endocytosis of Proteins

Protein	Internalization via Coated Pits and Vesicles	Fate of Internalized Protein
Transport Proteins		
LDL	Yes	Lysosomes
Yolk proteins	Yes	Yolk granules
Transferrin	Yes	Lysosomes
Transcobalamin II	?	Lysosomes
Protein Hormones		
Epidermal growth factor	Yes	Lysosomes
Insulin	?	Lysosomes, Golgi, nuclei
Chorionic gonadotropin	?	Lysosomes
Nerve growth factor	?	Nuclei
β-Melanotropin	?	Golgi, melanosomes
Others		
Asialoglycoproteins	?	Lysosomes
α-2-Macroglobulin	Yes	Lysosomes
Maternal immunoglobulins	Yes	Exocytic vesicles
Lysosomal enzymes	?	Lysosomes (not degraded)

Modified from Goldstein et al.[2] Reprinted with permission.

into cells, is one of the best understood of these systems. The LDL-receptor pathway illustrates each of the four characteristics of receptor-mediated endocytosis, and has served as a model with which to compare the various processes of receptor-mediated endocytosis.[1,2]

LDL-Receptor Pathway

Cholesterol is required for the normal function of the plasma membrane of all mammalian cells. Studies over the past 20 years have demonstrated that most of the cholesterol found in the membrane of nonhepatic cells is not synthesized locally, but rather is derived from cholesterol that is synthesized in the liver or absorbed from the intestine. The liver distributes endogenously synthesized or intestinally derived cholesterol to extrahepatic tissues by means of lipoproteins that are secreted into the plasma.[3] The main cholesterol-carrying lipoprotein in human plasma is LDL. LDL consists of a core of neutral lipid that is composed predominantly of cholesteryl esters. Surrounding this neutral lipid core is a polar coat composed of phosphilipid, free cholesterol, and a protein called apoprotein B. Body cells require free cholesterol for their membranes, and not cholesteryl esters. Hence, cells require a mechanism by which they can extract the cholesteryl esters from LDL and hydrolyze those esters, so as to generate free cholesterol for metabolic use.

The mechanism by which human cells take up and utilize the cholesteryl esters of LDL has been elucidated by studies of human fibroblasts in tissue culture.[3,4] Figure 3-1 shows the sequential steps in this LDL-uptake pathway as delineated biochemically and ultrastructurally in fibroblasts. When cultured human fibroblasts are in need of cholesterol, they synthesize a receptor that becomes localized to the plasma membrane. This receptor, which is believed to be a protein, specifically binds to the protein component of the lipoprotein. Once LDL binds to the receptor on the surface of these cells, the lipoprotein is internalized by adsorptive endocytosis and delivered to cellular lysosomes. Within the lysosomes, the protein component of LDL is hydrolyzed to amino acids, and the cholesteryl esters of LDL are exposed to an acid lipase that hydrolyzes the cholesteryl esters to generate free cholesterol. The free cholesterol is able to cross the lysosomal membrane to enter the cellular compartment where it is used by the cell for membrane synthesis.

Within the cell the cholesterol derived from the catabolism of LDL also regulates three metabolic events. (1) It suppresses the activity of 3-hydroxy-3-methylglutaryl CoA reductase (HMG CoA reductase), which is the rate-controlling enzyme in cholesterol biosynthesis. (2) The cholesterol activates an acyl-CoA:cholesterol acyltransferase (ACAT), which reesterifies the incoming cholesterol. The newly formed cholesteryl esters are stored within the cell as cytoplasmic lipid droplets. (3) The incoming cholesterol suppresses the synthesis of the LDL receptor. This feedback regulation of LDL receptor synthesis stops the uptake of LDL and prevents cells from becoming overloaded with cholesterol when exposed to LDL.[3,4]

The distribution of LDL receptors on the surface of fibroblasts has been studied by electron microscopy with the use of LDL coupled to the iron-containing, electron-dense protein ferritin[5-7] or by ^{125}I-LDL autoradiography.[8] Between 50% and 80% of LDL receptors in human fibroblasts are clustered over the 2% of the cell surface represented by coated pits. The membrane in these regions is indented and coated on its cytoplasmic surface by a bristle coat. LDL receptors are localized in coated pits, even when the cells are fixed with formaldehyde prior to incubation with the LDL-ferritin; hence, it is believed that receptor clustering occurs before the exposure to LDL.[6]

LDL-ferritin bound to coated pits at 4 C is rapidly internalized when the fibroblasts are warmed to 37 C.[5] In this process the coated pits invaginate to form coated endocytic vesicles. After five to ten minutes at 37 C, LDL-ferritin is seen in lysosomes as the result of their fusion with the incoming coated vesicles. The rapid sequence of events visualized in the electron micrographs precisely parallels biochemically derived data on the rapid uptake and degradation of LDL, labeled either in its protein moiety with ^{125}I or in its cholesteryl ester moiety with ^{3}H-cholesteryl linoleate.[3]

The inference that the LDL-receptor pathway is important in vivo is based on studies of human fibroblasts that show specific biochemical blocks in the receptor-mediated uptake of LDL. These naturally occurring

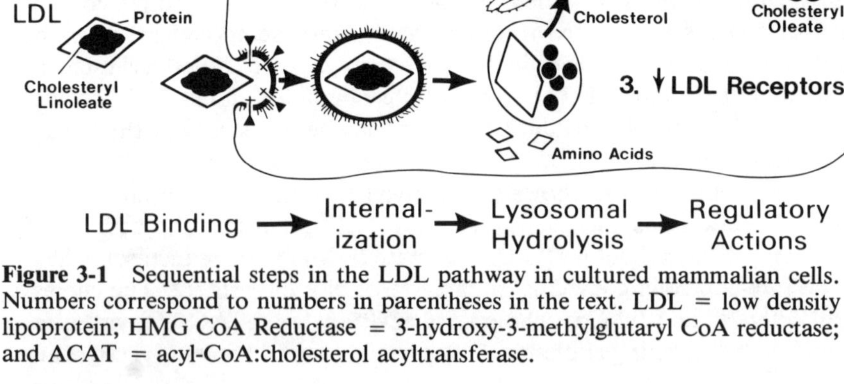

Figure 3-1 Sequential steps in the LDL pathway in cultured mammalian cells. Numbers correspond to numbers in parentheses in the text. LDL = low density lipoprotein; HMG CoA Reductase = 3-hydroxy-3-methylglutaryl CoA reductase; and ACAT = acyl-CoA:cholesterol acyltransferase.

mutant cells, which are cultured from humans with genetic disorders of lipoprotein metabolism, have been invaluable in detailing the structure and function of the LDL receptor.[3,4]

Receptor-Mediated Endocytosis of Proteins Other than LDL

Table 3-1 lists the protein ligands for which receptor-mediated endocytosis has now been demonstrated. The similarities and differences between the LDL-receptor pathway and each of these other systems have been discussed in detail in a recent review.[2] Several points deserve brief comment. First, in each of the systems listed in Table 3-1, the endocytosis is carried out by the same receptor that is believed to mediate the physiologic action of the protein. Second, for six of the systems, including the receptors for LDL, yolk proteins, transferrin, epidermal growth factor, α-2-macroglobulin, and maternal immunoglobulins (IgG), internalization has been shown by electron microscopy to occur in coated pits and coated vesicles.[5-14] Third, in those systems for which coated pits have been documented, the mechanism for receptor clustering into coated pits may vary. Some receptors (such as those for LDL and epidermal growth factor in human fibroblasts[5,6,9]) appear to spontaneously localize in coated pits as a result of a receptor-mediated mechanism, while other receptors (such as those for α-2-macroglobulin and epidermal growth factor in mouse 3T3 cells[10,14]) may require ligand binding in order to induce receptor migration to the pits. Finally, whereas the coated pit/coated vesicle may be a common portal of entry for receptor-bound proteins, the ultimate destination of the internalized proteins varies from system to system. In eight of the systems in Table 3-1, the internalized protein is delivered to lysosomes where it undergoes rapid degradation. On the other hand, other of the internalized proteins accumulate in cell structures without undergoing degradation, and some are transported intact in coated vesicles across the cell for discharge on the other side.

An important question raised by the above findings relates to the mechanism by which endocytic vesicles and their contents are directed to specific delivery sites within cells. Do proteins with different intracellular destinations enter cells in the same vesicles? If so, how are these proteins sorted and delivered to their separate destinations?

What Is the Function of Receptor-Mediated Endocytosis?

The need for a system by which receptor-bound proteins are carried into cells in a tightly controlled manner is obvious when cells deal with transport proteins such as LDL, transferrin, and transcobalamin II that must transfer their contents into the cell. In the case of protein hormones, receptor-mediated uptake and degradation may help to regulate the responses to hormones. If a hormone molecule is degraded during its action, then it, and in some cases its receptor, can act only once. Under these conditions, when a gland secretes a molecule of hormone, it sends a discrete packet of information to a target cell. The cell's response is dictated by the number of hormone molecules that bind to it per unit time, and not by the equilibrium level of binding. In fact, equilibrium binding may never be achieved ordinarily, because the hormone enters the cell and is degraded before it dissociates from its receptor. A strong target-cell response may be associated with a low plasma hormone level, owing to the rapid degradation that accompanies hormone binding. Conversely, resistance to a hormone, such as that which occurs when the number of receptors is reduced, may be associated with a high plasma hormone level owing to its slow uptake and degradation. Indeed, in clinical states of hormone resistance the circulating levels of peptide hormones do tend to be elevated.

But is degradation the only reason for receptor-mediated endocytosis of protein hormones? Does the endocytic uptake mediate any of the actions of the hormones? To date, no regulatory action of a protein hormone (with the possible exception of nerve growth factor) has been shown to require endocytosis of the protein. However, only a few studies have addressed the question of whether the long-term biologic effects of protein hormones require endocytic uptake. More systems are needed in which the entire sequence of hormone binding and hormone action can be studied by combined biochemical, morphologic, and genetic techniques in isolated cells.

REFERENCES

1. Brown MS, Goldstein JL: Receptor-mediated endocytosis: insights from the lipoprotein receptor system. *Proc Natl Acad Sci USA* (In press).

2. Goldstein JL, Anderson RGW, Brown MS: Coated pits, coated vesicles, and receptor-mediated endocytosis. *Nature* 279:679–685, 1979.

3. Goldstein JL, Brown MS: The low-density lipoprotein pathway and its relation to atherosclerosis. *Annu Rev Biochem* 46:897–930, 1977.

4. Brown MS, Goldstein JL: Receptor-mediated control of cholesterol metabolism. *Science* 191:150–154, 1976.

5. Anderson RGW, Brown MS, Goldstein JL: Role of the coated endocytic vesicle in the uptake of receptor-bound low density lipoprotein in human fibroblasts. *Cell* 10:351–364, 1977.

6. Anderson RGW, Goldstein JL, Brown MS: Localization of low density lipoprotein receptors on plasma membrane of normal human fibroblasts and their absence in cells from a familial hypercholesterolemia homozygote. *Proc Natl Acad Sci USA* 73:2434–2438, 1976.

7. Orci L, Carpentier J-L, Perrelet A, et al: Occurrence of low density lipoprotein receptors within large pits on the surface of human fibroblasts as demonstrated by freeze-etching. *Exp Cell Res* 113:1–13, 1978.

8. Carpentier J-L, Gorden P, Goldstein JL, et al: Binding and internalization

of ^{125}I-LDL in normal and mutant human fibroblasts: a quantitative autoradiographic study. *Exp Cell Res* (In press).

9. Gorden P, Carpentier J-L, Cohen S, et al: Epidermal growth factor: morphological demonstration of binding, internalization, and lysosomal association in human fibroblasts. *Proc Natl Acad Sci USA* 75:5025–5029, 1978.

10. Maxfield FR, Willingham MC, Davies PJA, et al: Amines inhibit the clustering of α-2-macroglobulin and EGF on the fibroblast cell surface. *Nature* 277:661–663, 1979.

11. Roth TF, Cutting JA, Atlas SB: Protein transport: a selective membrane mechanism. *J Supramol Struct* 4:527–548, 1976.

12. Roth TF, Porter KR: Yolk protein uptake in the oocyte of the mosquito *Aedes aegypti*. *J Cell Biol* 20:313–332, 1964.

13. Sullivan AL, Grasso JA, Weintraub LR: Micropinocytosis of transferrin by developing red cells: an electron-microscopic study utilizing ferritin-conjugated transferrin and ferritin-conjugated antibodies to transferrin. *Blood* 47:133–143, 1976.

14. Willingham MC, Maxfield FR, Pastan IH: α-2-Macroglobulin binding to the plasma membrane of cultured fibroblasts: diffuse binding followed by clustering in coated regions. *J Cell Biol* (In press).

4 Clathrin – The Major Structural Component of the Coated Vesicle System of Intracellular Transport

Barbara M.F. Pearse, PhD

Coated vesicles are found in most eukaryotic cells. They are characterized by the curious bristle coats they have on the cytoplasmic surfaces of their lipid membranes; their diameters vary between 500 and 1500 Å. Of their various functions, one of the most striking examples takes place in developing chicken oocytes, where coated vesicles can pinocytose up to 1 gm of yolk proteins per day in a single cell. Another example is the recycling of membrane after transmitter release in excised frog neuromuscular-junction preparations. These terminals can secrete hundreds of times more quanta of neurotransmitter than the number of vesicles they initially contain; yet they still look full of vesicles. Here, the replenishment of these secretory vesicles from the presynaptic membrane is mediated by coated vesicles. In fibroblasts, coated vesicles are responsible for the uptake of several macromolecules, including low density lipoprotein and epidermal growth factor, which ultimately reach lysosomes and are degraded (see Brown and Goldstein, Chapter 3). In other cells coated vesicles appear to bud from the Golgi apparatus and seem to be involved in secretion. In general, the function of coated vesicles apparently entails the transfer of membrane and selected macromolecules, between different membranous sites within the cell, and in a defined direction. However, much still has to be learned about their full role.

Biochemically, coated vesicles are characterized by the major structural protein of their coats—clathrin, a polypeptide of 180,000 daltons. They can be disrupted in vitro to give defined subunits containing several molecules of clathrin that, on suitable treatment, reassemble to form coat-like particles. Electron micrographs of tilted specimens of coated vesicles show that their coats are based on polyhedral lattices constructed from 12 pentagons (required by Euler's theorem) plus a variable number of hexagons. Among the smaller particles, three such structures have been identified, two containing 108 molecules of clathrin each and a third containing 84. The coats of larger particles are thought to be constructed on similar principles. Thus, clathrin forms flexible lattices ideally suited to participating in the formation of vesicles during their budding from membranes.

Clathrin from several different animal sources shows a high degree of conservation of amino acid sequence as shown by one-dimensional peptide mapping. Although the precise extent of this conservation must await detailed sequence studies, the picture which has emerged so far is reminiscent of actin, another "scaffold" protein. The reason that actin sequences are highly conserved is presumably because this molecule forms many associations with other proteins (eg, with actin, myosin, and tropomyosin), and this makes it difficult for its sequence to drift without the protein-protein contacts being disturbed. Perhaps the same constraints apply to clathrin, since it is evident that here, too, several interactions between a clathrin molecule and other proteins must exist. During a cycle of pinocytosis, this would include: 1) the binding of the receptor (protein) for the pinocytosed molecule by clathrin; 2) assembly of clathrin subunits to form a flexible cage; and 3) the association of clathrin with other unknown proteins that may be involved in the transport of the coated vesicle to another membrane, and its specific fusion with that membrane. Since the clathrin of coated vesicles seems to operate in a wide variety of contexts in different cells, where different receptors are involved and the vesicles shunted between different membranes, the case for associations of clathrin with many other proteins would seem to be strong.

Purified coated vesicles (eg, from chick oocytes) seem to lack specific contents and receptor molecules, and therefore represent only a restricted form of the intact cellular system of coated membranes. The quantity of clathrin isolated from lymphoma cells (E.L.4 cell line) suggests that about 1% of these cells' lipid could be in coated vesicles at any one moment. If coated vesicles have a half-life of only minutes, as estimated in fibroblasts, they could therefore be responsible for a considerable transfer of membrane components within cells. Future biochemical studies on this interesting cell organelle should be rewarding.

SUGGESTED READINGS

Crowther RA, Finch JT, Pearse BMF: On the structure of coated vesicles. *J Mol Biol* 103:785–798, 1976.

Pearse BMF: Coated vesicles from pig brain: purification and biochemical characterization. *J Mol Biol* 97:93–98, 1975.

Pearse BMF: Clathrin: a unique protein associated with intracellular transfer of membrane by coated vesicles. *Proc Natl Acad Sci USA* 73:1255–1259, 1976.

Pearse BMF: On the structural and functional components of coated vesicles. *J Mol Biol* 126:803–812, 1978.

5 Changes in Cell Membrane Fluidity and Cell Function Induced by Abnormalities of Plasma Lipoproteins

Richard A. Cooper, MD

Cholesterol and phospholipid are the two major lipids of cell surface membranes. Cholesterol is insoluble in water but is solubilized by phospholipids both in membranes and in plasma lipoproteins. Moreover, cholesterol is exchanged between membranes and lipoproteins. An equilibrium partition is established based on the amount of cholesterol relative to phospholipid (C/P) in these two compartments.

Increases in the C/P of red cell membranes have been studied under three conditions: 1) spontaneous increases in vivo have been observed in the spur red cells of patients with severe liver disease; 2) similar red cell changes in vivo have been induced by the administration of cholesterol-enriched diets to rodents and dogs; and 3) increases in membrane cholesterol have been induced in vitro by enriching the C/P of the lipoprotein environment with cholesterol-phospholipid dispersions (liposomes) having a C/P of > 1.0. In each case, there is a close relationship between the C/P of the plasma environment and the C/P of the red cell membrane. In vivo, the C/P molar ratio of red cell membranes ranges from a normal value of 0.9 to 1.0, to values which approach but do not reach 2.0. In vitro, this ratio approaches 3.0.

Changes in red cell membrane cholesterol content are associated with a change in membrane surface area of 0.22% for each 1% change in cholesterol. This is true throughout the range of membrane C/P from 0.5 to 2.0.

The cholesterol content of red cell membranes also influences membrane lipid fluidity, as assessed by the anisotropy of hydrophobic fluorescent probes such as diphenyl hexatriene (DPH). A close correlation exists between increases in red cell membrane C/P and decreases in membrane fluidity over the range of membrane C/P from 0.5 to 2.0; however, little further change in fluidity occurs when membrane C/P is increased to 2.0 to 3.0. This suggests that the cone angle in which DPH is free to move is maximally decreased at a membrane C/P of 2.0.

Cholesterol enrichment of red cell membranes is associated with the transformation of cell contour to one which is redundant and folded, and this is associated with a decrease in red cell filterability in vitro. Circulation in vivo in the presence of the spleen further modifies cell shape to a spiny, irregular (spur) form, and the survival of cholesterol-rich red cells is decreased in the presence of the spleen. Although active Na-K transport is not influenced by cholesterol enrichment of human red cells, several carrier-mediated transport pathways are inhibited. We have demonstrated this effect for the co-transport of Na + K and similar results have been obtained by others in studies of organic acid transport and the transport of small neutral molecules such as erythritol and glycerol.

Exposure of human platelets to an environment with a high C/P also leads to membrane enrichment. Brief exposure (four to five hours) to liposomes with a C/P of 2.0 selectively enriches the surface membrane system without affecting granule membranes. Platelets so modified aggregate in response to concentrations of epinephrine or adenosine diphosphate one to two orders of magnitude lower than required to aggregate normal platelets. Insel et al (see Suggested Readings) have shown that this is not due to a change in the number of surface receptors, thereby implying that cholesterol exerts its effect at some step distal to that of agonist binding.

Nucleated cells are also affected by the C/P of their environment. Thus, for example, lymphoblastoid cells growing as ascites tumors in cholesterol-fed mice become enriched with cholesterol. Similarly, the surface membrane of mouse splenic lymphocytes develops an increased C/P and a decreased fluidity when these cells are incubated with cholesterol-rich liposomes for 24 hours. These lymphocytes display a response to lectin mitogens, such as concanavalin A or phytohemagglutinin, which is greater in magnitude and earlier in time than is seen in normal lymphocytes following mitogenic stimulation.

Thus, the membrane C/P of many cells is sensitive to the C/P of the plasma environment. Changes in membrane C/P affect membrane fluidity and membrane surface area, and these changes are associated with altered membrane permeability, distorted cell shape, and abnormal cell function.

SUGGESTED READINGS

Cooper RA: Abnormalities of cell membrane fluidity in the pathogenesis of disease. *N Engl J Med* 297:371–377, 1977.

Cooper RA, Leslie MH, Fischkoff S, et al: Factors influencing the lipid composition and fluidity of red cell membranes in vitro: production of red cells possessing more than two cholesterols per phospholipid. *Biochemistry* 17:327–331, 1978.

Insel P, Nirenberg P, Turnbull J, et al: Relationships between membrane cholesterol, α-adrenergic receptors, and platelet function. *Biochemistry* 17:5269–5274, 1978.

6 Mechanisms of Capping Surface Antigens

Mark S. Bretscher, PhD

While the mechanism whereby cross-linked surface antigens cap is unknown, several hypotheses have been advanced which could provide a molecular mechanism for the process. Broadly speaking, these mechanisms can be grouped into three classes: those which invoke 1) the cytoskeletal apparatus as a "mini-muscle," which pulls the patches to the pole of the cell where the cap forms; 2) the continuous flow of membrane from the cell periphery to a site of resorption where the cap forms; or 3) a flow system superficially similar to (2), but invoking only the lipid molecules in the flow.

My purpose is to examine the advantages and disadvantages of these schemes. Much of the experimental evidence to which I shall refer has been collected in an admirable review by de Petris.[1] The process of capping is characterized by the following properties[2]:

1. Antigens must be cross-linked in order to cap. A cross-linking antibody that does not lead to extensive cross-linking does not lead to capping, whereas addition of a second-layer cross-linking anti-antibody does lead to capping. This

feature of the process is often observed – a common example would be the capping of histocompatibility antigens.

2. The process requires metabolic energy and warmth (ie, it is sensitive to DNP, N_3^-, and cold).
3. The process can be very fast: on lymphocytes, the first caps of cross-linked surface immunoglobulin are seen one to two minutes after the cells are warmed up to 37 C.
4. Most important, the process is polar in the sense that the cap forms on the cell surface in a region that bears a specific relationship to the cell interior. Thus, motile fibroblasts on a substratum form caps over the perinuclear region, and B lymphocytes form Ig caps on that end of the cell remote from the nucleus. It would be interesting to understand the mechanism of capping, so that we might shed light on intracellular processes that are, at present, otherwise inaccessible.
5. Effect of drugs. Colchicine, which is believed to lead to the disruption of microtubules, has no very obvious effect on capping. Cytochalasin B, which is believed to lead to the disruption of microfilaments, does inhibit capping, but very feebly. The surprising thing is that, when both drugs are added together, capping of surface immunoglobulin on B lymphocytes is abolished. This observation has, I believe, stimulated many to look for evidence of the involvement of microtubules and microfilaments in the process.

Cytoskeletal Mechanisms

The literature is full of evidence that implicates the filamentous molecules, actin fibres and microtubules, as providing the motive forces for capping. By and large that evidence is necessarily indirect; I do not wish to review it here because I suspect it may be misleading. Why?

First, I cannot envisage a simple molecular mechanism whereby the cytoskeletal apparatus inside the cell can recognize that an antigen has been cross-linked on the outside of the cell. The fact that alloantisera against histocompatibility antigens (which presumably yield H-2 dimers in the membranes) do not cap, yet on further cross-linking by a second layer do cap, indicates to me that here capping is the direct result of an extensive cross-linkage. This degree of cooperativity might not be so hard to explain if the cross-linked antigen complex made a sort of crystalline lattice on the cytoplasmic side of the membrane in which a large number of weak interactions with the cytoskeleton could add up to make a strong one. But there is no reason to believe that patches have a crystalline nature.

Second, the fact that apparently any antigen caps when properly cross-linked makes it hard to see how a specific molecular recognition could be the guiding force, unless every cappable membrane protein has a specific region in its cytoplasmic domain that is recognized by a component of the cytoskeletal system.

Third, there is the case of those molecules that have no cytoplasmic portion, ie, antigenic lipids. There are several cases now in which glycolipids or modified phospholipids cap, when cross-linked by two layers of antibody.[3,4] We have recently shown that Forssman glycolipid, when introduced into a variety of Forssman-negative cells, will cap.[5] This capping behaves just as described above with respect to the drugs dinitrophenol, cytochalasin, and colchicine. Although other interpretations are possible, the simplest interpretation of such experiments is that cross-linked lipid molecules cap. If this is so, it is hard to see how this could be directly effected by the cytoskeleton (as lipid molecules in the outer monolayer would be inaccessible to cytoplasmic components).

Finally, a point which is not often appreciated. Different antigens cap at quite different rates. If a mini-muscle "sliding-filament" mechanism were pulling the patch to the pole of the cell, I would have expected all antigens to cap at the same rate. They do not seem to do so.

It is for these reasons that I feel one should keep an open mind on the mechanism of capping.

Membrane Flow

The idea of a polar membrane flow came from the elegant experiments of Abercrombie and his colleagues, when they observed that carbon particles move rearward over the dorsal surface of chick heart fibroblasts.[6] This type of experiment showed that the particles move more or less directly away from the ruffling membrane at the leading edge of the cell, migrating toward that area of the cell surface above the perinuclear region. They do so at a roughly uniform rate. Abercrombie and his colleagues suggested that membrane is internalized in the region of the cell where the carbon particles accumulate, and is reinserted into the plasma membrane at the cell's leading edge. This cycle of membrane flow—from leading edge, along the surface of the cell (in which flow the carbon particles are caught), to the perinuclear region, then (in a vesicular form) into the cell, and through the cell cytoplasm back to the leading edge—requires that at any moment in time there will be a (small) pool of plasma membrane in transit inside the cell. Vasiliev and his colleagues, in a series of ingenious experiements, looked for this pool and could not detect it.[7] And I think they could have detected quite a small pool. On the face of it, these experiments deny the existence of membrane flow.

Lipid Flow

This is an idea[8] based in part on Abercrombie's proposal of membrane flow. But it also takes into account the fact that if there were a flow, the flow cycle would probably include some intracellular organelles (endoplasmic reticulum? lysosomes? Golgi apparatus?) and would also be concerned with the transport of newly synthesized proteins to the plasma membrane—a sort of conveyor belt. If that were so, it might be more convenient for the cell if the flow were restricted to the lipid matrix of the plasma membrane, in order to keep the protein specificity of the different membrane compartments separate.

In addition, the scheme takes into account the lateral diffusion of proteins in the flow. The essence of the idea, then, is as follows: Lipid vesicles are added to the ruffling membrane at the leading front of a fibroblast, and lipid is resorbed (presumably in a vesicular form) toward the rear of the cell. The endocytosed lipid is transferred back to the leading front via the cytoplasm and perhaps comes into contact with various organelles en route. There is thus a polar bulk-flow of lipid from front to back on the surfaces of a fibroblast. What happens to the plasma membrane proteins? They will tend to be swept along by the flow, but will tend to randomize by diffusion. Which wins?

This can be calculated in a simple case (Figure 6-1). If we consider a protein (having a diffusion coefficient of D) sitting in a sheet of membrane in which the lipid is flowing (at a rate of F), we can calculate the ratio of its concentration C_A/C_B at two points A and B separated by a distance L (along the flow direction). This turns out to be $C_A/C_B = e^{-FL/D}$. Substituting numbers we might take $L = 10\ \mu$ (the "length" of a fibroblast over which carbon particles travel), $F = 3\ \mu/\text{min}$ (the rate of movement of particles seen by Abercrombie et al[6]), and $D = 5 \times 10^{-9}\ \text{cm}^2/\text{sec}$ (the diffusion coefficient of rhodopsin measured by Poo and Cone[9]). Putting these numbers in the equation we get $C_A/C_B = 1/3$; that is, if these numbers were applicable to a fibroblast, then a protein sitting in the flow might not have a random distribution, but might be swept a bit by the flow such that the concentration of the protein at the fibroblast leading front might be only one third of that found over the perinuclear region. More importantly (since the numbers used above are rather approximate), what the equation shows is that C_A/C_B varies in an exponential fashion with respect to D. Thus, if proteins are cross-linked to form a patch, D would decrease and C_A/C_B would lower dramatically, such that the patch would be seen to cap.

The advantages of this scheme are that it is a simple mechanism, and that it would apply to all mobile molecules in the plasma membrane when they are cross-linked. It implies that the functions of microfilaments and microtubules in the process would be indirect (which is what the effects of drugs suggest). In addition, it would provide a simple explanation for why different antigens cap at different rates—the rate would simply depend on how easily a patch of a particular antigen could diffuse against the flow.

At this point I should like to make it clear that I do not envisage the flow as restricted solely to lipid: the endocytotic vesicles will contain some selected protein components in order to help form the vesicle. But for the sake of clarity, it is useful to think of it as a pure lipid flow.

How could one look for such a flow? I would argue that the observation that cross-linked lipid molecules cap demonstrates it. But this is not generally accepted. In order to show the difficulty of demonstrating such a slow flow, let me explain why it is that an experiment, often suggested to me, would not work. The suggestion is: if you could insert (how, I am not sure) a vesicle of fluorescent lipid into the membrane at the leading front of a fibroblast, couldn't you watch it move backwards in the flow? The reason why such an experiment would be unhelpful is because of the high rate of lipid diffusion. In one minute, when the flow would have moved the area of labeled lipid back 3 μ (in the absence of diffusion), diffusion would have redistributed the fluorescent lipid over all of the 30-μ-long cell. So nothing would be seen except a fluorescent cell.

Together with Whytock (unpublished data), I have looked for the effects of such a flow in two ways, both of them depending on the possibly uneven distribution of proteins on a mobile fibroblast. The two experiments were: 1) to fix moving cells in glutaraldehyde at 37 C, decorate them with ferritin-conjugated concanavalin A, embed them and cut sections along the length of the cell, and look for a gradient of ferritin and thus concanavalin A receptors; or 2) to freeze motile fibroblasts (growing on a gold block) without fixation, fracture them, and look for a gradient of particles in the fracture plane.

In neither case did we see any reproducible or convincing gradients, either of ferritin or of particles in the fracture plane. This means that either lipid flow is wrong, or that the numerical basis for looking for the particle gradient is wrong. For example, the sought gradient depends critically on D, and if D were twice as large as the number taken (say 10^{-8} cm^2/sec rather than 5×10^{-9} cm^2/sec), we would not expect to see any gradients.

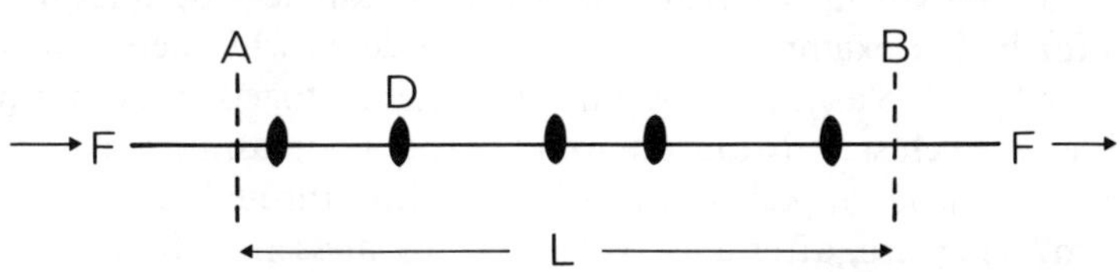

Figure 6-1 Lipid flows across a sheet of membrane in which proteins (●), having a diffusion coefficient of D, are embedded. At A and B there are molecular filters which prevent the protein from diffusing beyond those points, but which allow lipid to flow through them. The flow rate is F, and A and B are separated by a distance L. It then follows that the concentration of the protein at A (C_A) or B (C_B) is given by $C_A/C_B = e^{-FL/D}$.

My prejudice is that something along the lines of lipid flow will eventually explain capping. There are a few points which I believe are worth considering at this point.

First, there is already evidence that newly synthesized proteins first appear at the leading edge of a fibroblast, albeit on a nonmotile polynucleated giant HeLa cell. This was shown in a lovely experiment by Marcus,[10] in which the hemagglutinin of Newcastle disease virus (detected by its ability to bind erythrocytes) first appeared at the cell periphery of those cells infected with NDV.

Second, where, in a fibroblast, would the lipid flow into the cell? This point is made more acute by reconsidering the carbon particle experiments of Abercrombie and his colleagues.[6] They usually studied triangular fibroblasts. If we presume that the leading edge is the site of lipid insertion, and (for sake of argument) the opposite apex (tail of the cell) is the site of lipid resorption, then, because membrane must be conserved, the rate of lipid flow would increase as we get closer to the tail of the cell. In fact, the carbon particles move roughly at a uniform rate until they reach the region of the cell surface over the perinuclear region. This form of movement suggests that the site of resorption would be over the whole surface of the cell. That is, insertion of lipid at the leading front, coupled with its uniform retrieval over the cell's surface, would provide a decreasing rearward bulk flow of lipid on a triangular cell. The fact that extensive endocytosis occurs in what are usually regarded as nonpinocytotic fibroblastic cells (see Brown and Goldstein, Chapter 3, and Helenius, Chapter 10) encourages me in my views. This uptake occurs uniformly over the cell surface and is mediated by coated vesicles.

Finally, even if it should turn out that lipid flow does not explain the phenomenon of capping, the ideas outlined above will, I believe, often apply in cell biology. Cells are continually adding and removing membrane from one compartment to another, eg, from endoplasmic reticulum to Golgi. This transfer must be selective, as otherwise the membranes would all be the same. Of course, we do not know how large the flow rates are, nor to what extent they have a polar nature. But the chances are that they will have polar components to them, and then one can develop diffusion rates and so forth. For example, I have often wondered why there is a smooth and a rough endoplasmic reticulum, often joined together: ribosomes are sometimes seen clustered at just one end of the endoplasmic reticulum. Can it be that they have capped because of a lipid flow through the endoplasmic reticulum? They are, after all, cross-linked by messenger RNAs.

REFERENCES

1. de Petris S: Distribution and mobility of plasma membrane components on lymphocytes, in Nicolson GL, Poste G (eds): *Cell Surface Reviews,* vol 3. Amsterdam, North Holland, 1977, pp 643–728.

2. Taylor RB, Duffus PH, Raff MC, et al: Redistribution and pinocytosis of lymphocyte surface immunoglobulin molecules induced by anti-immunoglobulin antibody. *Nature* 233:225–229, 1971.

3. Craig SW, Cuatrecasas P: Mobility of cholera toxin receptors on rat lymphocyte membranes. *Proc Natl Acad Sci USA* 72:3844–3848, 1975.

4. Revesz T, Greaves M: Ligand-induced redistribution of lymphocyte membrane ganglioside GM1. *Nature* 257:103–106, 1975.

5. Stern PL, Bretscher MS: Capping of exogenous Forssman glycolipid on cells. *J Cell Biol* 82:829–833, 1979.

6. Abercrombie M, Heaysman JEM, Pegrum SM: The locomotion of fibroblasts in culture. III. *Exp Cell Res* 62:389–398, 1970.

7. Vasiliev JM, Gelfand IM, Domnina LV, et al: Active cell edge and movements of concanavalin A receptors of the surface of epithelial and fibrolastic cells. *Proc Natl Acad Sci USA* 73:4085, 1976.

8. Bretscher MS: Directed lipid flow in cell membranes. *Nature* 260:21–23, 1976.

9. Poo M, Cone RA: Lateral diffusion of rhodopsin in the photoreceptor membrane. *Nature* 247:438–441, 1974.

10. Marcus PI: Dynamics of surface modification in myxovirus-infected cells. *Cold Spring Harbor Symp Quant Biol* 27:351–365, 1962.

SECTION II
Membranes, Toxins, Viruses, and Parasites

7 Cholera and Related Enterotoxins
Tools for Membranologists

Richard A. Finkelstein, PhD

Cholera may be defined as a metabolic disturbance of the epithelial cells of the small bowel of man, caused by a potent, specialized form of exotoxin, the cholera enterotoxin, also known as choleragen, which results in a copious life-threatening secretory diarrhea. Cholera and the cholera enterotoxin serve as prototypes for an increasingly recognized variety of diarrheal diseases collectively known as enterotoxic enteropathies, of which diarrhea due to enterotoxigenic strains of *Escherichia coli* may be regarded as the most important example. (For a summary of background information and additional references, the reader is referred to recent reviews.[1-3]) Although this is presumably its primary function in nature, because of its specific biologic effect—the activation of adenylate cyclase—its broad

Studies in the author's laboratory cited herein were supported in part by US Public Health Service grants AI-08877 and (formerly) AI-10066, under the US-Japan Cooperative Medical Science Program, from the National Institute of Allergy and Infectious Diseases.

spectrum of responsive cells, its propensity to interact with a specific glycolipid, and its availability as a pure well-characterized protein, choleragen has also become useful as a research tool, or reagent, to those interested in membrane function and other biologic phenomena.

The pathway to the recognition that cholera is an enterotoxin-related disease was relatively straightforward, although like the action of the cholera toxin itself, it became evident only after a prolonged latent period—in this case, it was about 80 years. Robert Koch, who first isolated the causative agent of cholera, which he called "the Kommabacillus," postulated in the 1880s, but did not prove, that cholera was a toxicosis. Recently, our learned colleague, W.E. van Heyningen at Oxford, reflecting on "Why Was the Discovery of Cholera Toxin So Long Delayed?" (in preparation), criticized Koch on the grounds that Koch implied that the toxin may work systemically rather than locally, and that this misled subsequent investigators. I think rather that the blame, if any, should be put on the subsequent investigators for their failure to recognize that the test material should be administered by a relevant route, and should reproduce relevant manifestations in an appropriate experimental animal model; and in fact, it is surprising, indeed remarkable, how this simple concept, which was introduced abortively by some investigators around the time of Koch and after, was never exploited to the point of yielding useful or memorable results. Instead, many investigators focused on numerous toxicities associated with *Vibrio cholerae* or its culture fluids, which were capable of causing death, usually without diarrhea, in various experimental animals, and which were primarily associated with the lipopolysaccharide endotoxin of the cell wall of the vibrios.

In the 1950s, two groups in India independently revived models which had been introduced by earlier investigators, and De and Dutta and their associates were able to show in 1959 that cell-free products of cholera vibrios could elicit outpouring of fluid into the gut of rabbits, following enteral challenges. De used culture filtrates and found that occasionally, but not regularly, he could produce positive enterosorption into ligated ileal loops. Dutta, a pharmacologist, on the other hand used lysed vibrios and found that if he gave multiple doses, equivalent to the cytosol of approximately 10^{12} vibrios, he could induce diarrhea in suckling rabbits.

Having seen cholera patients in the Philippines in 1961, as had Koch, I too became convinced that cholera vibrios had to elaborate a toxin and, adopting Dutta's model, I found that experimental cholera could be produced reliably and reproducibly in infant rabbits given single peroral doses of dilutions of cholera culture filtrates grown under defined conditions. We next set about isolating the responsible factor, to which we gave the laboratory epithet "choleragen"[4] because at the outset we did not dare to presume that this was, in fact, *the* cholera enterotoxin. Relevance to cholera in humans was established on March 5, 1964, when a Thai physician-researcher, Benyajati,[5] found that such material evoked a 20-liter diarrhea over the course of two days in a volunteer who received material administered via a catheter in the small bowel. No other manifestations were produced.

The material was isolated to homogeneity by us, as reported in 1969.[6] As the available bioassays were extremely tedious and involved gastric lavage of individual infant rabbits for the purification, we gambled on the possibility that a single precipitin band, which remained in an antiserum raised against culture supernatants following adsorption with the live cholera vibrios, represented the specific toxin-antitoxin precipitate. This hypothesis was fortunately almost correct. On gel filtration, two peaks of antigen activity were detected, of which the first, the choleragen, had biologic activity in the infant rabbit model, and the second had essentially no biologic activity, but appeared to be antigenically identical to choleragen. We gave this material the name "choleragenoid."[6]

Subsequent events established that choleragenoid was, in fact, the B or binding region of the intact holotoxin, and that it consists of noncovalently associated 11,000-dalton B subunits, of which the number is believed to be 5 (range 4 to 6).[7,8] Choleragenoid, whose molecular weight is approximately 56,000 daltons, differs from choleragen by lacking the 28,000-dalton A region which is responsible for the biologic effects of the toxin. Thus, the molecular weight of the intact toxin is 84,000 daltons. Interestingly, on gels containing agarose, both proteins, choleragen and choleragenoid, comigrate and are retarded in their elutions. This observation became important later on. The A and B regions can be separated from each other by gel filtration under mild dissociating conditions and they can be reassembled to restore the biologic activity of the holotoxin.[9]

The A region is composed of two peptides, A_1 at 21,000 daltons and A_2 at 7000 daltons, joined by a disulfide bond at residue 5 of the A_2 peptide. The B chains also contain one internal disulfide loop each.[8] The complete sequence of the B chains, formula weight 11,500, has been established independently by Kurosky et al[10] and by Lai.[11] Interestingly, there are similarities in amino acid sequence between the B chains of choleragen and the β chains of certain mammalian hormones.[12] The A region has not yet been sequenced. Studies by Gill,[13] using a pigeon erythrocyte lysate assay, have established conclusively that it is the A_1 peptide that is responsible for activity of the toxin. No function has been determined for the A_2 peptide, and it is therefore presumed to provide the linkage between the B subunits and the A region. All cholera toxin isolated by routine procedures from culture supernatants is nicked, presumably by bacterial proteases, in the A region; however, if precautions are taken to avoid proteolysis, the A portion can be obtained intact, and this then requires proteolytic processing for full activity (D.M. Gill, personal communication).

Choleragen and choleragenoid have been crystallized.[14] The structure of choleragen is presently being attacked by x-ray crystallography.[15]

Studies by van Heyningen at Oxford,[16] subsequently confirmed and extended by others including Holmgren et al[17] and Bennett and Cuatrecasas,[18] have established that the receptor on host-cell membranes, with which the B or choleragenoid region interacts to allow penetration of the A_1 peptide, is a glycolipid, the G_{M1} ganglioside. The ubiquity of this ganglioside in mammalian cell membranes is probably responsible for the broad spectrum of biologic activity of the cholera toxin experimentally in cells with which the toxin never comes into contact in nature. Unresponsive cells have been rendered responsive by treatment with G_{M1} ganglioside.

The cholera toxin has thus become a valuable reagent for those studying cyclic AMP (cAMP)-mediated reactions. The list of such observations (summarized by Finkelstein,[3] and Bennett and Cuatrecasas[18]) is far too long to be covered adequately here.

The initial observations which led to our present understanding of the mode of action of choleragen were those of Field[19] and Vaughan et al,[20] who showed, respectively, 1) that agents that elevated cAMP levels in the small bowel stimulated secretion similar to that induced by choleragen; and 2) that the purified protein stimulated lipolysis in a cAMP-dependent, hormonally responsive, isolated cell system: rat epididymal fat cells.

More recent studies[13,21-24] have shown that the A region of the toxin acts in a manner somewhat similar to an "older" toxin, diphtheria toxin, in catalyzing the ADP-ribosylation of a target protein. In the case of diphtheria toxin, that protein is elongation factor 2, and its ADP-ribosylation results in the cessation of protein synthesis.[25] In the case of cholera toxin, the target is now believed to be the GTP-binding protein of adenylate cyclase, and the effect of its ADP-ribosylation is to lock adenylate cyclase in the "on" position. Results of Moss and Vaughan[26] suggest that arginine residues may be the actual substrate. While most studies have dealt with effects of the holotoxin or of its A subunit, recent studies by Moss et al[27] and Tosteson and Tosteson[28] have suggested that the B or choleragenoid region may not be entirely inert, in that evidence has been presented that both choleragen and choleragenoid can effect channel formation in liposomal model membranes containing G_{M1}.

In 1968/69, when we first discovered choleragenoid,[6] we predicted that if a *V cholerae* mutant could be isolated that produced only the choleragenoid region of the toxin, this mutant might be useful as an attenuated or avirulent living vaccine against cholera.

It should be mentioned at this point that vaccines composed of killed cholera vibrios inoculated parenterally have been in use as prophylaxis against cholera since the turn of the century.[2,29] However, it was not until the 1960s and early 1970s that scientifically controlled field studies established that conventional cholera vaccines offer only a limited degree of protection against cholera, and then only among the older populations of heavily endemic regions where cholera is primarily a pediatric disease. With the recognition that cholera was a toxin-mediated disease, it was considered that, as with diphtheria and tetanus, immunization with a toxoid might be protective. A toxoid, consisting of the cholera toxin detoxified with glutaraldehyde and administered parenterally, was evaluated and also found not to offer significant protection against cholera. The reason may well be that because cholera is a local disease restricted to the lumen and the surface of the intestine, the largely humoral antibody response generated by such a procedure cannot be expected to be effective. That immunity can occur was demonstrated conclusively by volunteer studies at the University of Maryland. Volunteers recovered from induced cholera were found to be solidly resistant to subsequent challenge with cholera vibrios.[30] In our efforts to obtain the desired choleragenoid-producing mutant, which could now be regarded as a A^-B^+ mutant, we isolated a hypotoxinogenic mutant, M13,[31] which was shown to be avirulent in laboratory animals and in man. This mutant, administered perorally to volunteers, was found to stimulate immunity to challenge with virulent cholera vibrios.[32] However, the immunity was not quite as solid as that generated by the disease itself. In addition, as there was some suggestive evidence that the mutant could revert to toxinogenicity, and as it could not effectively stimulate antitoxic immunity, M13 was not suitable for further evaluation.

We have recently isolated an A^-B^+ mutant, designated "Texas Star," and have shown that this mutant stimulates immunity against cholera in experimental animal models.[33] It is hoped that studies in humans, to be conducted soon, will, support our hypothesis that this live avirulent vaccine strain will produce solid immunity resembling that produced by the disease itself. In addition, following the recognition that cholera was a toxin-mediated disease, a number of investigators were stimulated to examine the possibility that other enteric bacteria, previously regarded as components of the normal flora, might also be able to cause diarrheal disease by means of toxins resembling the cholera enterotoxin. The result of these studies has been an expanding list of bacteria associated with diarrheal diseases, now known as enterotoxic enteropathies, which are actually far more important than cholera as causes of morbidity and mortality. In many instances, the enterotoxins have been shown to be immunologically related and functionally similar (in activating adenylate cyclase) to the cholera enterotoxin. The most prominent among these is diarrhea due to strains of *E coli,* which have become infected with a plasmid, an extrachromosomal genetic element that confers the ability to elaborate an enterotoxin or enterotoxins.

Thus, it is also hoped that by virtue of its ability to elaborate the cross-reactive cholera B antigen, this mutant may also stimulate some degree of immunity against the immunologically related enterotoxic enteropathies.

Even though it has been recognized for ten years that *E coli* can produce a cholera-like enterotoxin, and despite the availability of new methodology, the *E coli* heat-labeled enterotoxin (LT) has resisted purification and characterization in numerous laboratories around the world,[34] including our own, until now. The rationale has invariably been to fractionate crude preparations containing enterotoxic activity and to evaluate the biologic activity of the fractions, using one or more of the bioassays which have been applied with the cholera enterotoxin. These include whole animal models such as infant mice, isolated intestinal loops, and skin tests in rabbits, as well as assays employing cultured responsive cell lines such as Y-1 adrenal cells or Chinese hamster ovary cells. In each instance, the assays lacked sufficient precision to evaluate recovery of biologic activity. In almost every instance, separation was attempted on columns containing agarose, in part because this menstrum had been used successfully with choleragen.

In each instance, it was recognized that *E coli* elaborates substantially less toxin than cholera vibrios, and in each instance the biologic activity of isolated products was between one hundred- and one million-fold less than that of purified choleragen, in whatever assay was being used. The *coli* enterotoxins described in numerous preliminary reports ranged from 10,000 to greater than 1,000,000 daltons in molecular size.

After nearly ten years of effort in our own laboratory, we finally discarded bioassays as a means of recognizing active fractions and relied instead on our observations that sufficiently concentrated preparations of crude *E coli* toxin precipitated with antisera raised against cholera toxin. In fact, antisera against both the cholera A and B regions were reactive with *coli* enterotoxin.[35,36] This enabled Clements and myself to recognize that antigenic material, which had been applied to columns containing agarose, was no longer present in the eluted fractions, even when they were recombined and concentrated to the original volume. Either the antigen had been destroyed or it was still on the column. As agarose is a polymer of galactose—a component of G_{M1} ganglioside—we treated the columns with galactose, and the result was immediate success. A single-step procedure resulted in homogeneous *E coli* enterotoxin from three sources: cell-free supernatants, extracts of the bacterial cells, and lysates of the bacterial cells.

The resultant pure enterotoxin resembles choleragen in its subunit composition, although there are some differences.[37] In contrast to choleragen, the biologic activity of *coli* preparations (dependent on their source) benefits from mild proteolysis with trypsin, and with appropriate trypsin treatment, the activity approaches that of choleragen. In intestinal models such as the rabbit ileal loop, sufficient protease activity is present in the succus entericus to accomplish activation of the toxin.

The amino acid composition of LT resembles that of choleragen, although there are significant differences between the two toxins, as well as among the *E coli* preparations, depending on their source. In contrast to the cholera vibrios, or rather to the strain 569B widely used for choleragen production, the major source of *E coli* enterotoxin is the bacteria themselves.

However, that material from the cytosol of the bacteria themselves is the material that most benefits from proteolytic treatment, and this fact has eluded those investigators who have found the cytosol to be a relatively poor source of biologically active *E coli* toxin.

As with choleragen, *E coli* A and *E coli* B can be separated from each other by gel filtration under dissociating conditions. Partial amino acid sequence analysis reveals that of the first 20 amino acids in the B chain of both toxins, they differ by five one-base substitutions.[38] We have not been able to begin the sequence of the *coli* A chains.

We have earlier postulated that cholera enterotoxin may confer survival advantage on cholera vibrios, for which man is the only known natural host, and which survive poorly in the colon and in the external environment; thus, cholera enterotoxin, by causing diarrhea, can increase to 10^{12} per day the number of vibrios disseminated into the environment, and hence their likelihood of finding another human host. *E coli* does not suffer similar disadvantages, and therefore the acquisition of the capability of producing LT may be an accidental event from which *E coli* derives little advantage and for which it is not well equipped. Enterotoxigenic *E coli* may therefore be relatively recent and relatively inept pathogens. Based on this reasoning, they may have received the *tox* gene from *V cholerae,* rather than vice versa.

There is one other aspect that I would like to touch upon, which may be of interest to membranologists: that is, the mechanism by which the cholera vibrios adhere to the epithelial cells of the small bowel, which they must do in order to avert the otherwise highly effective host intestinal clearance mechanisms.[39] We have isolated from cholera culture supernatants a protein lectin-like factor of about 67,000 daltons which inhibits attachment of vibrios to tissue in vitro and in vivo, and we are interested in further delineation of the mechanism or mechanisms involved.[40]

Prospects for the Future

Waiting in the wings are enterotoxins from such diverse genera as *Salmonella; Shigella; Klebsiella-enterobacter; Pseudomonas; Aeromonas;* non-agglutinable vibrios (a heterogeneous group which some claim are related taxonomically to the true cholera vibrios); *Vibrio parahaemolyticus,* a halophilic marine vibrio associated with enteritis acquired by eating uncooked seafood; and others. For most of these, evidence has been presented suggesting an immunologic and/or functional relationship with the cholera enterotoxin. Either the products have been shown to cause similar reactions in the same model systems, or they have been shown to be neutralizable by cholera antisera and, in a smaller number of instances, both phenomena have been reported.[34]

Other classes of enterotoxins which are unrelated to the cholera enterotoxin are also waiting patiently to be discovered and characterized. Some are, in contrast to choleragen, small molecular weight, heat-stable factors, resembling the so-called ST or stable toxin from *E coli,* which appear to act (again in contrast to the cholera family of toxins) by activating host-cell guanylate cyclase immediately and transiently.[41]

We have thus far considered only enterotoxins, or enterotoxin-like activities, from gram-negative bacteria. The gram-positives, although generally less studied in this regard, may also be sources of enterotoxic activities. *Perfringens* enterotoxin is one such example, which has been isolated from an anaerobic spore-forming bacillus, *Clostridium perfringens,* associated with food poisoning. This enterotoxin is apparently distinct from the cholera family. Additionally, strains of *Bacillus cereus,* a gram-positive spore-former, have recently been associated with food poisoning, including gastroenteritis, but the responsible factor or factors have not yet been characterized. Most recently a cytotoxin-enterotoxin from a protozoan, *Entamoeba histolytica,* has been reported. I eliminated from this discussion the "classic enterotoxin" produced by staphylococci, responsible for food-borne intoxication, as these well-characterized toxins act at the level of the central nervous system to cause vomiting. Other factors, yet to be recognized, may be responsible for the diarrheal component of staphylococcal food poisoning.

So, to conclude, even though it is my hope, but not my expectation, that we may ultimately eliminate diarrheal diseases as causes of human misery and economic losses in animal husbandry, there will undoubtedly be a surfeit of products of use to, and to be studied by, alert membranologists in the decades ahead. We all could be very busy.

REFERENCES

1. Finkelstein RA: Cholera. *CRC Crit Rev Microbiol* 2:553–623, 1973.

2. Finkelstein RA: Immunology of cholera. *Curr Top Microbiol Immunol* 69:137–196, 1975.

3. Finkelstein RA: Progress in the study of cholera and related enterotoxins, in Bernheimer A (ed): *Mechanisms in Bacterial Toxinology.* New York, John Wiley & Sons Inc, 1976, pp 53–84.

4. Finkelstein RA, Norris HT, Dutta NK: Pathogenesis of experimental cholera in infant rabbits. I. Observations on the intraintestinal infection and experimental cholera produced with cell-free products. *J Infect Dis* 114:203–216, 1964.

5. Benyajati C: Experimental cholera in humans. *Br Med J* 1:140–142, 1966.

6. Finkelstein RA, LoSpalluto JJ: Pathogenesis of experimental cholera: preparation and isolation of choleragen and choleragenoid. *J Exp Med* 130: 185–202, 1969.

7. Gill DM: The arrangement of subunits in cholera toxin. *Biochemistry* 15:1242–1248, 1976.

8. Klapper DG, Finkelstein RA, Capra JD: Subunit structure and N-terminal

amino acid sequence of the three chains of cholera enterotoxin. *Immunochemistry* 13:605–611, 1976.

9. Finkelstein RA, Boesman M, Neoh SH, et al: Dissociation and recombination of the subunits of the cholera enterotoxin (choleragen). *J Immunol* 113: 145–150, 1974.

10. Kurosky A, Markel DE, Peterson JW: Covalent structure of the β chain of cholera enterotoxin. *J Biol Chem* 252:7257–7264, 1977.

11. Lai C-Y: Determination of the primary structure of the cholera toxin B subunit. *J Biol Chem* 252:7249–7256, 1977.

12. Kurosky A, Markel DE, Peterson JW, et al: Primary structure of cholera toxin β-chain: a glycoprotein hormone analog? *Science* 195:299–301, 1977.

13. Gill DM, Meren R: ADP-ribosylation of membrane proteins catalyzed by cholera toxin: basis of the activation of adenylate cyclase. *Proc Natl Acad Sci USA* 75:3050–3054, 1978.

14. Finkelstein RA, LoSpalluto JJ: Crystalline cholera toxin and toxoid. *Science* 175:529–530, 1972.

15. Sigler PB, Druyan ME, Kiefer HC, et al: Cholera toxin crystals suitable for x-ray diffraction. *Science* 197:1277–1279, 1977.

16. van Heyningen WE: Gangliosides as membrane receptors for tetanus toxin, cholera toxin and serotonin. *Nature* 249:415–417, 1974.

17. Holmgren J, Månsson J-E, Svennerholm L: Tissue receptor for cholera exotoxin: structural requirements of G_{M1} ganglioside in toxin binding and inactivation. *Med Biol* 52:229–233, 1974.

18. Bennett V, Cuatrecasas P: Cholera toxin: membrane gangliosides and activation of adenylate cyclase, in Cuatrecasas P (ed): *The Specificity and Action of Animal, Bacterial, and Plant Toxins. (Receptors and Recognition,* Series B, vol 1.) London, Chapman & Hall Ltd, 1977, pp 1–66.

19. Field M: Intestinal secretion: effect of cyclic AMP and its role in cholera. *N Engl J Med* 284:1137–1144, 1971.

20. Vaughan M, Pierce NF, Greenough WB III: Stimulation of glycerol production in fat cells by cholera toxin. *Nature* 226:658–659, 1970.

21. Gill DM: Seven toxic peptides that cross cell membranes, in Jeljaszewicz J, Wadström T (eds): *Bacterial Toxins and Cell Membranes.* New York, Academic Press Inc, 1978, pp 291–332.

22. Cassel D, Pfeuffer T: Mechanism of cholera toxin action: covalent modification of the guanyl nucleotide-binding protein of the adenylate cyclase system. *Proc Natl Acad Sci USA* 75:2669–2673, 1978.

23. Cassel D, Selinger Z: Mechanism of adenylate cyclase activation by cholera toxin: inhibition of GTP hydrolysis at the regulatory site. *Proc Natl Acad Sci USA* 74:3307–3311, 1977.

24. Cassel D, Selinger Z: Mechanism of adenylate cyclase activation through the β-adrenergic receptor: catecholamine-induced displacement of bound GDP by GTP. *Proc Natl Acad Sci USA* 75:4155–4159, 1978.

25. Collier RJ: Inhibition of protein synthesis by exotoxins from *Corynebacterium diphtheriae* and *Pseudomonas aeruginosa,* in Cuatrecasas P (ed): *The Specificity and Action of Animal, Bacterial, and Plant Toxins. (Receptors and Recognition,* Series B, vol 1.) London, Chapman & Hall Ltd, 1977, pp 67–98.

26. Moss J, Vaughan M: Isolation of an avian erythrocyte protein possessing ADP-ribosyltransferase activity and capable of activating adenylate cyclase. *Proc Natl Acad Sci USA* 75:3621–3624, 1978.

27. Moss J, Fishman PH, Richards RL, et al: Choleragen-mediated release of trapped glucose from liposomes containing ganglioside G_{M1}. *Proc Natl Acad Sci USA* 73:3480–3483, 1976.

28. Tosteson MT, Tosteson DC: Bilayers containing gangliosides develop channels when exposed to cholera toxin. *Nature* 275:142–144, 1978.

29. Finkelstein RA: Immunology of *Vibrio cholerae,* in Nahmias AJ, O'Reilly RJ (eds): *Immunology of Human Infections.* (Series, *Comprehensive Immunology;* Good RA [ed]) New York, Plenum Publishing Corporation. (In press).

30. Cash RA, Music SE, Libonati JP, et al: Response of man to infection with *Vibrio cholerae.* II. Protection from illness afforded by previous disease and vaccine. *J Infect Dis* 130:325–333, 1974.

31. Finkelstein RA, Vasil ML, Holmes RK: Studies on toxinogenesis in *Vibrio cholerae.* I. Isolation of mutants with altered toxinogenicity. *J Infect Dis* 129: 117–123, 1974.

32. Woodward WE, Gilman RH, Hornick RB, et al: Efficacy of a live oral cholera vaccine in human volunteers, in Regamey RH, Hennessen W, Hulse EC, et al (acting eds): *Developments in Biological Standardization, vol 33: International Symposium on Vaccination of Man and Animals by the Non-Parenteral Route.* Basel, Switzerland, S Karger AG, 1976.

33. Honda T, Finkelstein RA: Selection and characteristics of a *Vibrio cholerae* mutant lacking the A (ADP-ribosylating) portion of the cholera enterotoxin. *Proc Natl Acad Sci USA* 76:2052–2056, 1979.

34. Finkelstein RA: Laboratory production and isolation of enterotoxins and isolation of a candidate live vaccine for diarrheal disease, in *43rd Nobel Symposium, on Cholera and Related Diarrheas—Molecular Aspects of a Global Health Problem. Proceedings.* (Stockholm, August 1978.) Basel, Switzerland, S Karger AG. (In press).

35. Clements JD, Finkelstein RA: Immunological cross-reactivity between a heat-labile enterotoxin(s) of *Escherichia coli* and subunits of *Vibrio cholerae* enterotoxin. *Infect Immun* 21:1036–1039, 1978.

36. Clements JD, Finkelstein RA: Demonstration of shared and unique immunological determinants in enterotoxins from *Vibrio cholerae* and *Escherichia coli. Infect Immun* 22:709–713, 1978.

37. Clements JD, Finkelstein RA: Isolation and characterization of homogeneous heat-labile enterotoxins with high specific activity from *Escherichia coli* cultures. *Infect Immun* 24:760–769, 1979.

38. Clements JD, Finkelstein RA: Isolation and characterization of homogeneous heat-labile enterotoxins with high specific activity from *Escherichia coli* cultures. *Infect Immun* 24:760–769, 1979.

39. Nelson ET, Clements JD, Finkelstein RA: *Vibrio cholerae* adherence and colonization in experimental cholera: electron microscopic studies. *Infect Immun* 14:527–547, 1976.

40. Finkelstein RA, Arita M, Clements JD, et al: Isolation and purification of an adhesive factor ("cholera lectin") from *Vibrio cholerae,* in *Proceedings of the 13th Joint Conference on Cholera.* (DHEW Publ. No. (NIH) 78-1590.) Bethesda, Md, National Institutes of Health, 1978, pp 137–147.

41. Field M, Graf LH Jr, Laird WJ, et al: Heat-stable enterotoxin of *Escherichia coli: in vitro* effects on guanylate cyclase activity, cyclic GMP concentration, and ion transport in small intestine. *Proc Natl Acad Sci USA* 75:2800–2804, 1978.

8 The Mechanism of Action of Choleragen

Martha Vaughan, MD
Joel Moss, MD, PhD

Choleragen is the exotoxin of the *Vibrio cholerae* that is responsible for the clinical manifestations of cholera. In patients with cholera, the effects of the toxin are confined to the cells of the intestinal mucosa, but, in fact, choleragen can act on most tissues and usually mimics the effects of the hormones that can increase cAMP. The effects of choleragen on vertebrate cells result from the activation of adenylate cyclase which leads to the intracellular accumulation of cAMP. The action of choleragen differs in several ways, however, from that of the numerous hormones that activate adenylate cyclase. In intact cells, hormonal activation of adenylate cyclase is extremely rapid, but the action of choleragen is relatively slow; in many systems, there is a delay of 15 to 45 minutes before an effect is detected. Unlike the effects of the hormones that are rapidly reversed on removal of the agonist, activation by choleragen is not readily reversible and may persist for hours or days. Whereas receptors for a given hormone may occur on only a few types of cells, the vast majority of cells respond to choleragen. Because choleragen can activate adenylate cyclase from so many sources, it seems probable that by learning its mechanism of action we shall learn something very basic about the organization and regulation of adenylate cyclases in vertebrate cells. In the following paragraphs we have attempted to summarize and place in perspective some of our approaches to this problem. Several recent reviews[1-7] provide a more comprehensive consideration of the extensive choleragen literature.

Interaction of Choleragen with Gangliosides

Knowledge of the structure of choleragen is due largely to the work of Finkelstein[2] (see Chapter 7). Choleragen is a complex molecule (Table 8-1) made up of one A subunit, comprised of A_1 and A_2 peptides linked through a single disulfide bond, and five B subunits. The A_1 peptide, molecular weight about 23,500, is responsible for the activation of adenylate cyclase, as first shown by Gill and King.[8] The function of the small A_2 peptide, molecular weight 5500, has not been established, but it may be involved in binding the A subunit to the B complex. Work in several laboratories[1,3-7] has shown that the B subunits are responsible for choleragen binding to the cell surface through interaction with the oligosaccharide moiety of ganglioside G_{M1} in the plasma membrane.

Table 8-1
Choleragen: A_1–S–S–A_2 $(B)_5$

Peptide	MW*	Function
A_1	23,500	Activation of adenylate cyclase
A_2	5500	Unknown
B	11,600	Interaction with ganglioside G_{M1} on cell surface

Based on sources cited in references 1–7.
*Molecular weight

In collaboration with investigators at the National Institutes of Health and elsewhere, we have studied the interactions of choleragen or its isolated A and B subunits with gangliosides, using multilamellar liposomes, model membranes composed of cholesterol, dimyristoyl lecithin, and diacetylphosphate, into which pure gangliosides can be incorporated.[9,10] Choleragen binding was determined with ^{125}I-labeled toxin, and permeability changes were assessed by measuring the release of glucose from liposomes prepared with glucose trapped in the interior. From these studies, it was concluded that choleragen, in the absence of other proteins, interacts specifically with G_{M1}-liposomes and alters their permeability. It was further shown that the B subunit is responsible for the interaction with G_{M1} and for the resultant perturbations of the model membranes. In experiments with liposomes lacking gangliosides, it was demonstrated that the

A subunit can interact with lipid membranes when it is freed from the constraints present in the intact choleragen molecule.

We have also investigated the interaction with gangliosides by monitoring changes in the tryptophan fluorescence of the choleragen subunits, and have shown that G_{M1} (but not other gangliosides) specifically causes a so-called blue shift in the fluorescence spectrum of the B (but not the A) subunit, indicative of a ganglioside-induced conformational change in the protein that places the tryptophan residues in a more hydrophobic environment.[11] In addition, by using the isolated oligosaccharide moiety of G_{M1} separated from the lipid portion of the molecule, it has been possible to demonstrate directly that the B subunit of toxin recognizes and binds the oligosaccharide portion of G_{M1}, as had been inferred from the ganglioside specificity of the interaction.[12]

Van Heyningen and coworkers first showed that mixed gangliosides could bind choleragen and later demonstrated that binding occurs specifically to the monosialoganglioside G_{M1}.[5] There is now considerable evidence[1,3-7] that G_{M1} functions as a cell surface "receptor" for choleragen. To investigate the role of G_{M1} in choleragen action in intact cells, we used cultured mouse fibroblasts (NCTC 2071 cells), which can be grown in chemically defined medium. When grown without serum, the fibroblasts did not accumulate cAMP in response to choleragen.[13] On analysis of these cells, no G_{M1} was detected by the usual chemical procedures, by looking for radioactive G_{M1} in cells grown with ^{14}C-labeled galactose, or by looking for ^{3}H-labeled G_{M1} after treatment of cells with galactose oxidase followed by NaB^3H_4. The NCTC 2071 cells that lacked G_{M1} did not bind or respond to choleragen. They were able to take up exogenous G_{M1} and thereafter were capable of binding and responding to choleragen.[13] The fibroblasts were also able to take up other gangliosides from the medium, but when present in similar quantities these did not make the cells responsive to choleragen.[14] The length of the delay between the addition of choleragen and the onset of cAMP accumulation depended on the amount of G_{M1} taken up by the cells as well as the toxin concentration. Exposure of intact cells to galactose oxidase led to oxidation of G_{M1} on the cell surface, but, in cells that had bound choleragen, the G_{M1} was protected from the oxidase.[15] Thus, all evidence points to the conclusion that G_{M1} is required for choleragen activation of adenylate cyclase in intact cells.

Ganglioside G_{M1} is responsible (and necessary) for the specific binding of choleragen to the cell surface. It is not, however, required for the activation of adenylate cyclase by choleragen in broken cell systems, as was shown in experiments with the ganglioside-deficient NCTC 2071 cells.[13] It, therefore, apparently does not function as a "receptor" in the same sense as those receptors that are an obligatory part of the mechanisms through which β-adrenergic agonists and certain peptide hormones activate adenylate cyclase. In cell-free adenylate cyclase preparations, where G_{M1} is not necessary, the B subunit of the toxin is not required. As shown by

Gill,[16] the A_1 peptide (liberated from the A subunit by reduction of the disulfide bond that links it to A_2) activates adenylate cyclase in a process that is dependent on NAD.[9]

Role of NAD as Substrate in Choleragen-Catalyzed Reactions

In considering the possible role of NAD, we were influenced by the information on the mechanism of action of diphtheria toxin, which causes cell death by inhibiting protein synthesis. This effect is dependent on NAD and results from toxin-catalyzed transfer of ADP-ribose from NAD to elongation factor 2.[17] The products of the reaction are ADP-ribosyl-elongation factor 2, which cannot function in protein synthesis, and nicotinamide. No analogous protein acceptor for the postulated reaction catalyzed by choleragen was known. It had been shown, however, that in the absence of elongation factor 2 diphtheria toxin catalyzes an abortive reaction in which water serves as an acceptor; the products are ADP-ribose and nicotinamide.[18] To determine whether choleragen exhibited a similar NAD-glycohydrolase activity, we incubated it with NAD, labeled with ^{14}C in the nicotinamide portion, and readily established that choleragen catalyzed the production of ^{14}C-nicotinamide. Equimolar quantities of ADP-ribose accumulated during the reaction.[19]

To demonstrate ADP-ribosyltransferase activity, a model ADP-ribose acceptor was needed, since the natural acceptor, which was presumed to be a protein component of the adenylate cyclase system, had not been identified. A number of amino acids commonly present in proteins were tested for their ability to serve as ADP-ribose acceptors when incubated with choleragen and NAD. Arginine (both *D*- and *L*-), arginine methyl ester, and several other guanidino compounds including guanidine itself accelerated nicotinamide formation.[20] Citrulline and ornithine as well as a number of other amino acids including lysine, histidine, and serine were ineffective. To establish that arginine was, in fact, serving as an acceptor for ADP-ribose from NAD, the products of the reaction were chromatographically separated. It was found that nicotinamide and a new compound with the properties of ADP-ribosylarginine were formed in equimolar amounts.[20] All of our evidence supports the conclusion that the ADP-ribosyltransferase activity is an intrinsic property of the A subunit of choleragen, and it has recently been shown that the reduced and alkylated A_1 peptide is catalytically active.[21]

Choleragen can also catalyze the ADP-ribosylation of several purified proteins, eg, histones, lysozyme, and polyarginine.[22] Incubation of ^{14}C-ADP-ribosylated lysozyme with snake venom phosphodiesterase led to the release of essentially all of the radioactivity in the form of 5′-AMP, as would be expected with an ADP-ribosylated, but not a poly(ADP-ribosylated), protein. The amino acids in lysozyme and histone that are

ADP-ribosylated have not been identified. It is notable, however, that arginine (or arginine methyl ester) can inhibit this process, whereas amino acids that are not ADP-ribose acceptors,[20] and presumably do not interact with the catalytic site of choleragen, do not. All of these observations are consistent with the view that the acceptor amino acid in proteins that are ADP-ribosylated by choleragen is arginine. The stereospecificity of the choleragen-catalyzed ADP-ribosyltransferase reaction was investigated by Oppenheimer.[23] Using nuclear magnetic resonance spectroscopy, he showed that the product of the reaction of β-NAD with L-arginine is α-ADP-ribose-L-arginine, which anomerizes rapidly to a mixture of α- and β-forms. It is not known whether this anomerization occurs when the arginine moiety is part of a peptide chain.

In the present view of the mechanism of action of choleragen, the toxin first binds to the cell surface through a receptor that is, or contains, ganglioside G_{M1}; the B subunit of choleragen is the site of specific interaction with the oligosaccharide of G_{M1}. There is a delay between binding and adenylate cyclase activation. This probably represents the time in which the enzymatically active A_1 peptide is released from the holotoxin, and catalyzes the transfer of ADP-ribose from NAD to a protein that influences cyclase activity. To identify this specific protein requires demonstration of an alteration of function resulting from an ADP-ribosylation.

Several workers have shown that choleragen can catalyze the ADP-ribosylation of membrane proteins.[24-26] When ^{32}P-labeled NAD was included in the incubations, one predominantly labeled protein with an apparent molecular weight of about 42,000 was found.[24,25] After solubilization with detergent, it was bound to GTP-Sepharose, was eluted with GTP$_\gamma$S, and appeared to have the properties of a GTP-binding component of adenylate cyclase.[27] Cassel and Pfeuffer have proposed a model for the activation of adenylate cyclase by choleragen.[24] They suggest that GTP binds to adenylate cyclase through a GTP-binding protein and thereby converts it from an inactive to an active form. Hydrolysis of the bound GTP by a specific GTPase inactivates the adenylate cyclase. ADP-ribosylation of the GTP-binding protein catalyzed by choleragen reduces GTPase activity and thus prolongs the life of the active GTP-adenylate cyclase complex. Work in several laboratories is currently directed toward testing and extending or revising this hypothesis.

Similarities Between Choleragen and the Heat-Labile Enterotoxin of *Escherichia coli*

The effects of the heat-labile *E coli* toxin on cells, like those of choleragen, result from the NAD-dependent activation of adenylate cyclase.[28] Ganglioside G_{M1} serves as a cell surface "receptor" for both enterotoxins.[29,30] In addition, the two proteins exhibit some similarities in structure[31] and in immunologic reactivity.[32-34] It has recently been shown that a highly purified *E coli* toxin exhibits NAD-glycohydrolase activity and catalyzes the ADP-ribosylation of arginine, a number of other guanidino compounds, and several proteins.[35,36] Some characteristics of the *E coli* toxin and choleragen are compared in Table 8-2. It appears that the heat-labile *E coli* enterotoxin is catalytically equivalent to the A_1 peptide of choleragen, and causes the NAD-dependent activation of adenylate cyclase by catalyzing the ADP-ribosylation of either the cyclase itself or a regulatory component of the cyclase system.

Table 8-2
Properties of Choleragen, Heat-Labile *E coli* Enterotoxin, and the ADP-Ribosyltransferase from Turkey Erythrocytes

Property	Choleragen	Heat-Labile *E coli* Toxin	ADP-Ribosyl-transferase
Molecular weight	87,000*	~ 80,000	28,000
Subunits	Yes	Yes	?
DTT requirement	Yes	Yes	No
K_m for NAD	4 mM	8 mM	30 μM
Ability to use NADP	Very little	?	Yes
K_m for ArgOMet†	50 mM	240 mM	2 and 50 mM
Stereospecificity			
Substrate	β-NAD	β-NAD	β-NAD
Product with Arg	α-Anomer	α-Anomer	α-Anomer
Effectiveness of guanidino substrates	ArgOMet > D-Arg ~ L-Arg > guanidine	ArgOMet > D-Arg ~ L-Arg ~ guanidine	ArgOMet > D-Arg > L-Arg > guanidine
Effectiveness of protein substrates	PolyArg > histones ~ lysozyme	PolyArg ≫ histones ~ lysozyme	Histones > lysozyme ~ polyArg
Product with lysozyme	ADP-ribosyl-	ADP-ribosyl-	ADP-ribosyl-

Based on sources cited in text.
*Active subunit (A_1 peptide) = 23,500.
†Abbreviations are: ArgOMet = arginine methyl ester; Arg = arginine; PolyArg = polyarginine.

ADP-Ribosyltransferase from Turkey Erythrocytes

To determine whether vertebrate cells contain enzymes similar to the toxins that might function in adenylate cyclase activation, we looked for an enzyme that could catalyze the synthesis of ADP-ribosylarginine. Such an ADP-ribosyltransferase has been purified from the supernatant of turkey erythrocytes using conventional protein fractionation procedures.[14,37] The stereospecificity of the reaction with arginine is identical to that of the two toxin-catalyzed reactions.[37] The enzyme also exhibits NAD-glycohydrolase activity and catalyzes the ADP-ribosylation of histones, lysozyme, and polyarginine.[37] The product of the reaction with lysozyme is an ADP-ribosyl-, not a poly(ADP-ribosyl)-, protein.[14] In addition to catalyzing the same model reaction as choleragen and the *E coli* toxin, the turkey enzyme can bring about the NAD-dependent activation of adenylate cyclase.[14]

Several structural and enzymatic characteristics of the ADP-ribosyltransferase differ, however, from those of the two toxins (see Table 8-2). The intact choleragen molecule is much larger than the *E coli* toxin or the turkey enzyme, although the enzymatically active A_1 subunit has a molecular weight of less than 25,000. Both toxins require a thiol group for maximal activity, but the turkey enzyme does not. The ADP-ribosyltransferase has a much higher affinity for NAD, with a K_m of 30 μM, than do the toxins, and can use NADP in place of NAD somewhat more effectively than does choleragen. With all three enzymes the K_m for arginine in the model reaction is high, but the ADP-ribosyltransferase, in contrast to the toxins, exhibits two apparent K_m for arginine. The effectiveness of certain proteins as ADP-ribose acceptors also differs with the three enzymes. With the turkey enzyme, histone was a much better ADP-ribosyl acceptor than was polyarginine, whereas with the two toxins polyarginine was clearly more effective than histone.

Despite clear differences in properties and structure, all of our evidence supports the view that the turkey enzyme and the two toxins activate adenylate cyclase by catalyzing the ADP-ribosylation of a regulatory protein in the cyclase system. Whether endogenous ADP-ribosyltransferases play a role in the regulation of adenylate cyclase activity in intact cells remains to be determined.

REFERENCES

1. Bennett V, Cuatrecasas P: Mechanism of action of *Vibrio cholerae* enterotoxin. *J Membr Biol* 22:1–28, 1975.

2. Finkelstein RA: Cholera. *CRC Crit Rev Microbiol* 2:553–623, 1973.

3. Gill DM: Mechanism of action of cholera toxin. *Adv Cyclic Nucleotide Res* 8:85–118, 1977.

4. Moss J, Vaughan M: Activation of adenylate cyclase. *Annu Rev Biochem* 48:581–600, 1979.

5. van Heyningen WE: Gangliosides as membrane receptors for tetanus, toxin, cholera toxin and serotonin. *Nature* 249:415–417, 1974.

6. van Heyningen S: Cholera toxin. *Biol Rev* 52:509–549, 1977.

7. Vaughan M, Moss J: Mechanism of action of choleragen. *J Supramol Struct* 8:473–488, 1978.

8. Gill DM, King CA: The mechanism of action of cholera toxin in pigeon erythrocyte lysates. *J Biol Chem* 250:6424–6432, 1975.

9. Moss J, Fishman PH, Richards RL, et al: Choleragen-mediated release of trapped glucose from liposomes containing ganglioside G_{M1}. *Proc Natl Acad Sci USA* 73:3480–3483, 1976.

10. Moss J, Richards RL, Alving CR, et al: Effect of the A and B protomers of choleragen on release of trapped glucose from liposomes containing or lacking ganglioside G_{M1}. *J Biol Chem* 252:797–798, 1977.

11. Moss J, Osborne JC Jr, Fishman PH, et al: Effect of gangliosides and substrate analogues on the hydrolysis of nicotinamide adenine dinucleotide by choleragen. *Proc Natl Acad Sci USA* 74:74–78, 1978.

12. Fishman PH, Moss J, Osborne JC Jr: Interaction of choleragen with the oligosaccharide of ganglioside G_{M1}: evidence for multiple binding sites. *Biochemistry* 17:711–716, 1978.

13. Moss J, Fishman PH, Manganiello VC, et al: Functional incorporation of ganglioside into intact cells: induction of choleragen responsiveness. *Proc Natl Acad Sci USA* 73:1034–1037, 1976.

14. Fishman PH, Moss J, Vaughan M: Uptake and metabolism of gangliosides in transformed mouse fibroblasts. Relationship of ganglioside structure to choleragen response. *J Biol Chem* 251:4490–4494, 1976.

15. Moss J, Manganiello VC, Fishman PH: Enzymatic and chemical oxidation of gangliosides in cultured cells: effects of choleragen. *Biochemistry* 16:1876–1881, 1977.

16. Gill DM: Involvement of nicotinamide adenine dinucleotide in the action of cholera toxin *in vitro*. *Proc Natl Acad Sci USA* 72:2064–2068, 1975.

17. Honjo T, Nishizuka Y, Kato I, et al: Adenosine diphosphate ribosylation of aminoacyl transferase II and inhibition of protein synthesis by diphtheria toxin. *J Biol Chem* 246:4251–4260, 1971.

18. Kandel J, Collier RJ, Chung DW: Interaction of fragment A from diphtheria toxin with nicotinamide adenine dinucleotide. *J Biol Chem* 249:2088–2097, 1974.

19. Moss J, Manganiello VC, Vaughan M: Hydrolysis of nicotinamide adenine dinucleotide by choleragen and its A protomer: possible role in the activation of adenylate cyclase. *Proc Natl Acad Sci USA* 73:4424–4427, 1976.

20. Moss J, Vaughan M: Mechanism of action of choleragen. Evidence for ADP-ribosyltransferase activity with arginine as an acceptor. *J Biol Chem* 252:2455–2457, 1977.

21. Moss J, Stanley SJ, Lin MC: NAD glycohydrolase and ADP-ribosyltransferase activities are intrinsic to the A_1 peptide of choleragen. *J Biol Chem* (In press).

22. Moss J, Vaughan M: Isolation of an avian erythrocyte protein possessing ADP-ribosyltransferase activity and capable of activating adenylate cyclase. *Proc Natl Acad Sci USA* 75:3621–3624, 1978.

23. Oppenheimer NJ: Structural determination and stereospecificity of the choleragen-catalyzed reaction of NAD^+ with guanidines. *J Biol Chem* 253:4907–4910, 1978.

24. Cassel D, Pfeuffer T: Mechanism of cholera toxin action: covalent modification of the guanyl nucleotide-binding protein of the adenylate cyclase system. *Proc Natl Acad Sci USA* 75:2669–2673, 1978.

25. Gill DM, Meren R: ADP-ribosylation of membrane proteins catalysed by cholera toxin: basis of the activation of adenylate cyclase. *Proc Natl Acad Sci USA* 75:3050–3054, 1978.

26. Johnson GL, Kaslow HR, Bourne HR: Genetic evidence that cholera toxin substrates are regulatory components of adenylate cyclase. *J Biol Chem* 253:7120–7123, 1978.

27. Cassel D, Selinger Z: Mechanism of adenylate cyclase activation by cholera toxin: inhibition of GTP hydrolysis at the regulatory site. *Proc Natl Acad Sci USA* 74:3307–3311, 1977.

28. Gill DM, Evans DJ Jr, Evans DG: Mechanism of activation of adenylate cyclase in vitro by polymyxin-released heat-labile enterotoxin of *Escherichia coli. J Infect Dis* 133(suppl):S103–S107, 1976.

29. Holmgren J: Comparison of the tissue receptors for *Vibrio cholerae* and *Escherichia coli* enterotoxins by means of gangliosides and natural cholera toxoid. *Infect Immun* 8:851–858, 1973.

30. Moss J, Garrison S, Fishman PH, et al: Gangliosides sensitize unresponsive fibroblasts to *Escherichia coli* heat-labile enterotoxin. *J Clin Invest* 64:381–384, 1979.

31. Robertson DC, Kunkel SL, Gilligan PH: Purification and characterization of heat-labile enterotoxin produced by ENT$^+$ *E. coli:* application of hydrophobic chromatography and use of defined media, in *The 14th Joint Conference US-Japan Cooperative Medical Science Program Cholera Panel,* Karatsu, Japan, September 1978, 75 (abs).

32. Gyles CL, Barnum DA: A heat-labile enterotoxin from strains of *Escherichia coli* enteropathogenic for pigs. *J Infect Dis* 120:419–426, 1969.

33. Holmgren J, Söderlind O, Wadström T: Cross-reactivity between heat labile enterotoxins of *Vibrio cholerae* and *Escherichia coli* in neutralization tests in rabbit ileum and skin. *Acta Pathol Microbiol Scand [B]* 81:757–782, 1973.

34. Smith NW, Sack RB: Immunologic cross-reactions of enterotoxins from *Escherichia coli* and *Vibrio cholerae. J Infect Dis* 127:164–170, 1973.

35. Moss J, Garrison S, Oppenheimer NJ, et al: NAD-dependent ADP-ribosylation of arginine and proteins by *Escherichia coli* heat-labile enterotoxin. *J Biol Chem* 254:6270–6272, 1979.

36. Moss J, Richardson SH: Activation of adenylate cyclase by heat-labile *Escherichia coli* enterotoxin. Evidence for ADP-ribosyltransferase activity similar to that of choleragen. *J Clin Invest* 62:281–285, 1978.

37. Moss J, Stanley SJ, Oppenheimer NJ: Substrate specificity and partial purification of a stereospecific NAD- and guanidine-dependent ADP-ribosyltransferase from avian erythrocytes. *J Biol Chem* (In press).

9 Cyclic AMP Increase Mediated by Cholera Toxin in *Xenopus laevis* Oocytes Inhibits Late Events of Progesterone-Induced Maturation In Vitro

Michel Schorderet, PhD
Sabine Schorderet-Slatkine, PhD
Patrice Boquet, MD, PhD
François Godeau, MD
Etienne-Emile Baulieu, MD

A cascade of events seems to be involved in the meiosis of *Xenopus laevis* oocytes induced by progesterone in vitro.[1] Surface steroid receptors, membrane calcium translocation, translation-dependent protein synthesis, and a cytosolic factor called Maturation-Promoting Factor (MPF) are all implicated in this biologic phenomenon. However, the interrelationships among these various cell constituents and/or machinery are far from being completely understood at the present time. That cAMP might have some regulatory effects on the process was derived from the experiments of Maller and Krebs.[2] They have shown that the regulatory subunit of cAMP-dependent protein kinase (isolated from mammalian muscles) was able to mimic the action of progesterone, whereas the catalytic subunit inhibited it. Recently, we have shown that cholera toxin, applied in various experimental conditions, was a potent inhibitor of progesterone action,[3] and that the cAMP increase induced by the toxin appeared to be responsible for that inhibition.[4]

This work was partially supported by the "Délégation Générale à la Recherche Scientifique et Technique," by the Swiss National Science Foundation (Grants No. 3,210,077 and No. 3,327,078), and by the World Health Organization.

Materials and Methods

Collection of oocytes and induction of maturation in vitro After immersion anesthesia of frogs in 0.25% tricaine methane sulfonate (MS 222, Sandoz), the surgically removed ovaries were rinsed in Barth's medium and the oocytes were isolated and defolliculated by collagenase digestion.[5] They were rinsed five times in Barth's medium and selected under a dissecting microscope to obtain a uniform population of stage VI oocytes. Maturation was usually induced by continuous exposure to progesterone (1 to 10 μM) in Barth's medium at 20 to 22 C. Alternatively, maturation was provoked by injecting recipient oocytes (not exposed to progesterone) with approximately 50 nl of MPF-containing cytoplasm removed from mature oocytes. Maturation was scored by observing the frequency of germinal vesicle breakdown (Gvbd), as indicated by the formation of a well-defined white spot on the pigmented animal pole of the oocyte after five to eight hours of hormonal treatment. Pools of 30 to 50 oocytes taken from the same female were used for each type of experiment.

Cholera toxin Pure cholera toxin (Schwartz-Mann, USA), or its subunits A and B prepared from the toxin by acid urea gel filtration,[6] were used after appropriate dilution in Barth's medium (up to 10 pM). In microinjection experiments, 50 nl of toxin solution was injected into the oocytes. The final concentration of the toxin (or of its subunits) inside the oocyte was calculated, assuming an oocyte volume of 1 μl (up to 100 pM). In all experiments performed with both cholera toxin and progesterone, the hormone was added in Barth's medium after treatment of oocytes with the toxin alone.

cAMP measurements For each determination of cAMP levels, ten oocytes (in triplicate) were transferred into a glass homogenizer containing 200 μl of distilled water, heated at 100 C for ten minutes, then homogenized and centrifuged at 4 C. The measurements of cAMP were made on lyophilized aliquots of the supernatants (40 μl) by the saturation assay method.[7] The usual methodologic controls of identity and/or recovery were accurately completed.[4]

Results

Previous reports[3,4] characterize several features of the cAMP-dependent inhibition of progesterone-induced meiosis after an oocyte's exposure to or injected with choleragen or A and B subunits of the toxin:

1. Cholera toxin (1 to 50 pM) added into the incubation medium is able to inhibit in a dose-dependent manner (concentration needed for 50% inhibition of maturation, ED_{50} = 5 pM) the maturation induced by 10 μM progesterone.

2. The cholera toxin inhibition described above is irreversible. However, it is achieved only when the toxin has been applied at a definite time before progesterone exposure. The sooner the occurrence of Gvbd after exposure to progesterone, the earlier cholera toxin must be added before progesterone.

3. The entire holotoxin (10 to 100 pM) injected into the oocytes leads also to a dose-dependent inhibition of progesterone action. The time of injection (versus progesterone exposure) also varies from female to female.

4. The A subunit of cholera toxin is also effective inside the oocyte (when injected) but not outside (when incubated).

5. The B subunit is totally ineffective either inside or outside the oocyte in the range of effective concentrations of the A subunit.

6. The inhibitory effects of cholera toxin or the A subunit described above seem to be correlated with cAMP increases inside the oocyte, which are achieved after 5 to 7 hours of incubation in the presence of the holotoxin, or 3 to 5 hours after injection of the holotoxin or its A subunit.

7. The levels of cAMP in oocytes treated with cholera toxin alone are higher than those of oocytes treated with cholera toxin and progesterone. Progesterone strongly inhibits the formation of cAMP induced by cholera toxin, the inhibition being detectable at any time (from 30 seconds to several minutes) after progesterone addition to the medium. Nevertheless, the cAMP level of oocytes treated with both cholera toxin and progesterone during the maturation process is higher than the basal level (no cholera toxin, no progesterone).

8. Derivatives of cAMP that are more resistant to phosphodiesterase(s) than the parent compound show a transient but reversible inhibition of progesterone-induced maturation. They have to be injected into the oocytes when the MPF is virtually under formation.

9. MPF itself (cytoplasm transferred from mature oocytes) can progressively (up to 24 hours) reverse the cholera toxin inhibition after the injection of a sufficient amount of MPF in cholera toxin-treated oocytes. However, the reversal time of inhibition is very much shortened when MPF is injected in oocytes treated with both cholera toxin and progesterone.

10. In contrast with other known inhibitors of maturation such as gammexane or cycloheximide, the preferential synthesis of proteins detected during the course of maturational events by a double labeling technique is neither impaired nor modified by cholera toxin or its A subunit.

Discussion

It was shown that in various experimental conditions cholera toxin was able to inhibit the progesterone-induced maturation of *X laevis* oocytes.

Since the effectomer of cholera toxin (A subunit) is supposed to act inside the cell by stimulating irreversibly the activity of the enzyme adenylate cyclase,[8] it is inferred that cAMP may play a role in the process of amphibian meiosis,[4] as well as in mammalian meiosis.[9] Arguments in favor of such a hypothesis have already been provided by the experiments of Maller and Krebs,[2] who have found that the injection of the regulatory subunit of cAMP-dependent protein kinase led to oocyte maturation, whereas the process was inhibited by the injection of the catalytic subunit. Protein phosphorylation changes, on the other hand, occurred during the course of progesterone-induced meiosis,[10] and can be related to bursts of activity of various cAMP-dependent protein kinases.

The crucial point is, however, when and how cAMP can carry out its regulatory influence on this well-defined biologic process.[2] Fluctuations of cAMP levels have been recorded during the early hours of progesterone exposure in amphibian oocytes.[11] Our experimental results summarized in this report would also favor a role of cAMP as a time-delayed event during the process. More precisely, cAMP might influence the formation (or activation) of the physiologically defined MPF, whose physicochemical characteristics have been, up to now, very well ignored. Several lines of evidence seem to support this hypothesis: first, the cholera toxin inhibition is fully achieved only when the toxin is added to the medium an appreciable amount of time before progesterone treatment. This lag period varies with the individual female. Some oocytes will undergo maturation within 2 to 4 hours under constant progesterone exposure; therefore cholera toxin must be applied at least 120 minutes before the steroid in order to get 100% inhibition of Gvbd. In comparison, oocytes taken from another female will undergo maturation within 6 to 12 hours; then the lag period between cholera toxin and progesterone treatment can be less than 60 minutes. These and other findings[4] seem to imply that cholera toxin does not antagonize progesterone directly, but that it impairs a key event of the process which takes place some time after the beginning of progesterone treatment.

Second, cholera toxin-induced accumulation of cAMP within the oocyte is detectable after a lag period of 3 to 5 hours, depending on whether the toxin is injected or incubated. Thus, the inhibitory effects of cholera toxin seem again to operate at a particular time during the process, which appears to coincide with the period of MPF formation or activation. Third, the experiments of inhibition with injection into the oocytes of cAMP derivatives at various times during the maturational time course would also support the same proposal. Thus, without excluding a priori an early effect of cAMP during the process of maturation, as postulated by Maller and Krebs[2] and previously discussed,[4] we are more inclined to consider a predominant role of cAMP at a time-delayed event of maturation, ie, when MPF is formed or activated. How cAMP can now achieve its function at this step needs further studies, which are currently under way.

Another striking feature of our experiments is the constant and significant inhibition of the cholera toxin-induced accumulation of cAMP inside the oocytes under progesterone action. These results clearly indicate that the steroid can in some way modulate the active state of adenylate cyclase, thereby decreasing the formation of cAMP induced by the toxin. They are in accordance with those of Mulner et al,[12] who have also found that progesterone can diminish the cholera toxin-induced synthesis of cAMP from labeled ATP previously injected into the oocytes. They are, on the other hand, in contradiction with those of Maller et al,[13] who have recently shown that progesterone can enhance the cholera toxin-induced accumulation of cAMP in *X laevis* oocytes. Recent experiments performed by our group have, however, confirmed that progesterone inhibits the activity of adenylate cyclase, in a reproducible and very rapid manner, when the latter enzyme is stimulated by cholera toxin. The mechanism of the progesterone-mediated inhibition is possibly linked to permeability changes, calcium availability, and/or the state of oxidation-reduction of membrane enzymes or constituents. Experiments are actually under way to investigate some characteristics of this inhibition with cell-free preparations of *X laevis* oocytes.

Finally, it appears that *X laevis* oocytes can be accurately used to test the intracellular (and toxic) effects of bacterial toxins such as cholera enterotoxin[3,4] and diphtheria toxin[14] on living cells. The first toxin was found to be ineffective on the preferential synthesis of proteins during the course of maturational events, as measured by a double labeling technique,[15] in contrast with diphtheria toxin,[14] as well as with other known inhibitors of maturation such as cycloheximide[16] or gammexane.[17] The size of oocytes (giant cells of 1 mm diameter), and their survival in vitro before and after microinjection of various soluble proteins, make them particularly useful for biologic, pharmacologic, and/or toxicologic studies.

Conclusions

1. Cholera enterotoxin (holotoxin or A subunit) inhibits the progesterone-induced maturation of *X laevis* oocytes in vitro.
2. This inhibition is linked to cAMP increase within the cell.
3. This inhibition is not linked to some alteration(s) of the synthesis of preferential proteins detected during the biologic process.
4. This inhibition (cAMP-dependent) appears to be related to a time-delayed event of the biologic process, namely to the step of Maturation Promoting Factor (MPF) formation or activation.
5. A regulatory role for cAMP in the meiotic process of amphibians and mammals is not unlikely.
6. Progesterone constantly decreases the formation of cAMP induced

by cholera enterotoxin, and it is supposed that it modulates the active state of adenylate cyclase.

7. *X laevis* oocytes are giant living cells which are available for testing the intracellular effect(s) of bacterial toxins.

The technical assistance of Miss G. Allenbach and the secretarial collaboration of Mrs N. Collet are gratefully acknowledged.

REFERENCES

1. Baulieu EE, Godeau JF, Schorderet M, et al: Steroid induced meiotic division in *Xenopus laevis* oocytes: surface and calcium. *Nature* 275:593–598, 1978.
2. Maller JL, Krebs EG: Progesterone-stimulated meiotic division in *Xenopus* oocytes. Induction by regulatory subunit and inhibition by catalytic subunit of adenosine 3′:5′-monophosphate-dependent protein kinase. *J Biol Chem* 252: 1712–1718, 1977.
3. Godeau F, Boquet P, Schorderet M, et al: Inhibition par l'entéro toxine de *Vibrio cholerae* de la réinitiation méiotique de l'ovocyte de *Xenopus laevis* induite *in vitro* par la progestérone. *CR Acad Sci (Paris)* 286:685–688, 1978.
4. Schorderet-Slatkine S, Schorderet M, Boquet P, et al: Progesterone-induced meiosis in *Xenopus laevis* oocytes: a role for cAMP at the "Maturation-Promoting-Factor" level. *Cell* 15:1269–1275, 1978.
5. Schorderet-Slatkine S, Drury KC: Progesterone-induced maturation in oocytes of *Xenopus laevis.* Appearance of a maturation promoting factor in enucleated oocytes. *Cell Differ* 2:247–254, 1973.
6. Finkelstein RA, Boesman M, Neoh SH, et al: Dissociation and recombination of the subunits of the choleraenterotoxin (choleragen). *J Immunol* 113:145–150, 1974.
7. Brown BL, Albano JDM, Ekins RP, et al: A simple and sensitive saturation assay method for the measurement of adenosine 3′:5′-cyclic monophosphate. *Biochem J* 121:561–562, 1971.
8. Gill DM: Mechanism of action of cholera toxin. *Adv Cyclic Nucleotide Res* 8:85–118, 1977.
9. Wassarman PM, Josefowicz WJ, Letourneau GE: Meiotic maturation of mouse oocyte *in vitro:* inhibition of maturation at specific stages of nuclear progression. *J Cell Sci* 22:531–545, 1976.
10. Maller JL, Wu M, Gerhart JC: Changes in protein phosphorylation accompanying maturation of *Xenopus laevis* oocytes. *Dev Biol* 58:295–312, 1977.
11. Speaker MG, Butcher FR: Cyclic nucleotide fluctuation during steroid induced meiotic maturation of frog oocytes. *Nature* 267:848–850, 1977.
12. Mulner O, Huchon D, Thibier C, et al: Cyclic AMP synthesis in *Xenopus laevis* oocytes. Inhibition by progesterone. *Biochim Biophys Acta* 582:179–184, 1979.
13. Maller JL, Butcher FR, Krebs EG: Early effect of progesterone on levels of cyclic adenosine 3′:5′-monophosphate in *Xenopus* oocytes. *J Biol Chem* 254: 579–582, 1979.
14. Godeau F, Boquet P, Schorderet-Slatkine S, et al: *Xenopus laevis* oocytes: a novel system for the study of microbial toxins. *Exp Cell Res* (In press).

15. Schorderet-Slatkine S, Baulieu EE: Induction by progesterone and a "maturation-promoting factor" of soluble proteins in *Xenopus laevis* oocytes *in vitro*. *Ann NY Acad Sci* 286:421–433, 1977.

16. Schorderet-Slatkine S: Action of progesterone and related steroids on oocyte maturation in *Xenopus laevis*. An *in vitro* study. *Cell Differ* 1:179–189, 1972.

17. Schorderet-Slatkine S, Schorderet M, Baulieu EE: Inhibition of the progesterone-dependent induction of meiosis by gammexane in *Xenopus laevis* oocytes. *Biochem Biophys Res Commun* 77:496–502, 1977.

10 Entry of Semliki Forest Virus into the Host Cell

Ari Helenius, PhD

To infect a cell, viruses must transfer their genome from the extracellular space into the cytosol. The T-even bacteriophages have a complicated organ, the tail with baseplate and fibers, which is responsible for the binding of the virus to the cell surface. It also punctures the bacterial cell wall and injects the DNA into the cell. Eucaryotic viruses do not have such a complicated instrument. In spite of intensive study, it is not yet known how the nucleoprotein of these viruses enters into the cell.[1-4]

We have studied this problem using a simple membrane-containing virus, the Semliki Forest virus (SFV). SFV is well suited for entry studies. It is simple in structure, the virus particles are highly uniform, and the ratio of particles to infective units is low. Moreover, the composition, structure, biosynthesis, and replication cycle have been characterized in detail.[5] The SFV consists of a spherical nucleocapsid with a single-stranded RNA molecule and 240 copies of a single protein species (the capsid protein, apparent molecular weight 30×10^3). The nucleocapsid is surrounded by a lipid bilayer membrane with 240 glycoprotein spikes. Each spike consists of three glycoproteins: E1, E2, and E3, with apparent molecular weights of

48×10^3, 51×10^3, and 10×10^3, respectively. The SFV infects a large variety of tissue culture cells from different species. The cells used in our studies were mainly human and mouse lymphoblastoid cells (for binding studies) and BHK-21 cells (for entry studies).

Figure 10-1 schematically illustrates the route of entry suggested by our results. In this discussion I shall briefly describe each of the steps indicated by the numbers in the figure. The scheme is based on results that have been published elsewhere in detail.[6–8]

1. Attachment of SFV to the cell surface The virus attaches to the cell surface through its glycoprotein spikes. The number of high affinity binding sites on different cells varies from less than 10^3 on human red blood cells to 50×10^3 on BHK-21 cells. Half-saturation of the high affinity sites occurs at a 10^{-11} to 10^{-10} M virus concentration. The binding occurs all over the cell but in certain cells, such as the BHK-21 cells, we observed preferential binding to microvilli. Some viruses bound to coated regions of the cell (so-called coated pits; see Brown and Goldstein, Chapter 3). Using isolated spike glycoprotein we have obtained a clue as to the identity of the receptors in the plasma membrane. With a variety of techniques it can be shown that the spike glycoprotein has an affinity for the major histocompatibility antigens on the surface of mouse and human lymphoblasts (H-2 and HLA antigens). These are highly conserved polymorphic surface glycoproteins with a two-chain structure (heavy chain 44,000, and β_2-microglobulin 12,000). These antigens (or homologous glycoproteins) are present on virtually all nucleated cells. This might explain the wide host-cell specificity of SFV.

The binding of SFV to the cell membrane does not lead to fusion between the viral membrane and the plasma membrane of the cell. We could not observe fusion using thin section electron microscopy, as has been observed with Sendai and other paramyxo viruses known to fuse with the plasma membrane. Nor could we detect any insertion of SFV spike glycoproteins into the plasma membrane by complement-dependent cytotoxicity assays using anti-spike glycoprotein serum.[9]

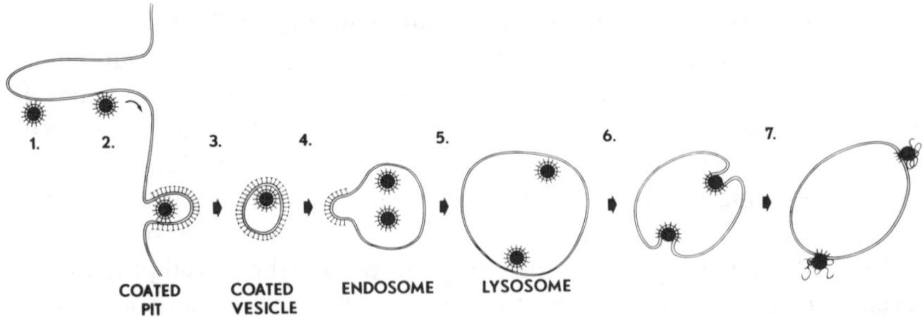

Figure 10-1 Schematic representation of the SFV entry into BHK-21 cells. For details see text.

2. Lateral displacement of SFV on the cell surface The virus is taken up into the cell by endocytosis. Using carrier-free ^{35}S-methionine virus at high and at low multiplicity, we could show that up to 90% of the cell-bound virus was taken up intact into the cytoplasm. On BHK cells we have to assume lateral movement of the virus from the microvilli to regions where the endocytosis occurs. Endocytosis in coated vesicles did not occur at the microvilli, but was frequently observed at their base.

3. Endocytosis in coated vesicles By thin section electron microscopy, we observed two ways by which the SFV was being endocytosed. Most viruses were taken up in coated vesicles (Figure 10-2). The coated vesicles had a diameter of 130 to 200 nm. They were formed through pinching-off from coated pits at the cell membrane. As a rule they contained only a single virus particle. By synchronizing the entry using a temperature shift from 0 to 37 C, it was found that the formation and passage of the coated vesicles into the cells were very rapid (5 to 20 seconds). The coated-vesicle uptake resembled in many ways that previously described for the uptake of serum low density lipoprotein.[10] Occasional viruses appeared to enter in small uncoated endocytotic vesicles, but the importance of this route of entry is not yet clear.

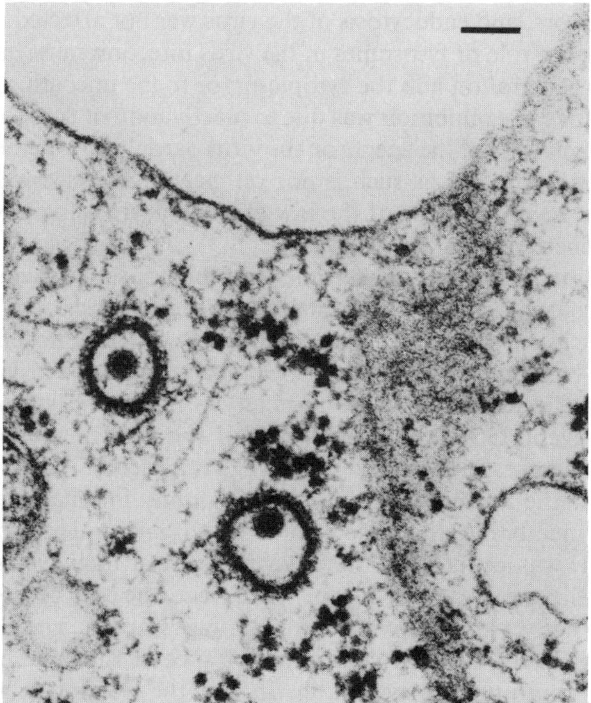

Figure 10-2 Thin section electron microscopy of SFV in coated vesicles. Bar = 0.1 μ.

4. Delivery into intracellular vacuoles Within 30 to 60 seconds after a temperature shift we found virus particles in larger uncoated or partially coated vacuoles in the cytoplasm. In contrast to the coated vesicles these contained several viruses. They were not lysosomes, as they could not be stained for the lysosomal enzyme markers (acid phosphatase and cathepsin D).

5. Entry into the lysosomes From the vacuoles, at least part of the SFV found its way into the lysosomal compartment. This was demonstrated using the lysosome-specific staining procedures. Part of the endocytosed viral protein was, in fact, degraded in the cell beginning about 30 minutes after uptake. A role for lysosomes in the actual route of productive infection was indicated by our studies on the effects of lysosomal inhibitors (so-called lysosomotropic agents) on SFV infection. We used NH_4Cl, amantadine, butylamine, trimethylamine, and chloroquine. These agents are weak bases and they accumulate in the lysosomes, owing to the low lysosomal pH. They increase the intralysosomal pH and inhibit many of the lysosomal acid hydrolases.[11] This eventually leads to disorders in the catabolism of cellular and endocytosed material.

We found that the lysosomotropic agents inhibited SFV infection. It could be shown that the inhibition took place within the first five minutes after virus entry into the cell, and it was reversible after removal of the agents from the medium. The virus bound to cells normally in the presence of the inhibitors, and endocytosis of the virus was not affected. Our results suggest that the role of lysosomes in the virus infection must be connected to the virus penetration into the cytoplasm, or to the uncoating of the viral RNA. Whether the inhibition was due to inactivation of some essential enzyme, a direct effect of the agents on the virus particle, or whether it was the result of the rise in pH as such is not yet clear. Indirect evidence speaks strongly, however, in favor of the low lysosomal pH being a requirement for virus penetration in the cytoplasm.

6. Fusion of the viral membrane with the lysosomal membrane In order to reconstruct what happens when the virus particle enters the low pH milieu in the lysosomes, we have used an in vitro system. Phospholipid- and cholesterol-containing liposomes were prepared and mixed with viruses. The pH was decreased to values normally occurring in lysosomes.[11] We found by several techniques that fusion of the viral membrane with the lysosomal membrane occurred with high efficiency. This resulted in the entry of the nucleocapsid into the liposomal internal space. Fusion of the viral membrane could also be induced with natural membranes, provided the pH was low enough. When viruses were artificially fused with the plasma membrane of cells by low pH treatment, the cells became infected.

We do not yet know how the low pH triggers fusion. The phenomenon is, however, likely to be of central importance in the final entry of the nucleocapsid into the cell cytoplasm. In vivo the reaction is probably induced by the low pH within the lysosomes, and since the fusion occurs from inside the lysosomes, it results in the release of the nucleocapsid into the cytoplasm. The inhibitory action of lysosomotropic agents is explained by the elevation of lysosomal pH.[12]

7. Uncoating, protein synthesis, and RNA replication Little is known about the final stages in the entry process. These include the uncoating of the RNA, the initial synthesis of nonstructural proteins, and the onset of RNA replication. The viral RNA polymerase is known, however, to be membrane-bound and associated with intracellular vacuoles which sediment in the lysosomal-mitochondrial organelle fraction. It is possible that the incoming viral genome remains attached to the lysosomes, and that this is the site for the early synthetic steps.

Our tentative scheme for SFV entry suggests that SFV is taken up passively by the normal cellular endocytotic apparatus.[13] The actual membrane penetration occurs first in the lysosomes, and this involves a membrane fusion event. It is too early to speculate about the generality of this mechanism. Most types of animal viruses are known to be endocytosed. Whether this is connected with the route of productive infection is a question of dispute. A survey of older literature on various different membrane-containing animal viruses shows that several compounds, now known to be lysosomotropic, are inhibitory to some early step of infection. This suggests a lysosomal involvement in the entry. It has previously been suggested that reovirus (a membrane-free DNA virus) enters through lysosomes.[14] Quite aside from the virologic aspects, our results may serve as a starting point for quantitative studies on endocytosis in tissue culture cells. It may help us to isolate the endocytotic coated vesicles. Studies on the route of virus movement inside the cell may reveal interesting aspects of the control of intracellular transport. Furthermore, the fusion of the viral membrane with liposomes and biologic membranes may serve as a simple model system to unravel the molecular mechanism of fusion.

REFERENCES

1. Dales S: Early events in cell-animal virus interactions. *Bacteriol Rev* 37:103–135, 1973.

2. Gallaher WR, Howe C: Identification of receptors for animal viruses. *Immunol Commun* 5:535–552, 1976.

3. Longberg-Holm K, Philipson L: Early interaction between animal virus and cells. *Monogr Virol* 9:1–149, 1974.

4. Meager A, Hughes RC: Virus receptors, in Cuatrecasas P, Greaves MF (eds): *Receptors and Recognition,* Series A, vol 4. London, Chapman & Hall Ltd, 1977, pp 143–187.

5. Simons K, Garoff H, Helenius A, et al: The structure and assembly of the membrane of Semliki Forest virus, in Pullman B (ed): *Frontiers of Physiochemical Biology.* New York, Academic Press Inc, 1978, pp 387–407.

6. Fries E, Helenius A: Binding of Semliki Forest virus and its isolated glycoprotein to cells. *Eur J Biochem* 97:213–220, 1979.

7. Helenius A, Kartenbeck J, Simons K, et al: On the entry of Semliki Forest virus into BHK-21 cells. *J Cell Biol* 84:404–420, 1979.

8. Helenius A, Morein B, Fries E, et al: Human (HLA-A and -B) and murine (H2-K and -D) histocompatibility antigens are cell surface receptors for Semliki Forest virus. *Proc Natl Acad Sci USA* 75:3846–3850, 1978.

9. Fan DP, Sefton BM: The entry into host cells of Sindbis virus, Vesicular stomatitis virus and Sendai virus. *Cell* 15:985–992, 1978.

10. Anderson RGW, Brown MS, Goldstein JL: Role of the coated endocytotic vesicle in the uptake of receptor bound low density lipoprotein in human fibroblast. *Cell* 10:351–364, 1977.

11. De Duve C, De Barsy T, Poole B, et al: Lysosomotropic agents. *Biochem Pharmacol* 23:2495–2531, 1974.

12. Ohkuma S, Poole B: Fluorescence probe measurement of the intralysosomal pH in living cells and the perturbation of pH by varying agents. *Proc Natl Acad Sci USA* 75:3327–3331, 1978.

13. Goldstein JL, Anderson RGW, Brown MS: Coated pits, coated vesicles and receptor mediated endocytosis. *Nature* 279:679–685, 1979.

14. Silverstein SC, Dales S: The penetration of reovirus RNA and initiation of its genetic function in L-strain fibroblasts. *J Cell Biol* 36:197–230, 1968.

11 The Association of Major Histocompatibility Complex Gene Products with Schistosomula of *Schistosoma mansoni*

Alan Sher, PhD*

In the process of adapting to the host environment, extracellular parasites have developed highly sophisticated surface structures. In addition to acting as an interface between the physiologic milieus of host and parasite, these structures play a vital role in protecting the parasitic organism against destruction by the immune system.

A well-studied example of such a surface structure is the tegumental membrane of the schistosome, a trematode parasite of man and other vertebrate hosts. This organism develops in the tissues of its snail intermediate host and is released into fresh water as an infective larva known as a cercaria. When the cercaria encounters a susceptible definitive host (eg, man), it burrows into the skin and undergoes a rapid transformation into a schistosomulum, a larval stage now physiologically adapted for life in the

The work described here was funded by grants from the Edna McConnell Clark Foundation, the Rockefeller Foundation, and the US-Japan Program of the National Institute of Allergy and Infectious Diseases.

*Present address: Laboratory of Parasitic Disease, National Institutes of Health, Bethesda, MD 20205.

vertebrate environment. During this remarkable transformation process, the parasite sheds its locomotive forked tail, sloughs off its external glycocalyx, and adds a second lipid bilayer to its tegument, producing a double outer membrane,[1] a surface structure common to all blood flukes.[2]

Immediately after host penetration the schistosomula undergo another fascinating transition, involving their susceptibility to immune damage. As described in the *Schistosoma mansoni*/mouse experimental model, within a 24-hour period after their transformation from cercariae the parasites become refractory to antibody-dependent killing mediated by passively transferred immune serum.[3] In the mouse, this insusceptibility to antibody-dependent damage is maintained throughout the remainder of the development of the worm, as well as its extended adult life in the host.

A number of structural and physiologic changes have been documented to occur in developing schistosomula, during this critical period when they lose their susceptibility to immune damage. Perhaps the most intriguing change is the surface acquisition of host molecules. The phenomenon of host molecule acquisition by schistosomes was first described by Smithers and colleagues, who studied the rejection of adult worms of xenogeneic origin by monkeys immunized against blood cells of the donor species.[4] By means of these worm transfer experiments and mixed agglutination techniques, they were able to demonstrate that adult worms bear on their surface immunochemically defined molecules derived from their donor host. Smithers and his colleagues speculated that these acquired host molecules might act as a disguise, protecting the surface of the older parasite from recognition by host antibody.[4] In support of this hypothesis, it was later shown that as developing schistosomula become refractory to immune damage they lose the capacity to bind host antibodies.[5,6] In the mouse, host molecules were shown to be acquired during the same transition period.[7] By means of experiments in which schistosomula were cocultured in the presence of human blood from individuals with different blood group types, it was later shown that the A, B, H, and Lewis blood group glycolipids are acquired by the surface of developing larvae.[8] Since, in the same experiments, glycoprotein blood group antigens (eg, M, N, rhesus) were not transferred to the parasite tegument, it was hypothesized that schistosomes acquire host molecules principally in the form of glycolipids.[8]

In our laboratory we have employed a somewhat different approach to studying the immunochemical identity of the host molecules acquired by schistosomes. Using the mouse system, we have asked whether five-day-old lung-stage schistosomula express genetically defined alloantigens on their surface.

The general protocol used in these experiments is outlined in Figure 11-1. Briefly, mice of different inbred (or inbred congenic or recombinant) strains are infected by injecting them intravenously with freshly transformed schistosomula. Five days later the parasites are recovered from the lungs of the animals, washed thoroughly, and tested for their reactivity with alloantisera, prepared by cross-immunization of mice of each of the donor strains or with other antisera with defined allospecificity. The reaction of the antisera with the worms is monitored by the technique of indirect immunofluorescence (Figure 11-1).

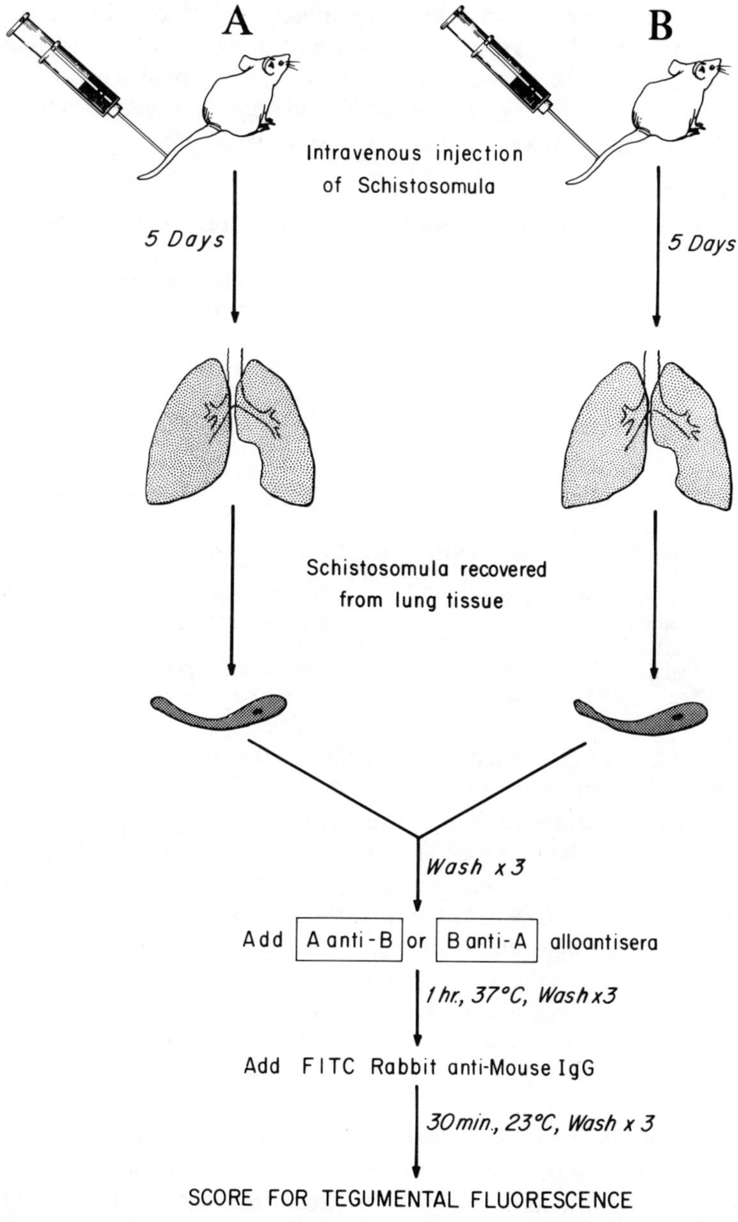

Figure 11-1 Protocol for detection of murine alloantigens on schistosomula.

Using this methodology in conjunction with broad-specificity alloantisera produced by cross-immunizing distantly related inbred mice, we were able to establish that lung-stage schistosomula do indeed express murine alloantigens on their tegumental membranes (Table 11-1).[9] Furthermore, by employing as worm donors congenic mouse strains differing in H-2 haplotype we were able to demonstrate (Table 11-1) that among these alloantigens acquired by the larvae are the products of the major histocompatibility complex (MHC). However, it was also apparent from the data that other alloantigens not encoded by the murine MHC are also expressed on the worm surface. We refer to these MHC and non-MHC gene products as "major" and "minor" host alloantigens, respectively.[9,10]

Table 11-1
Surface Expression of Murine Alloantigens by Lung-Stage Schistosomula*

		Immunofluorescent Staining of Worms with Antisera:	
Worm Donor	H-2 Haplotype	*C57 anti-C3H*	*C3H anti-C57*
C3H/HE	k	+	–
C57BL/6	b	–	+
B10/Sn	b	–	+
B10.BR	k	+	+

*The reaction of anti-C3H antisera with worms from C3H but not C57BL donors, and conversely the reaction of anti-C57 antisera with C57BL but not C3H worms, establishes the presence of alloantigens on the parasites. The reaction of anti-C3H antisera with B10.Br (H-2^k) worms but not with B10/SN (H-2^b) worms demonstrates the presence of MHC-encoded substances on the larvae. In addition, because anti-C57 antisera react with H-2^k (B10.Br) worms, it is evident that other non-MHC ("minor") alloantigens are also present on the parasite surface. From Sher et al.[9] Data reproduced with permission of Rockefeller University Press.

The presence of MHC-encoded molecules on the surface of lung-stage parasites was confirmed by the use of highly specific antisera directed against products of defined regions of that genetic locus. In this manner we were able to establish the presence of K and I region products on the tegument of the larvae.[9] More recently D region products have also been identified on the worms (B.F. Hall and A. Sher, unpublished data, 1979). In addition to lung-stage schistosomula, MHC gene products were also found on the surface of other developmental stages of the parasite in mice: 1- and 2-day-old schistosomula recovered from skin,[9] and 20-day-old worms obtained by hepatic perfusion (L. Hall and A. Sher, unpublished data, 1979).

Although products of the K, I, and D regions of the MHC could be identified on the worm surface, other mouse cell-surface alloantigens and host proteins could not be detected by the same methods (Table 11-2). The K, I, and D region products of the MHC have been characterized as integral membrane glycoproteins (ie, glycoproteins which span the lipid bilayer of the host cell membrane).[11] For this reason, our recent experiments (A. Sher, unpublished data, 1979), in which we have been unable to detect T-200 on the worm surface, are of particular interest. T-200 is a membrane glycoprotein originally identified on mouse lymphocytes by Trowbridge and his colleagues.[12] The molecule was later shown to have a broad tissue distribution.[13] Indeed the monoclonal rat anti-T-200 reagent (supplied to us by Dr Trowbridge) reacts strongly with purified mouse lymphocytes, macrophages, neutrophils, and eosinophils (S. James, unpublished data, 1979), but fails to stain lung-stage schistosomula. In addition to their broad tissue distribution, T-200 and the K, I, and D MHC molecules share another important biologic property. T-200, like the K, I, and D gene products, has been shown to be an integral membrane glycoprotein (I. Trowbridge and B. Omary, personal communication). Moreover, it appears to be closely associated with H-2 proteins in the lymphocyte membrane as judged by cocapping experiments.[14] Therefore, the apparent absence of T-200 on the surface of MHC-product–bearing schistosomula (Table 11-2) is an important argument for the selective association of K, I, and D molecules with the parasite surface. Nevertheless, as discussed earlier, other minor alloantigens have been detected on the surface of lung-stage schistosomula (see Table 11-1). The identity and structural properties of these non-MHC host molecules have not been characterized.

Table 11-2
Host Molecules Associated with the Surface of Lung-Stage Schistosomula Recovered from Mice

Molecules Detected	Molecules Not Detected*
K,† IA,† D,† MHC products	C3
	Ig
Unidentified minor alloantigens	Albumin
	Thy-1
	Ly-1
	H-Y
	T-200†
	β_2-Microglobulin

*Positive immunofluorescence controls were performed to show that each of the antisera employed were reactive with mouse tissue.
†Integral membrane glycoproteins.

A selectivity in the association of MHC gene products with schistosomula also appears to occur at the level of the molecules themselves. The K and D MHC gene products each consist of a heavy chain and a light chain. The heavy chain subunit (molecular weight 44,000) bears the H-2 serologic polymorphism and spans the membrane of the host cell. The light chain (molecular weight 12,000), which has been shown to be identical to plasma β_2-microglobulin, is noncovalently associated with the heavy chain and is situated on the external surface of the lipid bilayer of the cell.[11] Employing a high titer anti-mouse β_2-microglobulin antiserum (provided by Dr Tanagaki), we have been unable to detect β_2-microglobulin on the surface of schistosomula bearing K and D MHC-encoded host molecules, despite the fact that the antiserum produces strong immunofluorescence reactions with lymphocytes obtained from the mice serving as worm donors. This finding suggests either that K and D product heavy chains are transferred to the parasite surface without their light chains, or that the intact H-2 molecule is situated in the worm tegument in a manner in which the light chain is buried and not accessible to antibody.

It is apparent from the results discussed above that contrary to the previous hypothesis,[8] glycolipids are not the only host molecules which can associate with the schistosome tegument. Clearly, integral membrane glycoproteins are also acquired by the parasites in a manner that appears to a certain degree to be selective. What is the origin of these molecules? How are they associated with the tegument of the parasite? What is the mechanism by which these molecules are selectively transferred from host to parasite?

At the moment we are a long way from having definitive answers to these questions. However, we do have several important clues. For instance, it is evident from several lines of investigation that MHC gene products are associated with the schistosome tegument in an unusually stable fashion. Thus, in our immunofluorescence experiments on lung-stage schistosomula we never observe capping, shedding, or patching of MHC antigens. In addition, we are still able to detect MHC antigens of donor origin on lung-stage larvae after three days of in vitro culture and as late as 87 hours after re-infection of allogeneic recipients.[9] That the acquired host material is intimately associated with the parasite surface is also suggested by transmission electron microscopy, which shows that the tegument of lung-stage schistosomula is uniform in appearance with no evidence of adherent host cellular debris.[15]

The origin of schistosome-associated MHC products appears to be cellular rather than humoral. We base this hypothesis on the observation that schistosomula acquire MHC antigens when cocultured with mouse lymphocytes, but fail to do so when cultured in mouse serum.[9] The question of whether contact between cells and schistosomula is required for MHC antigen uptake has not been resolved. Recently, Caulfield and his colleagues have reported that when human granulocytes bind to schistosomula in vitro by means of antibody- or complement-dependent adherence reactions, fusion events between the outer leaflets of the cells and the worms are observed.[16] As discussed below, these cell-parasite fusions, if they occur in vivo, could explain the uptake of integral membrane glycoproteins by schistosomula.[16]

If, as the data suggest, the MHC products present on schistosomula are derived from host cells, one can conceive of three alternate mechanisms by which integral membrane proteins of this type could be transferred from cell to parasite. These models of host molecule transfer are diagrammed schematically in Figure 11-2. In the first model, integral proteins are acquired by the parasites as a consequence of cell membrane turnover. Turnover of MHC antigens by cultured cells has been reported by numerous investigators. In most of these studies, the MHC molecules were found to be rapidly degraded into small molecules after release into the supernatant.[17] However, Koch and Smith have observed that P815 tissue culture cells spontaneously shed microvilli in which actin and H-2 are the major components.[18] The molecules exist together in the microvillous structures as a noncovalently bound protein complex. Conceivably the schistosomulum could adsorb this exfoliated material on its tegument and thus acquire H-2 in a selective fashion. If this mechanism were operative, one would predict that the H-2 antigens present on schistosomes might be associated with actin molecules.

A second mechanism by which MHC K and D antigens could be transferred from cell to schistosome membranes would involve the proteolytic excision of the heavy chain of the molecule from the cell lipid bilayer and the binding of the excised polypeptide chain to the worm surface (Figure 11-2). The enzymatic cleavage of the H-2 molecule could conceivably be mediated by a tegument-associated or secreted worm protease. This model predicts that the H-2 heavy chains associated with the worm surface will be found to be smaller in size than the 44,000 molecular weight component associated with host cell membranes.

The third and last model of host molecule transfer outlined in Figure 11-2 is the one suggested by Caulfield et al[16] on the basis of their observation of fusions between granulocytes and schistosomula, already mentioned

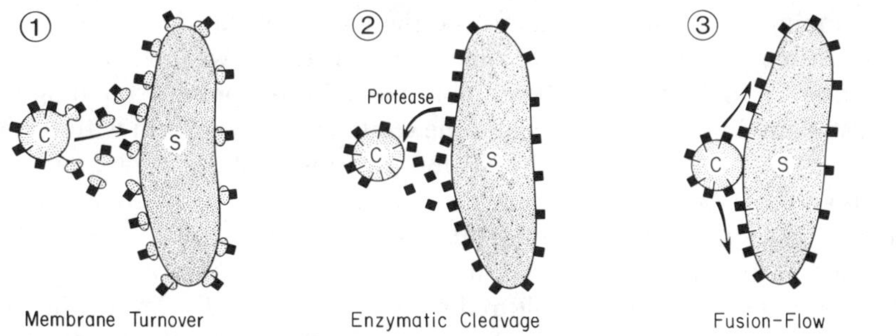

Figure 11-2 Mechanisms by which integral membrane proteins could be transferred from host cells to the schistosomulum surface. C = cell; S = schistosomulum. See text for description.

above. According to this model, after the formation of cell-parasite junctions, H-2 molecules would flow from the lipid bilayer of the cell across into the outer leaflet of the worm tegument. If this mechanism were operative one might expect the H-2 heavy chains of host cells and schistosomula to be equal or approximately equal in molecular size. However, the model fails to explain the partial selectivity of MHC acquisition by the parasite, and in particular the apparent absence on the tegument of T-200, which like H-2 is a common integral membrane protein.

At present the data do not distinguish between the three hypothetical mechanisms of MHC molecule transfer outlined above. Clearly, more information is needed concerning the molecular size of the worm-bound MHC products, the possible requirement for cell-parasite contact in host molecule transfer, and the specificity of the MHC antigen-schistosomulum association. Relevant to the last question is the recent demonstration of immunoglobulin Fc receptors on newly transformed schistosomula by Torpier and coworkers.[19] These investigators demonstrated that the worm Fc receptors also bind human β_2-microglobulin, presumably because of its homology with the third domain of the IgG heavy chain, and speculated that the same receptors might be responsible for the binding of H-2 via its β_2-microglobulin subunit to the parasite surface. This hypothesis is consistent with our inability to detect β_2-microglobulin on the surface of lung-stage schistosomula; if the light chains of the intact H-2 molecule were buried in tegumental Fc receptors one might not expect them to be accessible to antibody binding.

Although the elucidation of the mechanism by which MHC molecules are transferred from host cells to schistosomes is a fascinating problem in membrane chemistry, the critical question raised by the association of MHC gene products with the parasite concerns the biologic significance of the phenomenon. As originally proposed by Smithers and his colleagues,[4] host molecules might serve an important biologic function in protecting the surface of the schistosome from antibody-dependent immune killing. Recent data suggest, however, that the acquisition of host molecules by developing schistosomula cannot be the sole explanation of their loss in susceptibility to immunologic damage. For instance, we have shown that it is possible to bypass the host molecule disguise present on lung-stage schistosomula by surface-labeling the tegument of the worm with trinitrophenyl (TNP) groups. These TNP-lung worms bind the same quantity of anti-TNP antibody per unit surface area as TNP-early schistosomula, yet are not as susceptible to anti-TNP antibody-dependent killing as TNP-labeled early schistosomula.[20] Similarly, a number of investigators have shown that schistosomula cultured in vitro in chemically defined (ie, host-molecule–free) medium lose their susceptibility to antibody-dependent damage[21,22] (A. Dessein, J. Samuelson, A.E. Butterworth, et al, unpublished data, 1979). These findings suggest that a change in the surface of the schistosomulum, independent of its acquisition of host molecules, is responsible for the parasite's survival in the face of immunologic attack.

If MHC gene products, as acquired host molecules, are not important in disguising and thereby protecting the surface of the schistosome from immune damage, what is the functional significance of their association with the worm surface? MHC gene products have been shown to play an important role in immune recognition by governing the cell-cell interactions involved in lytic and regulatory immunologic events.[23,24] This consideration led us to explore the possibility that schistosome-associated MHC molecules might be involved in the recognition of the parasite by T lymphocytes and in particular by that class of T cell, the cytotoxic lymphocyte (T_C), which mediates the lysis of cellular targets. We were able to demonstrate that T_C generated against MHC products do indeed recognize these antigens on lung-stage schistosomula, as judged by their specific adherence to parasites recovered from inbred mice with the appropriate MHC haplotype.[25] Furthermore, in a subsequent study[10] we were able to demonstrate that schistosome-associated MHC molecules restrict the recognition of other antigens (ie, TNP groups or minor alloantigens) on the worm surface by T_C generated against the same antigens. Although in these studies T_C were found to adhere to schistosomula in an MHC-dependent manner, the interaction was observed not to result in damage to the parasite targets.[25] Thus, worm-associated MHC molecules are apparently not important in promoting the destruction of schistosomula by lymphocytic effector cells. The possibility that the molecules may play some alternative immunoregulatory function with regard to the induction or suppression of responses directed against the parasite is currently under investigation in our laboratory.

I am grateful to my colleagues, J. Caulfield, M. Vadas, D.J. Pratt, J. Lazdins, S. James, and G. Moser, for their helpful discussions and contributions to this work. I also thank I. Trowbridge and N. Tanagaki for their generosity in supplying antisera for our experiments. Finally, I thank Dr John David for his continued advice and support.

REFERENCES

1. Hockley DJ, McClaren DJ: *Schistosoma mansoni:* changes in the outer membrane of the tegument during development from cercaria to adult worm. *Int J Parasitol* 3:13–25, 1973.

2. McLaren DJ, Hockley DJ: Blood flukes have a double outer membrane. *Nature* 269:147–149, 1977.

3. Sher A: Immunity against *Schistosoma mansoni* in the mouse. *Am J Trop Med Hyg* 26:20–28, 1977.

4. Smithers SR, Terry RJ, Hockley DJ: Host antigens in schistosomula. *Proc R Soc Lond* 171:483–494, 1969.

5. Goldring OL, Sher A, Smithers SR, et al: Host antigen and parasite antigens of murine *Schistosoma mansoni. Trans R Soc Trop Med Hyg* 71:144–148, 1977.

6. McLaren DJ, Clegg JA, Smithers SR: Acquisition of host antigens by young *Schistosoma mansoni* in mice: correlation with failure to bind antibody *in vitro. Parasitology* 269:147–149, 1975.

7. Imohiosen EAE, Sher A, von Lichtenberg F: Early developmental changes of the schistosomula of *Schistosoma mansoni in vitro* and in mouse lung. *Parasitology* 76:317–326, 1978.

8. Goldring OL, Clegg JA, Smithers SR: Acquisition of human blood group antigens by *Schistosoma mansoni. Clin Exp Immunol* 26:181–187, 1976.

9. Sher A, Hall BF, Vadas MA: Acquisition of murine major histocompatibility complex gene products by schistosomula of *Schistosoma mansoni. J Exp Med* 148:46–50, 1978.

10. Vadas MA, Butterworth AE, Burakoff S, et al: Major histocompatibility complex products restrict the adherence of cytolytic T lymphocytes to minor histocompatibility antigens or to trinitrophenyl determinants on schistosomula of *Schistosoma mansoni. Proc Natl Acad Sci USA* 76:1982–1985, 1979.

11. Vitetta ES, Capra DJ: The protein products of the murine 17th chromosome: genetics and structure. *Adv Immunol* 26:147–193, 1979.

12. Trowbridge IS, Nilson-Hamilton M, Hamilton RT, et al: Preliminary characterization of two thymus dependent xenoantigens from mouse lymphocytes. *Biochem J* 163:211–217, 1977.

13. Trowbridge IS: Interspecies spleen-myeloma hybrid producing monoclonal antibodies against mouse lymphocyte surface glycoprotein, T-200. *J Exp Med* 148: 313–323, 1978.

14. Bourguignon LY, Hyman R, Trowbridge I, et al: Participation of histocompatibility antigens in capping of molecularly independent cell surface components by their specific antibodies. *Proc Natl Acad Sci USA* 75:2406–2410, 1978.

15. Clegg JA: The schistosome surface in relation to parasitism, in Taylor AE, Muller R (eds): *Functional Aspects of Parasite Surfaces,* Symposium No. 10 of the British Society for Parasitology. Oxford, Blackwell Scientific Publications Ltd, 1972.

16. Caulfield JC, Butterworth AE, Korman G, et al: Antibody and complement mediated junctions between *Schistosoma mansoni* and polymorphonuclear leukocytes. *Fed Proc* 38:908, 1979.

17. Cresswell P, Springer T, Strominger JL, et al: Immunological identity of the small subunit of HL-A antigens and β_2-microglobulin and its turnover on the cell membrane. *Proc Natl Acad Sci USA* 71:2123–2127, 1974.

18. Koch GL, Smith MJ: An association between actin and the major histocompatibility antigen H-2. *Nature* 273:274–278, 1978.

19. Torpier G, Capron A, Ouassi M: Receptor for IgG(Fc) and human β_2-microglobulin on *S. mansoni* schistosomula. *Nature* 278:449, 1979.

20. Moser G, Hall BF, Sher A: Trinitrophenyl groups as target antigens in the antibody dependent killing of schistosomula of *Schistosoma mansoni. Fed Proc* 38:1275, 1979.

21. Dean DA: Decreased binding of cytotoxic antibody by developing *Schistosoma mansoni.* Evidence for a surface change independent of host antigen adsorption and membrane turnover. *J Parasitol* 63:418–426, 1977.

22. Tavares CAP, Soares RC, Coelho PMZ, et al: *Schistosoma mansoni:* evidence for a role of serum factors in protecting artificially transformed schistosomula against antibody mediated killing *in vitro. Parasitology* 77:225–233, 1978.

23. Klein J: The major histocompatibility complex of the mouse. *Science* 203:516–521, 1979.

24. Paul WE, Benacerraf B: Functional specificity of thymus dependent lymphocytes. *Science* 195:1293–1300, 1977.

25. Butterworth AE, Vadas MA, Martz E, et al: Cytolytic T lymphocytes recognize alloantigens on schistosomula of *Schistosoma mansoni* but fail to induce damage. *J Immunol* 122:1314–1321, 1979.

SECTION III
Membranes and Superoxide

12 Chemical and Biological Aspects of the Superoxide Radical and of Superoxide Dismutases

Irwin Fridovich, PhD

The aerobic lifestyle provides enormous advantages when compared with an anaerobic existence. The aerobic oxidation of glucose yields approximately 15 times more energy than does its anaerobic conversion to lactate. Oxygen can also be used to good advantage in facilitating specific metabolic interconversions, such as hydroxylations. Yet, molecular oxygen is not an unmixed blessing. It is a peculiar oxidant in that its reduction to water, which requires four electrons, proceeds most readily by a stepwise pathway that involves dangerously reactive intermediates. Oxygen thus constitutes both a benefit and a threat to living things, and those organisms that have availed themselves of its benefits have, of necessity, developed a series of defenses against its dangers. The first intermediate encountered during the stepwise or univalent·pathway of oxygen reduction is the superoxide radical (O_2^-). A family of enzymes, called superoxide dismutases (SODs), is the countermeasure that has been evolved to minimize the potential toxicity of this radical. In this essay we shall focus upon the chemistry and biology of the superoxide radical and of its enzymic scavengers, the superoxide dismutases.

The spin restriction Molecular oxygen, in the ground state, is paramagnetic because it contains two unpaired electrons in parallel spin states. The direct divalent reduction of this molecule, by insertion of a pair of electrons, would result in the disallowed situation of two electrons with parallel spins occupying the same orbital. This is the spin restriction, and it makes molecular oxygen much less reactive than we would otherwise expect it to be. There are three ways to circumvent this spin restriction. The first depends on electronic excitation. This moves one of the electrons from the ground state orbital to a more energetic orbital and inverts its spin state. An input of a minimum of 23 kcal/mole thus converts the triplet ground state to a very reactive singlet excited state, designated $^1\Delta g$. The second involves ligating the oxygen to a transition metal that contains unpaired electron spins of its own. Such a complex can accept a pair of electrons without violating the need to conserve angular momentum. The third avoids the spin restriction by adding electrons to oxygen one at a time. This univalent pathway necessarily involves the production of intermediates, the first of which is O_2^-. The spin restriction and the means of circumventing it have been presented in greater detail elsewhere.[1-5] For our purposes it is sufficient that the univalent pathway of oxygen reduction is commonplace because of the spin restriction, and that O_2^- therefore is frequently encountered in oxidation reactions. We need to explore some of the properties of O_2^-.

pKa and spontaneous dismutation The superoxide anion, $O_2^-\cdot$, is the conjugate base of a weak acid, the perhydroxyl radical, $HO_2\cdot$, whose pKa is 4.8.[6] Its lifetime in protic solvents, such as water, is limited by a spontaneous dismutation reaction whose products are H_2O_2 and O_2. $HO_2\cdot$ reacts rapidly with itself and with O_2^-. In contrast, O_2^- does not readily react with itself, because of mutual electrostatic repulsion and because of the difficulty of producing O_2^- in water. The pertinent reactions and their rate constants are[6,7]:

$$(1)\quad HO_2\cdot + HO_2\cdot \rightarrow H_2O_2 + O_2 \qquad k = 7.6 \times 10^5\ M^{-1}\ sec^{-1}$$

$$(2)\quad HO_2\cdot + O_2^- + H^+ \rightarrow H_2O_2 + O_2 \qquad k = 8.5 \times 10^7\ M^{-1}\ sec^{-1}$$

$$(3)\quad O_2^- + O_2^- + 2H^+ \rightarrow H_2O_2 + O_2 \qquad k \leqslant 0.3\ M^{-1}\ sec^{-1}$$

As might be expected, reaction (3) is catalyzed by transition metal cations, and the spontaneous rate is difficult to measure because of the effects of trace metal contaminants. For that reason Behar et al[6] gave the rate constant as being less than 100 M^{-1} sec^{-1}. More recently a rate constant of 6 M^{-1} sec^{-1} has been reported,[8] and Bielski and Allen, through the use of EDTA to trap trace metals, have measured a value of 0.4 M^{-1} sec^{-1}.[7]

Optical and EPR spectra HO_2 absorbs at 230 nm where E_m = 1220 M^{-1} cm^{-1},[6] and O_2^- absorbs maximally at 240 nm where E_m = 2000 M^{-1} cm^{-1}.[9] The EPR spectrum of $HO_2\cdot$ has been observed from aqueous solutions at g = 2.014 with a line width of 27 G.[10–13] The EPR spectrum of O_2^- has been observed in frozen gasses but not in free solution, because interaction with the solvent would provide an efficient pathway for spin relaxation.[14] Rapid freezing of aqueous solutions has permitted detection of an EPR spectrum for O_2^- where $g_\perp$ = 2.00 and $g_\parallel$ = 2.08 to 2.09.[15] This methodology was applied to the detection of enzymically generated O_2^-.[16,17]

Preparation of O_2^- Superoxide can be prepared by the univalent reduction of O_2 or the univalent oxidation of H_2O_2. The spontaneous dismutation reactions, already discussed above, limit the lifetime of this radical in protic solvents, so that relatively stable solutions can be prepared only in nonprotic solvents. Cathodic reduction has been used to prepare stable solutions of soluble salts of O_2^-, such as the tetrabutyl ammonium salt, in such nonprotic solvents as dimethyl sulfoxide.[18–21] This method has been used by several workers interested in exploring the reactivities of O_2^-.[22–25] Potassium superoxide is commercially available, but its utility is limited by poor solubility in nonprotic solvents. This has been overcome through the use of crown ethers to solubilize this salt.[26] The cathodic reduction of oxygen is a univalent process even in aqueous solution.[27] Coating the cathode with a hydrophobic surfactant facilitates the escape of the O_2^- from the vicinity of the cathode.[8,28,29] Cathodic reduction of O_2 could thus serve as a continuous source of O_2^- even in aqueous solutions. Solid tetraalkyl ammonium superoxide has been prepared by reacting solid KO_2 with the tetraalkyl ammonium hydroxide, after which the desired product was extracted in liquid NH_3 and recovered by evaporating the NH_3.[30] Hydrated electrons produced during photolysis[31] or radiolysis[32–34] of water react rapidly with dissolved O_2, giving O_2^-. All of these methods are potentially useful in preparing O_2.

Oxidation-reduction potential Measuring the redox potential of a reactive free radical is not a straightforward undertaking. It is not surprising, therefore, that the literature holds discrepant values for this constant. The present consensus is that E_0 for O_2^-/O_2 is in the range of -0.27 to -0.33 V.[8,28,35–37] O_2^- is thus a good reductant, comparable to the reduced pyridino coenzymes. It is also a strong oxidant. Thus E_0 (H_2O_2/O_2^-) = $+1.71$ V.[35]

The reactivity of O_2^- Our knowledge of the chemical reactivity of O_2^- is in its infancy. This is probably so because until recently O_2^- was of interest only to a small group of radiation chemists. The realization that O_2^- is generated in biologic systems has led many investigators to consider its reactivities. The results of this new interest are just beginning to appear in the literature. O_2^- rapidly reduces ferricytochrome c,[38-41] methemoglobin,[42] nitroblue tetrazolium,[43] and tetranitromethane.[44] Indeed these reductions of cytochrome c and of nitroblue tetrazolium have been made the bases of assays for SOD.[23,43] O_2^- acts as an oxidant toward numerous substances including oxyhemoglobin,[42] catechols,[45] catecholamines,[23,46,47] pyrogallol,[48] sulfite,[49] and ascorbate.[50] With the catecholamines and pyrogallol, O_2^- functions as a chain carrier in a free-radical chain reaction. The enzymes which scavenge O_2^- then prevent chain propagation and act to inhibit these autoxidations; this has also been made the basis for convenient assays of these enzymes.[46-48] The reactivity of O_2^- to nitrones has been used in spin trapping,[51,52] and its reactivity with alkyl halides appears to provide a useful synthetic route for preparing dialkyl peroxides.[53,54]

Reaction of O_2^- with H_2O_2 In 1970 we observed[55] that methional was converted to ethylene when exposed to the aerobic xanthine oxidase reaction. Since this enzymic reaction was known to produce both O_2^- and H_2O_2, and since the production of ethylene was inhibited by SOD or by catalase, it appeared that both O_2^- and H_2O_2 were essential reactions. Furthermore there was a lag in ethylene production which could be eliminated by exogenous H_2O_2. Haber and Weiss, in 1934,[56] had proposed the reaction:

$$(4)\qquad O_2^- + H_2O_2 \rightarrow OH^- + OH\cdot + O_2$$

This was invoked to explain the properties of the ethylene production from methional.[55] Support for this view was gained from the observation that benzoate or ethanol, which rapidly scavenge $OH\cdot$ but not O_2^- or H_2O_2, also inhibited ethylene production. We noted that "the striking conclusion that so reactive a radical as $OH\cdot$ can be generated in significant amounts in an aqueous enzymatic system appears to be well supported by the data presented."[55] Since then the Haber-Weiss reaction has been repeatedly invoked,[46,57-71] and an organic analog of it has been described.[30] Furthermore, pulse radiolysis has recently been used to show that methional does react rapidly with $OH\cdot$ but not with O_2^-.[72] Yet there are indications that the Haber-Weiss reaction, as written above, has no reality.[73] How then explain the many cases[46,55,57-71] where a reaction was inhibited by either SOD or catalase or by scavengers of $OH\cdot$? One possibility is that metal (Me) catalysis is required, so that the Haber-Weiss reaction is really the sum of the two separate reactions:

$$(5)\qquad Me(II) + H_2O_2 \rightarrow Me(III) + OH^- + OH\cdot$$

$$(6)\qquad Me(III) + O_2^- \rightarrow Me(II) + O_2$$

In that case, the overall process could only be observed in reaction mixtures containing the requisite metal catalysts. McCord and Day have indeed observed that iron salts are potent catalysts of the Haber-Weiss reaction, even in the presence of EDTA.[74]

Enzymatic dismutation of O_2^- The superoxide dismutases catalyze the reaction:

$$(7)\ O_2^- + O_2^- + 2H^+ \rightarrow H_2O_2 + O_2$$

The rate of reaction of SOD with O_2^- is $2 \times 10^9\ M^{-1}\ sec^{-1}$,[75-79] which appears to be close to the diffusion limit. The catalytic efficiency of the SOD from bovine erythrocytes is not affected by varying the pH from 5.5 to 9.5. All of the superoxide dismutases thus far described are metallo-enzymes (E = Me) whose mechanism can be minimally described by the reactions:

$$(8)\ \rightarrow\ E\text{-}Me(n) + O_2^- \qquad E\text{-}Me(n-1) + O_2$$

$$(9)\ \rightarrow\ E\text{-}Me(n-1) + H^+ \qquad EH^+\text{-}Me(n-1)$$

$$(10)\ \rightarrow\ EH^+\text{-}Me(n-1) + O_2^- \qquad E\text{-}Me(n) + HO_2^-$$

In the case of the bovine erythrocyte SOD, which contains Zn(II) and Cu(II) joined by an imidazolate bridge,[80] it has been proposed[81] that the imidazolate bridge is released from the copper during its reduction by O_2^- (reaction 8), and that this bridging ligand then protonates (reaction 9), and donates a proton to the second O_2^-, which reoxidizes the Cu(I) (reaction 10).

Physicochemical properties of the enzymes Superoxide dismutases containing copper and zinc, manganese, or iron have been described. Their physicochemical properties have been reviewed recently.[82] The x-ray diffraction analysis that was so successful with the Cu-Zn-SOD[80] is now being applied to the other types of SODs and a preliminary report on the Fe-SOD from *Pseudomonas ovalis* has appeared.[83] In all cases it has been possible to reversibly remove the metal prosthetic group, and to observe that activity was lost with the metal and was regained during reconstitution. This is true for the Cu-Zn-SOD from bovine erythrocytes,[23,84] the Mn-SOD from *Escherichia coli*[85] and from *Bacillus stearothermophilus*,[86] and the Fe-SOD from *P ovalis*.[87] In no case has it proven possible to substitute for the native active-site metal without loss of activity. The zinc in the bovine erythrocyte enzyme can be replaced by several other divalent metals; the copper is the metal which participates in the redox cycle of this enzyme, and the role of the zinc in this case is primarily structural.[88] It is clear that the active sites of these structurally diverse, yet functionally similar enzymes have been precisely tuned to the properties of the metals they are found to contain.

Is O_2^- produced in living systems? Because of the spin restriction that hinders the divalent reduction, the univalent pathway of oxygen reduction is favored. O_2^- is thus a common intermediate of oxygen reduction. Numerous substances, of biologic relevance, have been found to autoxidize and in so doing to generate O_2^-. Among those which have been studied are the reduced flavins,[38,43,89] hydroquinones,[89] catecholamines,[47] tetrahydropterins,[90-92] hemeproteins,[93-97] and reduced ferredoxins.[98-103] Several oxidative enzymes have been shown to produce substantial amounts of O_2^-. The best characterized of these enzymes is the xanthine oxidase from bovine milk.[23,49,104] Proceeding from simpler to more complex systems we next encounter subcellular organelles. The production of O_2^- by submitochondrial particles[105–111] and by fragments of chloroplasts[50,51,59,112–125] has been well documented. Indeed whole cells, of a specialized type, have been shown to produce O_2^- and to release it to the suspending medium. The activation of phagocytes has long been associated with a burst of cyanide-insensitive respiration. Phagocytic engulfment by these cells is independent of oxygen, but phagocytic killing of engulfed microorganisms is dependent on this extra oxygen consumption. It is now known that O_2^- is the primary product of this phagocytic oxygen reduction.[126–133] The responsible enzyme is a membrane-bound NADPH oxidase which has been called superoxide synthetase.[134] It is thus clear that the superoxide radical ($O_2^{-}\bullet$) is produced in respiring cells.

Is O_2^- cytotoxic? When *E coli* are exposed to a flux of O_2^-, produced photochemically or by the aerobic action of xanthine oxidase on xanthine, they are killed, and SOD prevents this lethality.[135] A similar observation was made using *Staphylococcus epidermidis.*[57] It seems likely that an early report[136] of the antibacterial action of the xanthine oxidase reaction was actually due to the effects of O_2^-. Ionizing radiation is known to generate O_2^- in aerated aqueous systems, and the ability of SOD to diminish the damage caused by such radiation is another indication of the toxicity of O_2^-.[71,137–144] Submitochondrial particles are known to produce O_2^- in the presence of an electron source plus O_2. Incubation of such particles in the presence of NADH plus O_2 causes a gradual decline in respiratory capacity, and this decline is prevented by SOD or by such noncatalytic scavengers of O_2^- as cytochrome c, tiron (3,4-dihydro-*m*-benzenadenylsulfonic acid), or Mn(II)-EDTA.[111] Catalase and other proteins had no effect. It is clear that the submitochondrial particles were being damaged by autogenous O_2^-. A somewhat similar observation has been made at the level of whole cells. Activated neutrophils produce O_2^- and H_2O_2, and a suspension of such cells suffers a loss of viability that could be prevented by SOD or catalase.[145] Viability was assessed in terms of the integrity of the limiting membrane, as measured by exclusion of trypan blue. It seems clear that O_2^- is cytotoxic. Actually, the most convincing support for this contention comes from experiments in which the intracellular level of SOD was manipulated; this will be presented shortly.

Is the dismutation of O_2^- the true activity of SOD? The catalytic scavenging of O_2^- was discovered before the identity of the various superoxide dismutases was known.[23,49] The proteins now known as superoxide dismutases were then isolated, using this catalytic activity as the sole criterion of purification. These enzymes react with O_2^- at a diffusion-limited rate. Their catalytic activity toward O_2^- exceeds that of any other enzymes toward their respective substrates. SODs have not yet been found to exhibit any other substantial catalytic activity. It was proposed that SODs could act upon singlet oxygen, but that has been shown to be untrue. Polyacrylamide gel electrophoresis of extracts of various tissues or of microorganisms, followed by activity staining, reveals discrete bands of SOD.[43] Each of these bands can now be correlated with specific SODs that have been purified to homogeneity. One can thus fully account for the SOD activity of diverse biologic extracts in terms of the known SODs. It seems reasonable to conclude that the dismutation of O_2^- is the true or biologic function of the superoxide dismutases.

What is the biologic importance of SOD? If SOD catalytically scavenges O_2^- (and as stated above, that is its only known activity), and if SOD can be shown to protect against the toxicity of oxygen, then we must conclude that O_2^- is an important agent of oxygen toxicity and that SOD is the biologic defense against O_2^-. There are several lines of evidence which indicate that SOD does protect against the toxicity of oxygen. These will be presented in a historical sequence.

A. Surveys The production of O_2^- can only occur in the presence of oxygen. It follows that respiring cells face the threat of O_2^- and need SOD, whereas sensitive obligate anaerobes do not. The earliest attempts to probe the biologic importance of SOD were therefore based on surveys of diverse microorganisms. It was found that actively respiring organisms contained the highest levels of SOD, while aerotolerant organisms contained less, and sensitive anaerobes contained none.[146] One aerotolerant organism *(Lactobacillus plantarum)* appeared anomalous in that it contained no detectable SOD. Studies of oxygen uptake then demonstrated that this strain of *L plantarum* was unusual, in that it consumed no oxygen[61,146]; that being the case, it could not produce O_2^- and would not need SOD. Several anaerobes have since been shown to contain SOD, but at levels significantly below those found in aerobes.[147,148] A recent survey of anaerobes isolated from clinical material has demonstrated a positive correlation between SOD content and oxygen tolerance.[149]

B. Inductions If the substrate for SOD is O_2^- and O_2^- can only be produced in the presence of oxygen, then facultative organisms might show an induction of SOD upon oxygenation. This has been shown to occur in *Streptococcus faecalis,*[150] *E coli,*[150,151] and *Saccharomyces cerevisiae.*[152] In all cases the cells that contained high levels of SOD, by virtue of induction by oxygen, were thereby rendered more resistant to the lethality of hyperbaric oxygen.[61,150-152] One could object that exposure to oxygen might have caused changes in addition to the induction of SOD, and that one of these other changes might actually have been responsible for the acquisition of oxygen tolerance. An organism that failed to induce SOD in response to oxygenation, and that also failed to gain resistance to hyperbaric oxygen from growth in air, would constitute a useful control that could turn aside this objection. *Bacillus subtilis* exhibited these properties. It responded to oxygenation by induction of catalase, but not of SOD, and growth under oxygen did not make it more resistant to hyperbaric oxygen.[151]

E coli K12 was also examined by comparing cells grown in deep, still liquid culture with those grown under vigorous aeration,[152] and no significant differences in SOD content were noted. In this respect *E coli* K12 appeared to respond to oxygen in a way that was qualitatively different from that seen with *E coli* B. Recent studies[153] have shown that *E coli* K12 does induce SOD in response to oxygen, but that it is so exquisitely responsive that the undetectably low level of oxygen, available in deep, still liquid culture, is sufficient to call forth nearly maximal induction. It is furthermore clear that oxygen induces catalase and peroxidase, as well as SOD, in this organism, and that the induction of SOD is actually an induction of the Mn-SOD and the hybrid SOD, while the Fe-SOD remains constant.[153] Induction of these enzymes correlates with resistance to oxygen; when this induction was prevented by puromycin, the cells were killed by exposure to air and suffered morphologic damage that was evident under the electron microscope. Inductions of SOD by oxygen have also been seen in *Psuedomonas ovalis,*[154] in *Photobacter leiognathi,*[155] and in phagocytes.[156]

Induction of SOD and parallel acquisition of oxygen tolerance has also been seen in whole rats. Exposure of young rats to 85% oxygen causes an adaptation, such that subsequent exposure to 100% oxygen is well tolerated. In contrast, exposure to 100% oxygen, without prior adaptation, is uniformly fatal. It has been found that this adaptation correlates with an increase in lung SOD.[157]

C. Oxygen-dependent antibiotics There are several antibiotics whose lethality is enhanced by oxygen. The first of these to be carefully studied was streptonigrin.[158,159] It is a paraquinone which can be readily reduced to a hydroquinone. Hydroquinones are known to autoxidize with the production of O_2^-. Thus streptonigrin could divert electron flow within a cell from its normal channel and conduct it to oxygen via a univalent pathway, producing O_2^-. Studies in vitro have shown that streptonigrin, coupled with a source of electrons, could generate radicals that attack DNA.[160-162] More recently SOD and catalase have been shown to prevent the formation of single-strand breaks in DNA caused by this in vitro streptonigrin system.[58] If streptonigrin really acts as an extra source of O_2^- and if this accounts for the oxygen enhancement of its lethality, then cells with a high level of SOD, by virtue of induction, should thereby be rendered resistant to streptonigrin lethality. This has been shown to be the case for *E coli* B,[151] *E coli* K12,[153] and *Chlorella.*[163] Streptonigrin is not the only antibiotic which behaves in this way. Bleomycin[164] and mitomycin C[68,165] also produce O_2^- by redox cycling; the same seems to apply to alloxan.[46]

D. Mutants The least equivocal means of demonstrating the biologic function of SOD would be to prepare mutants with defects in these enzymes. Comparisons of these mutants with the parental strain should then expose the effect of these enzymes. There is an oxygen-resistant mutant of *Chlorella sorokiniana* that is also resistant to streptonigrin. This mutant has 3.5 times more SOD than the wild type.[163] Mutants of *E coli* K12 were selected that had temperature-sensitive intolerance for oxygen. One of these mutants, designated TS-2, was evaluated and was found to have a parallel, temperature-sensitive decline in SOD level.[166] Recently other oxygen-intolerant mutants have been isolated that have defects in SOD.[167]

Conclusions O_2^- is produced in respiring cells and is an important agent of oxygen toxicity. Superoxide dismutases do catalytically scavenge O_2^- with great efficiency. They provide protection against oxygen toxicity and against the oxygen enhancement of the lethality of antibiotics, such as streptonigrin and mitomycin C, which can, by redox cycling within the cell, cause enhanced production of O_2^-. Superoxide dismutases also diminish the oxygen enhancement of radiation lethality and protect engulfed bacteria against phagocytic kill. There can be no doubt that our understanding of many phenomena, related to the aerobic lifestyle, has been greatly increased by an appreciation of the common occurrence of superoxide radicals and of the importance of superoxide dismutases.

REFERENCES

1. Collman JP: Patterns of organometallic reactions related to homogeneous catalysis. *Accounts Chem Res* 1:136–143, 1968.
2. Hamilton GA: Mechanisms of two- and four-electron oxidations catalyzed by some metalloenzymes. *Adv Enzymol* 32:55–96, 1969.
3. Hamilton GA: The proton in biological redox reactions, in Kaiser ET, Kezdy FJ (eds): *Progress in Bioorganic Chemistry,* vol I. New York, John Wiley & Sons Inc, 1971, pp 83–157.
4. Hamilton GA: Chemical models and mechanisms for oxygenases, in Hayaishi O (ed): *Molecular Mechanisms of Oxygen Activation.* New York, Academic Press Inc, 1974, pp 405–451.
5. Taube H: Mechanisms of oxidation with oxygen. *J Gen Physiol* 49(part 2):29–52, 1965.
6. Behar D, Czapski G, Rabani J, et al: Acid dissociation-constant and decay kinetics of perhydroxyl radical. *J Phys Chem* 74:3209–3213, 1970.
7. Bielski BHJ, Allen AO: Mechanism of the disproportionation of superoxide radicals. *J Phys Chem* 81:1048–1050, 1977.
8. Divisek J, Kastening B: Electrochemical generation and reactivity of the superoxide ion in aqueous solutions. *J Electroanal Chem* 65:603–621, 1975.
9. Behar D, Czapski G: Ozonide and hydroperoxy radicals in flash photolysis of hydrogenperoxide. *Isr J Chem* 8:699–708, 1970.
10. Bielski BHJ, Saito E: The activation energy for the disproportionation of the HO_2 radical in acid solutions. *J Phys Chem* 66:2266–2268, 1962.

11. Czapski G, Levanon H, Samuni H: ESR studies of uncomplexed and complexed HO_2 radical formed in the reaction of H_2O_2 with Ce^{4+}, Fe^{2+} and Ti^{3+} ions. *Isr J Chem* 7:375–386, 1969.

12. Saito E, Bielski BHJ: The electron paramagnetic resonance spectrum of the HO_2 radical in aqueous solution. *J Am Chem Soc* 83:4467–4468, 1961.

13. Samuni A, Czapski G: ESR study of a complex formation between HO_2 radical and peroxy-vanadium (V) ion. *Isr J Chem* 8:563–573, 1970.

14. Czapski G: Radiation chemistry of oxygenated aqueous solutions. *Annu Rev Phys Chem* 22:171–208, 1971.

15. Nilsson R, Pick FM, Bray RC, et al: ESR evidence for O_2^- as a long-lived transient in irradiated oxygenated alkaline aqueous solutions. *Acta Chem Scand* 23:2554–2556, 1969.

16. Bray RC, Pick FM, Samuel D: Oxygen-17 hyperfine splitting in the electron paramagnetic resonance spectrum of enzymically generated superoxide. *Eur J Biochem* 15:352–355, 1970.

17. Knowles PF, Gibson JF, Pick FM, et al: Electron-spin-resonance evidence for enzymic reduction of oxygen to a free radical, the superoxide ion. *Biochem J* 111:53–58, 1969.

18. Goolsby AD, Sawyer DT: The electrochemical reduction of superoxide ion and oxidation of hydroxide ion in dimethyl sulfoxide. *Anal Chem* 40:83–86, 1968.

19. McElroy AD, Hashman JS: Synthesis of tetramethylammonium superoxide. *Inorg Chem* 3:1798, 1964.

20. Peover ME, White BS: Electrolytic reduction of oxygen in arotic solvents: the superoxide ion. *Electrochim Acta* 11:1061–1067, 1966.

21. Sawyer DT, Roberts JL Jr: Electrochemistry of oxygen and superoxide ion in DI-methylsulfoxide at platinum, gold and mercury electrodes. *J Electroanal Chem* 12:90–101, 1966.

22. Fee JA, Hildenbrand PG: On the development of a well-defined source of superoxide ion for studies with biological systems. *FEBS Lett* 39:79–82, 1974.

23. McCord JM, Fridovich I: Superoxide dismutase—an enzymic function for erythrocuprein (hemocuprein). *J Biol Chem* 244:6049–6055, 1969.

24. Merritt MV, Sawyer DT: Electrochemical studies of the reactivity of superoxide ion with several alkyl halides in dimethyl sulfoxide. *J Org Chem* 35:2157–2159, 1970.

25. Odajima T, Yamazaki I: Myeloperoxidase of the leukocyte of normal blood. III. The reaction of ferric myeloperoxidase with superoxide anion. *Biochim Biophys Acta* 284:355–359, 1972.

26. Valentine JS, Curtis AB: A convenient preparation of solutions of superoxide anion and the reaction of superoxide anion with a copper(II) complex. *J Am Chem Soc* 97:224–226, 1975.

27. Forman HJ, Fridovich I: Electrolytic univalent reduction of oxygen in aqueous solution demonstrated with superoxide dismutase. *Science* 175:339, 1972.

28. Chevalet J, Rouelle F, Gierst I, et al: Electrogeneration and some properties of the superoxide ion in aqueous solutions. *J Electroanal Chem* 39:201–216, 1972.

29. Kastening B, Kazemifard G: Elektrochemische Reduktion von Sauerstoff zum Superoxide-anion in wässriger Lösung. *Ber Bunsenges Phys Chem* 74:551–556, 1970.

30. Peters JW, Foote CS: Chemistry of superoxide ion. II. Reaction with hydroperoxides. *J Am Chem Soc* 98:873–875, 1976.

31. McCord JM, Fridovich I: Production of O_2^- in photolyzed water demonstrated through the use of superoxide dismutase. *Photochem Photobiol* 17:115–121, 1973.

32. Keene JP: The absorption spectrum and some reaction constants of the hydrated electron. *Radiat Res* 22:1–13, 1964.

33. Gordon S, Hart EJ, Matheson MS, et al: Reaction constants of the hydrated electron. *J Am Chem Soc* 85:1375–1377, 1963.

34. Sweet JP, Thomas JK: Absolute rate constants for H atom reactions in aqueous solutions. *J Phys Chem* 68:1363–1368, 1964.

35. Ilan YA, Czapski G, Meisel D: The one-electron transfer redox potentials of free radicals. I. The oxygen/superoxide system. *Biochim Biophys Acta* 430: 209–224, 1976.

36. Muirwood P: The redox potential of the system oxygen-superoxide. *FEBS Lett* 44:22–24, 1974.

37. Sawada Y, Iyanagi T, Yamazaki I: Relation with redox potentials and rate constants in reactions coupled with the system oxygen-superoxide. *Biochemistry* 14:3761–3764, 1975.

38. Ballou D, Palmer G, Massey V: Direct demonstration of superoxide anion production during the oxidation of reduced flavin and of its catalytic decomposition by erythrocuprein. *Biochem Biophys Res Commun* 36:898–904, 1969.

39. Butler J, Jayson GG, Swallow AJ: The reaction between the superoxide anion radical and cytochrome *c*. *Biochim Biophys Acta* 408:215–222, 1975.

40. Land EJ, Swallow AJ: One-electron reactions in biochemical systems as studied by pulse radiolysis. *Arch Biochem Biophys* 145:365–372, 1971.

41. Simic MG, Tabu IA, Tocci J, et al: Free radical reduction of ferricytochrome-*c*. *Biochem Biophys Res Commun* 62:161–167, 1975.

42. Sutton HC, Roberts PB, Winterbourn CC: The rate of reaction of superoxide radical ion with oxyhaemoglobin and methaemoglobin. *Biochem J* 155: 503–510, 1976.

43. Beauchamp C, Fridovich I: Superoxide dismutase: improved assays and an assay applicable to polyacrylamide gels. *Anal Biochem* 44:276–287, 1971.

44. Rabani J, Mulac WA, Matheson MS: The pulse radiolysis of aqueous tetranitromethane. I. Rate constants and the extinction coefficient of e_{aq}^-. II. Oxygenated solutions. *J Phys Chem* 69:53–70, 1965.

45. Miller RW, Rapp U: The oxidation of catechols by reduced flavins and dehydrogenases: an electron spin resonance study of the kinetics and initial products of oxidation. *J Biol Chem* 248:6084–6090, 1973.

46. Cohen G, Heikkila R: The generation of hydrogen peroxide, superoxide radical, and hydroxyl radical by 6-hydroxydopamine, dialuric acid, and related cytotoxic agents. *J Biol Chem* 249:2447–2452, 1974.

47. Misra HP, Fridovich I: The role of superoxide anion in the autoxidation of epinephrine and a simple assay for superoxide dismutase. *J Biol Chem* 247: 3170–3175, 1972.

48. Marklund S, Marklund G: Involvement of the superoxide anion radical in the autoxidation of pyrogallol and a convenient assay for superoxide dismutase. *Eur J Biochem* 47:469–474, 1974.

49. McCord JM, Fridovich I: The reduction of cytochrome *c* by milk xanthine oxidase. *J Biol Chem* 243:5753–5760, 1968.

50. Greenstock CL, Miller RW: The oxidation of tiron by superoxide anion: kinetics of the reaction in aqueous solution and in chloroplasts. *Biochim Biophys Acta* 396:11–16, 1975.

51. Harbour JR, Bolton JR: Superoxide formation in spinach chloroplasts: electron spin resonance detection by spin trapping. *Biochem Biophys Res Commun* 64:803–807, 1975.

52. Harbour JR, Chow V, Bolton JR: An electron spin resonance study of the spin adducts of OH and HO_2 radicals with nitrones in the ultraviolet photolysis of aqueous hydrogen peroxide solutions. *Can J Chem* 52:3549–3553, 1974.

53. Johnson RA, Nidy EG: Superoxide chemistry. A convenient synthesis of dialkyl peroxides. *J Org Chem* 40:1680–1681, 1975.

54. San Filippo J Jr, Chern C-I, Valentine JS: The reaction of superoxide with alkyl halides and tosylates. *J Org Chem* 40:1678–1680, 1975.

55. Beauchamp C, Fridovich I: A mechanism for the production of ethylene from methional: the generation of hydroxyl radical by xanthine oxidase. *J Biol Chem* 245:4641–4646, 1970.

56. Haber F, Weiss, J: The catalytic decomposition of hydrogen peroxide by iron salts. *Proc R Soc Lond* A147:332–351, 1934.

57. Babior B, Curnutte JT, Kipnes RS: Biological defense mechanisms. Evidence for the participation of superoxide in bacterial killing by xanthine oxidase. *J Lab Clin Med* 85:235–244, 1975.

58. Cone R, Hasan SK, Lown JW, et al: The mechanism of the degradation of DNA by streptonigrin. *Can J Biochem* 54:219–223, 1976.

59. Elstner EF, Konze JR: Light-dependent ethylene production by isolated chloroplasts. *FEBS Lett* 45:18–21, 1974.

60. Goscin SA, Fridovich I: The role of superoxide radical in a nonenzymatic hydroxylation. *Arch Biochem Biophys* 153:778–783, 1972.

61. Gregory EM, Fridovich I: Oxygen metabolism in *Lactobacillus plantarum*. *J Bacteriol* 117:166–169, 1974.

62. Halliwell B, Ahluwalia S: Hydroxylation of *p*-coumaric acid by horseradish peroxidase. *Biochem J* 153:513–518, 1976.

63. Heikkila RE, Winston B, Cohen G, et al: Alloxan-induced diabetes—evidence for hydroxyl radical as a cytotoxic intermediate. *Biochem Pharmacol* 25:1085–1092, 1976.

64. Hodgson EK, Fridovich I: The mechanism of the activity-dependent luminescence of xanthine oxidase. *Arch Biochem Biophys* 172:202–205, 1976.

65. Kellogg EW III, Fridovich I: Superoxide, hydrogen peroxide and singlet oxygen in lipid peroxidation by a xanthine oxidase system. *J Biol Chem* 250: 8812–8817, 1975.

66. Kon S, Schwimmer S: Degradation of plant and other polysaccharides catalyzed by xanthine oxidase and its inhibition by catalase and superoxide dismutase. *Fed Proc* 35:1553, 1976.

67. Konze JR, Elstner EF: Pyridoxalphosphate-dependent ethylene production from methionine by isolated chloroplasts. *FEBS Lett* 66:8–11, 1976.

68. Lown JW, Begleiter A, Johnson D, et al: Studies related to antitumor antibiotics. Part V. Reactions of mitomycin *c* with DNA examined by ethidium fluorescence assay. *Can J Biochem* 54:110–119, 1976.

69. Mieyal JJ, Ackerman RS, Blumer JL, et al: Characterization of enzyme-like activity of human hemoglobin: properties of the hemoglobin-P-450 reductase-coupled aniline hydroxylase system. *J Biol Chem* 251:3436–3441, 1976.

70. Okita R, Bidlack WR, Hochstein P: The participation of cytochrome b_5 reductase and hydroxyl radicals in microsomal ethanol oxidation (MEOS). *Fed Proc* 33:1256, 1974.

71. Van Hemmen JJ, Meuling WJA: Inactivation of biologically active DNA by γ-ray-induced superoxide radicals and their dismutation products singlet molecular oxygen and hydrogen peroxide. *Biochim Biophys Acta* 402:133–141, 1975.

72. Bors W, Lengfelder E, Saran M, et al: Reactions of oxygen radical species with methional: a pulse radiolysis study. *Biochem Biophys Res Commun* 70:81–87, 1976.

73. McClune GJ, Fee JA: Stopped flow spectrophotometric observation of superoxide dismutation in aqueous solution. *FEBS Lett* 67:294–298, 1976.

74. McCord JM, Day ED Jr: Superoxide-dependent production of hydroxyl radical catalyzed by iron-EDTA complex. *FEBS Lett* 86:139–142, 1978.

75. Forman HJ, Fridovich I: Superoxide dismutases: a comparison of rate constants. *Arch Biochem Biophys* 159:396–400, 1973.

76. Klug D, Fridovich I, Rabani J: Pulse radiolytic investigations of superoxide catalyzed disproportionation. Mechanism for bovine superoxide dismutase. *J Am Chem Soc* 95:2786–2790, 1973.

77. Klug D, Rabani J, Fridovich I: A direct demonstration of the catalytic action of superoxide dismutase through the use of pulse radiolysis. *J Biol Chem* 247:4839–4842, 1972.

78. Pick M, Rabani J, Yost F, et al: The catalytic mechanism of the manganese-containing superoxide dismutase of *Escherichia coli* studied by pulse radiolysis. *J Am Chem Soc* 96:7329–7333, 1974.

79. Rotilio G, Bray RC, Fielden EM: A pulse radiolysis study of superoxide dismutase. *Biochim Biophys Acta* 268:605–609, 1972.

80. Richardson JS, Thomas KA, Rubin BH, et al: Crystal structure of bovine Cu,Zn superoxide dismutase at 3 Å resolution: chain tracing and metal ligands. *Proc Natl Acad Sci USA* 72:1349–1353, 1975.

81. Hodgson EK, Fridovich I: The interaction of bovine erythrocyte superoxide dismutase with hydrogen peroxide: inactivation of the enzyme. *Biochemistry* 14:5294–5299, 1975.

82. Fridovich I: Superoxide dismutases. *Annu Rev Biochem* 44:147–159, 1975.

83. Yamakura F, Suzuki K, Mitsui Y: Crystallization and preliminary crystal data of iron-containing superoxide dismutase from *Pseudomonas ovalis*. *J Biol Chem* 251:4792–4793, 1976.

84. Beem KM, Rich WE, Rajagopalan KV: Total reconstitution of copper-zinc superoxide dismutase. *J Biol Chem* 249:7298–7305, 1974.

85. Ose DE, Fridovich I: Superoxide dismutase: reversible removal of manganese and its substitution by cobalt, nickel or zinc. *J Biol Chem* 251:1217–1218, 1976.

86. Brock CJ, Harris JI, Sato S: Superoxide dismutase from *Bacillus stearothermophilus*. Preparation of stable apoprotein and reconstitution of fully active Mn enzyme. *J Mol Biol* 107:175–178, 1976.

87. Yamakura F, Suzuki K: Reconstitution of iron-superoxide dismutase. *Biochem Biophys Res Commun* 72:1108–1115, 1976.

88. Forman HJ, Fridovich I: On the stability of bovine superoxide dismutase. The effects of metals. *J Biol Chem* 248:2645–2649, 1973.

89. Misra HP, Fridovich I: The univalent reduction of oxygen by reduced flavins and quinones. *J Biol Chem* 247:188–192, 1972.

90. Fisher DB, Kaufman S: Tetrahydropterin oxidation without hydroxylation catalyzed by rat liver phenylalanine hydroxylase. *J Biol Chem* 248:4300–4304, 1973.

91. Heikkila RE, Cohen G: The generation of the superoxide radical by 2-amino-4-hydroxy-6,7-dimethyl-5,6,7,8-tetrahydropteridine ($DMPH_4$). *Experientia* 31:169–170, 1975.

92. Nishikimi M: The generation of superoxide anion in the reaction of tetrahydropteridines with molecular oxygen. *Arch Biochem Biophys* 166:273–279, 1975.

93. Brunori M, Falcioni G, Fioretti E, et al: Formation of superoxide in the autoxidation of the isolated α and β chains of human hemoglobin and its involvement in hemichrome precipitation. *Eur J Biochem* 53:99–104, 1975.

94. Caughey WS, Barlow CH, Maxwell JC, et al: Part I. Hemeproteins: liquid phenomena. Reactions of oxygen with hemoglobin, cytochrome *c* oxidase and other hemeproteins. *Ann NY Acad Sci* 244:1–9, 1975.

95. Lynch RE, Lee GR, Cartwright GE: Inhibition by superoxide dismutase of methemoglobin formation from oxyhemoglobin. *J Biol Chem* 251:1015–1019, 1976.

96. Misra HP, Fridovich I: The generation of superoxide radical during the autoxidation of hemoglobin. *J Biol Chem* 247:6960–6962, 1972.

97. Winterbourn CC, McGrath BM, Carrell RW: Reactions involving superoxide and normal and unstable haemoglobins. *Biochem J* 155:493–502, 1976.

98. Allen JF: A two-step mechanism for the photosynthetic reduction of oxygen by ferredoxin. *Biochem Biophys Res Commun* 66:36–43, 1975.

99. Misra HP, Fridovich I: The generation of superoxide radical during the autoxidation of ferredoxins. *J Biol Chem* 246:6886–6890, 1971.

100. Nakamura S: Initiation of sulfite oxidation by spinach ferredoxin-NADP reductase and ferredoxin system: a model experiment on the superoxide anion radical production by metalloflavoproteins. *Biochem Biophys Res Commun* 41:177–183, 1970.

101. Nakamura S, Kimura T: Studies on aggregated multienzyme systems: stimulation of oxygen uptake of ferredoxin-nicotinamide adenine dinucleotide phosphate reductase-ferredoxin complex by cytochrome *c*. *J Biol Chem* 247: 6462–6468, 1972.

102. Nilsson R, Pick FM, Bray RC: EPR studies on reduction of oxygen to superoxide by some biochemical systems. *Biochim Biophys Acta* 192:145–148, 1969.

103. Orme-Johnson WM, Beinert H: On the formation of the superoxide anion radical during the reaction of reduced iron-sulfur proteins with oxygen. *Biochem Biophys Res Commun* 36:905–911, 1969.

104. Fridovich I: Quantitative aspects of the production of superoxide anion radical by milk xanthine oxidase. *J Biol Chem* 245:4053–4057, 1970.

105. Azzi A, Montecucco C, Richter C: The use of acetylated ferricytochrome *c* for the detection of superoxide radicals produced in biological membranes. *Biochem Biophys Res Commun* 65:597–603, 1975.

106. Boveris A, Cadenas E: Mitochondrial production of superoxide anions and its relationship to the antimycin insensitive respiration. *FEBS Lett* 54:311–314, 1975.

107. Boveris A, Cadenas E, Stoppani AOM: Role of ubiquinone in the mitochondrial generation of hydrogen peroxide. *Biochem J* 156:435–444, 1976.

108. Forman HJ, Kennedy JA: Role of superoxide radical in mitochondrial dehydrogenase reactions. *Biochem Biophys Res Commun* 60:1044–1050, 1974.

109. Forman HJ, Kennedy J: Superoxide production and electron transport in mitochondrial oxidation of dihydroorotic acid. *J Biol Chem* 250:4322–4326, 1975.

110. Loschen G, Azzi A, Richler C, et al: Superoxide radicals as precursors of mitochondrial hydrogen peroxide. *FEBS Lett* 42:68–72, 1974.

111. Tyler DD: A protective function of superoxide dismutase during respiratory chain activity. *Biochim Biophys Acta* 396:335–346, 1975.

112. Allen JF, Hall DO: Superoxide reduction as a mechanism of ascorbate-stimulated oxygen uptake by isolated chloroplasts. *Biochem Biophys Res Commun* 52:856–862, 1973.

113. Allen JF, Hall DO: The relationship of oxygen uptake to electron transport in photosystem I of isolated chloroplasts: the role of superoxide and ascorbate. *Biochem Biophys Res Commun* 58:579–585, 1974.

114. Asada K, Kiso K: Initiation of aerobic oxidation of sulfite by illuminated spinach chloroplasts. *Eur J Biochem* 33:253–257, 1973.

115. Asada K, Kiso K: The photo-oxidation of epinephrine by spinach chloroplasts and its inhibition by superoxide dismutase: evidence for the formation of superoxide radicals in chloroplasts. *Agric Biol Chem* 37:453–454, 1973.

116. Asada K, Kiso K, Yoskikawa K: Univalent reduction of molecular oxygen by spinach chloroplasts on illumination. *J Biol Chem* 249:2175–2181, 1974.

117. Elstner EF, Heupel A: Involvement of the superoxide free radical ion in photosynthetic oxygen reduction. *Z Naturforsch* 29C:559–563, 1974.

118. Elstner EF, Heupel A: Lamellar superoxide dismutase of isolated chloroplasts. *Planta* 123:145–154, 1975.

119. Elstner EF, Stoffer C, Heupel A: Determination of superoxide free radical ion and hydrogen peroxide as products of photosynthetic oxygen reduction. *Z Naturforsch* 30C:53–57, 1975.

120. Elstner EF, Wildner GF, Heupel A: Oxygen activation by isolated chloroplasts from *Euglena gracilis*. *Arch Biochem Biophys* 173:623–630, 1976.

121. Epel BL, Neuman J: The mechanism of the oxidation of ascorbate and Mn^{2+} by chloroplasts: the role of the radical superoxide. *Biochim Biophys Acta* 325:520–529, 1973.

122. Halliwell B: Hydroxylation of *p*-coumaric acid by illuminated chloroplasts: the role of superoxide. *Eur J Biochem* 55:355–360, 1975.

123. Kono Y, Takahashi M, Asada K: Oxidation of manganous pyrophosphate by superoxide radicals and illuminated spinach chloroplasts. *Arch Biochem Biophys* 174:454–462, 1976.

124. Miller RW, Macdowell FDH: The tiron free radical as a sensitive indicator of chloroplastic photoautoxidation. *Biochim Biophys Acta* 387:176–187, 1975.

125. Schmid R: Deactivation of superoxide dismutase on EDTA-treated chloroplasts. *FEBS Lett* 60:98–102, 1975.

126. Babior BM, Kipnes RS, Curnutte JT: Biological defense mechanisms: the production by leukocytes of superoxide, a potential bactericidal agent. *J Clin Invest* 52:741–744, 1973.

127. Curnutte JT, Babior BM: Biological defense mechanisms: the effect of bacteria and serum on superoxide production by granulocytes. *J Clin Invest* 53:1662–1672, 1974.

128. Drath DB, Karnovsky ML: Superoxide production by phagocytic leukocytes. *J Exp Med* 141:257–262, 1975.

129. Johnston RB, Keele BB, Misra HP, et al: The role of superoxide anion generation in phagocytic bactericidal activity: studies with normal and chronic granulomatous disease leukocytes. *J Clin Invest* 55:1357–1372, 1975.

130. Johnston RB, Lehmeyer JE, Guthrie LA: Generation of superoxide anion and chemiluminescence by human monocytes during phagocytosis and on contact with surface-bound immunoglobulin G. *J Exp Med* 143:1551–1556, 1976.

131. Nakagawa A, Minakami S: Generation of superoxide anions by leucocytes treated with cytochalasin E. *Biochem Biophys Res Commun* 64:760–767, 1975.

132. Salin ML, McCord JM: Superoxide dismutase in polymorphonuclear leukocytes. *J Clin Invest* 54:1005–1009, 1974.

133. Weening RS, Wever R, Roos D: Quantitative aspects of the production of superoxide radicals by phagocytizing human granulocytes. *J Lab Clin Med* 85:245–252, 1975.

134. Babior BM, Curnutte JT, McMurrich BJ: The particulate superoxide-forming system from human neutrophils: properties of the system and further evidence supporting its participation in the respiratory burst. *J Clin Invest* 58:989–996, 1976.

135. Gregory EM, Yost FJ Jr, Fridovich I: Superoxide dismutase of *Escherichia coli:* intracellular localization and functions. *J Bacteriol* 115:987–991, 1973.

136. Green DE, Pauli R: The antibacterial action of the xanthine oxidase system. *Proc Soc Exp Biol Med* 54:148–150, 1943.

137. Lavelle F, Michelson AM, Dimitrijevic L: Biological protection by superoxide dismutase. *Biochem Biophys Res Commun* 55:350–357, 1973.

138. Michelson AM, Buckingham ME: Effects of superoxide radicals on myoblast growth and differentiation. *Biochem Biophys Res Commun* 58:1079–1086, 1974.

139. Misra HP, Fridovich I: Superoxide dismutase and the oxygen enhancement of radiation lethality. *Arch Biochem Biophys* 176:577–581, 1976.

140. Nordenson I, Beckman G, Beckman L: The effect of superoxide dismutase and catalase on radiation-induced chromosome breaks. *Hereditas* 82:125–126, 1976.

141. Petkau A, Chelack WS: Radioprotective effects of superoxide dismutase on model phospholipid membranes. *Biochim Biophys Acta* 433:445–456, 1976.

142. Petkau A, Chelack WS, Pleskach SD: Protection of post-irradiated mice by superoxide dismutase. *Int J Radiat Biol* 29:297–299, 1976.

143. Petkau A, Chelack WS, Pleskach SD, et al: Radioprotection of mice by superoxide dismutase. *Biochem Biophys Res Commun* 65:886–893, 1975.

144. Petkau A, Kelly K, Chelack WS, et al: Radioprotection of bone marrow stem cells by superoxide dismutase. *Biochem Biophys Res Commun* 67:1167–1174, 1975.

145. Salin ML, McCord JM: Free radicals and inflammation: protection of phagocytosing leukocytes by superoxide dismutase. *J Clin Invest* 56:1319–1323, 1975.

146. McCord JM, Keele BB Jr, Fridovich I: An enzyme-based theory of obligate anaerobiosis: the physiological function of superoxide dismutase. *Proc Natl Acad Sci USA* 68:1024–1027, 1971.

147. Hewitt J, Morris JG: Superoxide dismutase in some obligately anaerobic bacteria. *FEBS Lett* 50:315–318, 1975.

148. Lindmark DG, Müller M: Superoxide dismutase in the anaerobic flagellates, *Trichomonas foetus* and *Monocercomonas* sp. *J Biol Chem* 249:4634–4637, 1974.

149. Tally FP, Goldin BR, Jacobus NV, et al: Superoxide dismutase in anaerobic bacteria of clinical significance. *Infect Immun* 16:20–25, 1977.

150. Gregory EM, Fridovich I: Induction of superoxide dismutase by molecular oxygen. *J Bacteriol* 114:543–548, 1973.

151. Gregory EM, Fridovich I: Oxygen toxicity and the superoxide dismutase. *J Bacteriol* 114:1193–1197, 1973.

152. Gregory EM, Goscin SA, Fridovich I: Superoxide dismutase and oxygen toxicity in a eukaryote. *J Bacteriol* 117:456–460, 1974.

153. Hassan HM, Fridovich I: Enzymatic defenses against the toxicity of oxygen and of streptonigrin in *Escherichia coli* K12. *J Bacteriol* 129:1574–1583, 1977.

154. Yamakura F: Purification, crystallization and properties of iron-containing superoxide dismutase from *Pseudomonas ovalis*. *Biochim Biophys Acta* 422:280–294, 1976.

155. Puget K, Michelson AM: Isolation of a new copper-containing superoxide dismutase bacteriocuprein. *Biochem Biophys Res Commun* 58:830–838, 1974.

156. Rister M, Baehner RL: Induction of superoxide dismutase (SOD) activity *in vivo* by oxygen (O_2) in polymorphonuclear leukocytes (PMNs) and alveolar macrophages (AM). *Blood* 46:1016, 1975.

157. Crapo J, Tierney D: Superoxide dismutase and oxygen toxicity. *Clin Res* 21:222, 1973.

158. Hochstein P, Laszlo J, Miller D: A unique, dicoumarol-sensitive, non-phosphorylating oxidation of DPNH and TPNH catalyzed by streptonigrin. *Biochem Biophys Res Commun* 19:289–295, 1965.

159. White HL, White JR: Lethal action and metabolic effects of streptonigrin on *Escherichia coli*. *Mol Pharmacol* 4:549–565, 1968.

160. White JR, Dearman HH: Generation of free radicals from phenazine methosulfate, streptonigrin, and rubiflavin in bacterial suspensions. *Proc Natl Acad Sci USA* 54:887–891, 1965.

161. White JR, Vaughan TO, Yeh W-S: Superoxide radical in the mechanism of action of streptonigrin. *Fed Proc* 30:1145, 1971.

162. White HL, White JR: Interaction of streptonigrin with DNA in vitro. *Biochim Biophys Acta* 123:648–651, 1966.

163. Pulich WM: Resistance to high oxygen tension, streptonigrin, and ultraviolet irradiation in the green alga *Chlorella sorokiniana* strain *ORS. J Cell Biol* 62:904–907, 1974.

164. Ishida R, Takahashi T: Increased DNA chain breakage by combined action of bleomycin and superoxide radical. *Biochem Biophys Res Commun* 66:1432–1438, 1975.

165. Handa K, Sato S: Generation of free radicals of quinone group-containing anti-cancer chemicals in NADPH-microsome system as evidenced by initiation of sulfite oxidation. *Gan* 66:43–47, 1975.

166. McCord JM, Beauchamp CO, Goscin S, et al: Superoxide and superoxide dismutase, in King TE, Mason HS, Morrison M (eds): *Oxidases and Related Redox Systems, Proceedings of the 2nd International Symposium 1971.* Baltimore, Md, University Park Press, 1973, pp 51–76.

167. Hassan HM, Fridovich I: Superoxide, hydrogen peroxide and oxygen tolerance of oxygen-sensitive mutants of *Escherichia coli. Rev Infect Dis* 1:357–369, 1979.

13 Superoxide and Superoxide Dismutases in *Escherichia coli*—Threat and Defense

H. Moustafa Hassan, PhD

Oxygen is exceedingly useful in energy-yielding metabolism and in biosynthesis, yet all living forms are prone to oxygen toxicity. The toxicity of oxygen is due to the reactivities of the intermediates of oxygen reduction (O_2^- and H_2O_2) and to reactive entities that can be generated therefrom.[1-6] It is obvious therefore that the first and the best defense against oxygen toxicity would be avoidance of the generation of these reactive intermediates. Indeed, this is the strategy employed by most aerobic organisms, where most of the oxygen consumed is reduced to water by the cytochrome oxidase system, which accomplishes the tetravalent reduction of oxygen, without the release of any free reactive intermediates. Nevertheless, a small but significant amount of both O_2^- and H_2O_2 are normal products of the biologic reduction of oxygen,[5-7] and the second line of defense against oxygen toxicity utilizes enzymes capable of scavenging both O_2^- and H_2O_2. The superoxide radical is eliminated by the superoxide dismutases (SODs), while the hydrogen peroxide is removed by hydroperoxidases.

The superoxide theory of oxygen toxicity states that O_2^- is an important agent of oxygen toxicity and that superoxide dismutases are the defense against this radical. The physiologic function of SOD has been thoroughly explored both in procaryotes[8–17] and eucaryotes,[18–20] and these studies have provided strong support for the superoxide theory of oxygen toxicity. This review will deal with studies carried out with *Escherichia coli* to demonstrate how this organism, a facultative anaerobe, responds to the threat of oxygen toxicity by inducing the protective enzyme, superoxide dismutase.

Induction of SOD in *E coli* in Response to Oxygen

Aerobically grown *E coli* have three distinct isozymes of SOD,[10] one of which contains manganese,[21] the second contains iron,[22] and the third contains iron, but appears to be a hybrid containing one subunit from each of the other isozymes.[23] Anaerobically grown cells, in contrast, have only the iron-containing SOD (Figure 13-1). Thus, it appears that the manganese-containing enzyme is synthesized only in the presence of oxygen, while the iron-containing enzyme is synthesized in both the absence and presence of oxygen.

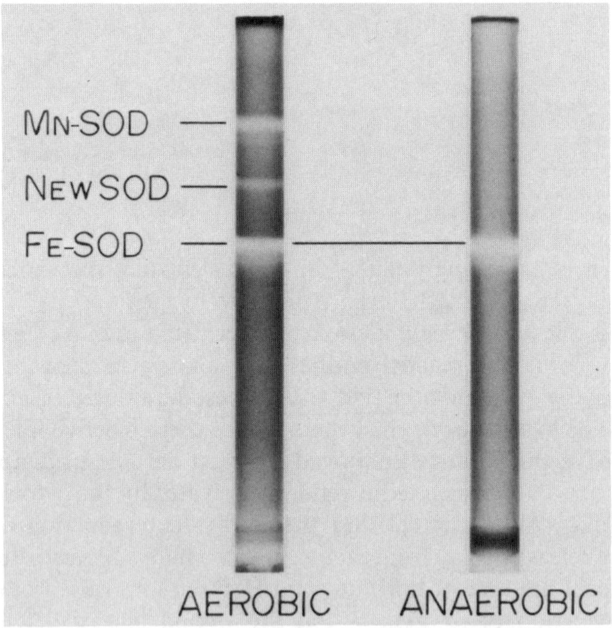

Figure 13-1 Responses of SOD isozymes of *E coli* to oxygen exposure. *E coli* K12 were grown aerobically (left gel) or anaerobically (right gel) in trypticase soy and yeast (TSY) media to late logarithmic phase. Dialyzed cell-free extracts were prepared and applied onto 10% polyacrylamide gels (250 μg of protein per gel), and the gels were developed and stained for SOD activity.[24] New SOD is the same as hybrid-SOD.[23]

Exposure of anaerobically grown *E coli* to air causes rapid synthesis (induction) of the manganese-containing and the hybrid SODs (Figure 13-2). During the course of three hours at 37 C in the presence of air, the levels of Mn- and hybrid-SOD continued to rise linearly and showed no indication of leveling off.[10] This clearly indicates that the derepression or induction mechanism requires several generations for full expression. This might explain the wisdom for *E coli* having a constitutive Fe-SOD, as a compromise between the expense of maintaining a large standing army of defensive enzyme, and the damage manifested by the delay normally inherent in the synthesis of optimum levels of Mn-SOD upon exposure to O_2.

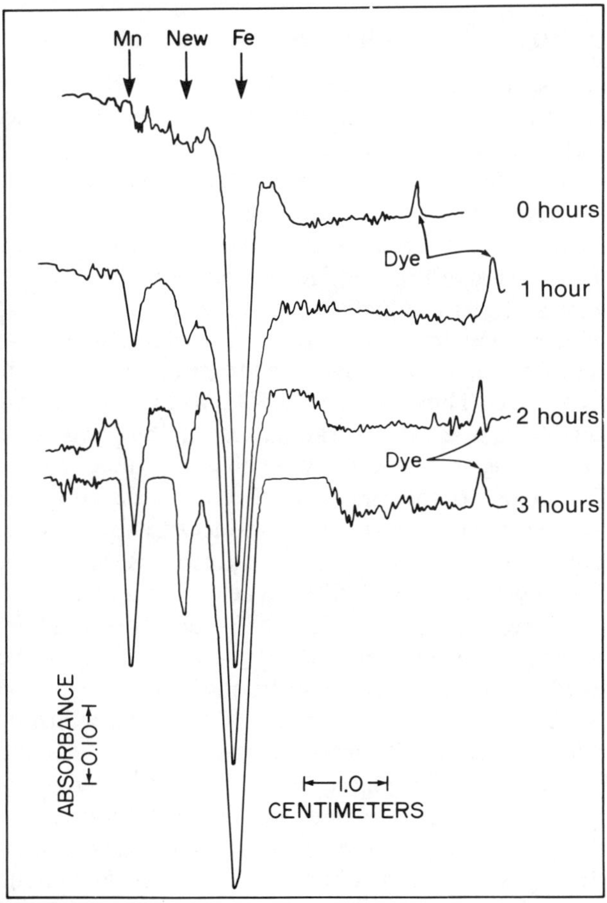

Figure 13-2 Densitometric scans of polyacrylamide gel electropherograms stained for SOD activity, showing induction of SOD by oxygen. *E coli* K12 were grown anaerobically in TSY medium. At zero time the culture was shifted to aerobic conditions and at intervals after the shift, portions of the dialyzed cell-free extracts were applied onto 10% gels. The gels were stained for activity as in Figure 13-1, and scanned at 560 nm. From Hassan and Fridovich.[10] Reprinted with permission.

We anticipated that exposure of anaerobically grown cells to air, while preventing the induction of Mn-SOD, would damage the cells and cause loss of viability. Indeed, this was the case.[10] Puromycin was used to prevent the synthesis of Mn-SOD during exposure to air, and controls showed that puromycin was not toxic to the anaerobically grown cells in the absence of O_2, or to aerobically grown cells in the presence of O_2. Electron micrographs of thin sections showed damage to cytoplasmic membranes and cell breakage.[10] Gaps were also seen at the poles of the cells. Exposure to concentrations of oxygen higher than 20% elicits increased synthesis of SOD. This has been seen in both procaryotes[8,9] and eucaryotes.[18-20]

Synthesis of SOD in *E coli* in Response to Metabolic Changes

Effects of specific growth rates When the specific growth rate of *E coli* was manipulated by changing the dilution rate in a glucose-limited chemostat, while maintaining constant and abundant aeration, the concentration of superoxide dismutase varied in proportion to the specific growth rate and to the rate of respiration.[11] If we assume that a constant fraction of the total oxygen metabolized by the cells is reduced via the univalent pathway, then low growth rate and low respiration rate must be associated with a slow intracellular production of O_2^- and a correspondingly low level of SOD, which suffices to scavenge this low level of O_2^-. Increased growth rates would be accompanied by increased respiration, O_2^- production, and higher levels of SOD. Thus, the level of SOD within the bacterium is in line with its growth rate and with its rate of respiration. This correlation was especially impressive, since the cells were all grown under constant oxygenation; and it provided the first clue that the inducer for SOD is not molecular oxygen per se, but rather a product of its metabolism.

Nutritional effects *E coli* grown on glucose as the sole source of carbon contained lower levels of SOD than cells grown on succinate, lactate, or on trypticase-peptone (TP).[12,13] Table 13-1 demonstrates this effect. The level of SOD was not a simple function of growth rate, as was the case in the glucose-limited chemostat culture mentioned above, since growth was slightly slower in the TP medium than in the nutrient broth plus 0.5% glucose medium; yet the SOD content was two times higher with the former medium. Furthermore, cells placed in minimal media containing succinate or lactate grew at much slower rates and contained higher levels of SOD than did cells grown on glucose. A kinetic study of SOD synthesis, during a growth cycle on TSY medium, which contained a limited about of glucose (0.25%), revealed the complexity of this glucose effect.[13] The data clearly showed that SOD activity declined early in the growth cycle, when the decline in pH suggested a fermentative metabolism, and then increased again, when the increase in pH indicated a reliance on catabolism of acids and nitrogenous compounds. When cells were shifted from a glucose-minimal medium to a rich medium (TSY) the cells' content of SOD decreased in the early phase, while glucose was being utilized, and then increased abruptly as the glucose was exhausted and the cells began to utilize other components of the medium, via a more oxidative type of metabolism (Figure 13-3). When the TSY medium was prepared free of glucose, SOD synthesis increased at the outset.[13]

Table 13-1
Effect of Different Growth Media on the Levels of SOD in *E coli* B

Medium	Generation Time (min)	SOD (U/mg)
Glucose-minimal	46	6–8
Nutrient broth + 0.5% glucose	39	7–9
Minimal salts + 0.5% succinate	160	11–13
Minimal salts + 0.5% lactate	105	15–16
Trypticase-peptone (TP)	42	16–18
Trypticase soy + yeast extract (TSY)	32	21–24

The cells were grown in the different media at 37 C on a rotary shaker (200 rpm) and samples were removed at different intervals for growth and SOD assays. The values reported for SOD were taken at peak values (ie, late logarithmic phase of growth). Reproduced from Hassan and Fridovich[13] with permission.

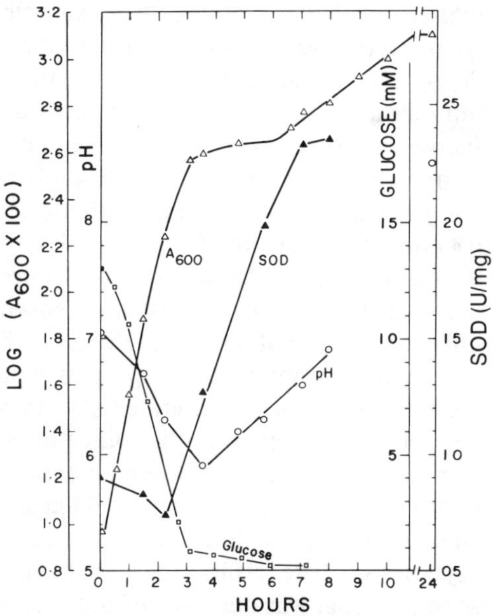

Figure 13-3 Kinetics of growth and SOD biosynthesis. The growth of *E coli* B was initiated in the TSY medium with an inoculum from an early stationary phase culture growing in glucose-minimal medium. From Hassan and Fridovich.[13] Reprinted with permission.

Suppression of the synthesis of SOD by glucose could have been due to catabolite repression, or alternately to diminution of the level of the true inducer (perhaps O_2^-) during the metabolism of glucose. The effect of glucose was not due to a classic type of catabolite repression, since adding cyclic AMP had no effect on the synthesis of SOD. Furthermore, α-methylglucoside, which mimics glucose in causing catabolite repression, was without effect on the synthesis of SOD in TSY medium devoid of glucose.[13] These studies led us to conclude that the rate of O_2^- production was low during glucose utilization, but increased during the oxidative phase, and that the cells modulated the level of SOD to meet their changing needs for the scavenging of O_2^-. Strong support for this conclusion was obtained through studies on the effects of paraquat and other redox active compounds on SOD biosynthesis in *E coli.*[14-17]

Induction of SOD in *E coli* in Response to Increases in Intracellular O_2^- by Redox Active Compounds

If the depression of SOD biosynthesis by glucose was actually a secondary consequence of a decreased production of O_2^-, during glucose fermentation, then any compound that could increase the production of O_2^- inside the cells should induce the synthesis of SOD. Paraquat (methyl viologen), a widely used herbicide, is an excellent example of such a compound. It is easily reduced to a stable cation radical that reacts very rapidly with oxygen to yield O_2^- plus paraquat.[25,26] We found that paraquat causes a rapid and profound induction of SOD in *E coli* grown in TSY medium and did so while glucose was still present.[13] This is illustrated in Figure 13-4.

E coli B, growing in TSY medium, was remarkably resistant to paraquat. Thus, 0.1 mM was without effect while 1 mM decreased the specific growth rate by 20% and 10 mM by only 50%. This high tolerance to paraquat was due to the protection afforded by SOD synthesized in response to the paraquat. The toxic effect of paraquat was only seen when the cells were prevented from achieving rapid synthesis of SOD, either by growing them in minimal media or by adding inhibitors of protein synthesis. The toxic effect of paraquat is due to its ability to generate O_2^-, and is seen only in the presence of both oxygen and a metabolizable source of electrons.[15,27]

Addition of 0.5 mM paraquat to an aerobic exponentially growing culture of *E coli* resulted, after a brief lag, in a rapid linear increase in SOD. Removal of paraquat from the growth medium resulted in a sharp decline in the rate of SOD synthesis. These results indicated that increased synthesis of SOD in response to paraquat was due to true induction and ruled out the possibility of selection by paraquat of some hyperproducers of SOD.[14]

Paraquat subverts electron flow from the normal cytochrome chain. This was seen as an increase in the cyanide-resistant respiration, without any change in total respiration. This increase in cyanide-resistant respiration was taken to represent an increase in the rate of O_2^- production. Disc gel electrophoresis, activity staining, and linear scanning densitometry demonstrated that paraquat specifically induced the manganese- but not the iron-containing SOD.[14] The increase in the biosynthesis of the Mn-SOD by paraquat was prevented by inhibitors of transcription or translation, but not by inhibitors of replication.[17] Thus, the increase in Mn-SOD biosynthesis in response to increases in intracellular O_2^-, as caused by paraquat, was due to de novo enzyme synthesis activated or derepressed at the level of transcription, and O_2^-, or a unique product of it, was the effector. Extracellularly generated O_2^- did not cause induction of SOD, and there is evidence that O_2^- does not cross the *E coli* cell envelope.[28]

Paraquat is not the only compound able to increase the rate of intracellular production of O_2^- and to induce the synthesis of Mn-SOD. Indeed, several redox active compounds were found to behave like paraquat.[17] These include pyocyanine, phenazine methosulfate, streptonigrin, juglone, plumbagin, menadione, methylene blue, and azure C. All of them caused an increase in cyanide-resistant respiration and an induction of Mn-SOD in *E coli*.[16,17] Furthermore, addition of cyanide or azide, to block cytochrome oxidase, resulted in increased synthesis of Mn-SOD in *E coli*.

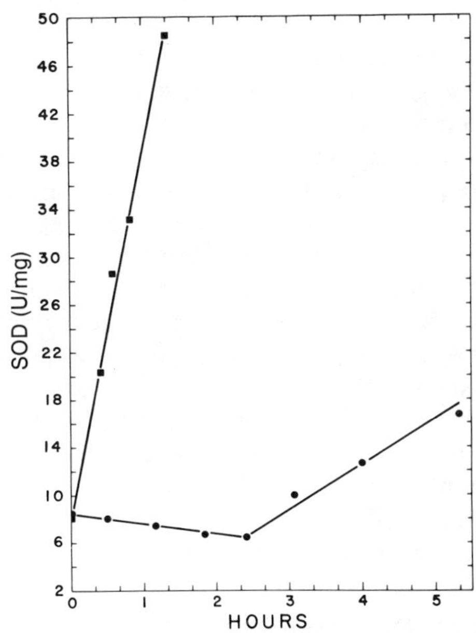

Figure 13-4 Effect of methyl viologen on the kinetics of SOD biosynthesis in *E coli* B. An inoculum from a late logarithmic phase culture growing in glucose-minimal medium was transferred into: •–• TSY medium; or ■-■ TSY medium containing 1 mM methyl viologen. From Hassan and Fridovich.[13] Reprinted with permission.

This indicated that one or more components of the normal electron transport chain were autoxidizable and able to generate O_2^-, and blocking the cytochrome oxidase by CN^- or N_3^- increased the steady-state concentration of the reduced form of such carriers, thus increasing the rate of reduction of oxygen to O_2^-, with the subsequent induction of Mn-SOD.[16,17]

Obligate Anaerobiosis in *E coli* in Response to Mutational Losses of SOD and Catalase

The physiologic roles of SOD and catalase in protecting *E coli* against the toxic effects of oxygen were further explored by genetic manipulations.[29,30] Mutants of *E coli,* deficient in SOD and/or catalase, were isolated and found to be ultrasensitive to oxygen toxicity. One mutant was temperature-sensitive with respect to oxygen tolerance, and was found to be temperature-sensitive with respect to SOD biosynthesis.[30]

Recently, we isolated oxygen-sensitive mutants of *E coli* and found that they fall into two groups.[29] The first group of mutants had lost the ability to induce both catalase and peroxidase, when exposed to oxygen, but retained their ability to induce Mn-SOD. The second group of mutants had lost the ability to induce all three enzymes. Oxygen was lethal to both groups. However, when a member of the first group of mutants was prevented from inducing Mn-SOD by adding puromycin, its susceptibility to the lethal effect of oxygen was greatly increased. These results indicated that enzymic scavenging of both O_2^- and H_2O_2 provides the ultimate defense against oxygen toxicity.

The mutants were found to revert to wild-type phenotype (oxygen tolerance) at a frequency of 10^{-5} to 10^{-4}, which is much higher than the normal spontaneous rate expected for point mutations. This might be due to endogenous mutagenesis caused by the reactivities of the unscavenged O_2^- and H_2O_2. The revertants were found to have regained the ability to induce the activity of the missing enzymes, except for one revertant from the first group, that became oxygen-tolerant without regaining the activities of catalase and peroxidase. Further characterization of this revertant revealed that it had lost the ability to respire, and therefore could dispense with catalase and peroxidase activities, since it could not longer generate O_2^- or H_2O_2. This revertant mutant is physiologically similar to the strain of *Lactobacillus plantarum* that was found to be oxygen-indifferent and was devoid of catalase and SOD activities.[31,32]

Correlations between the Level of SOD in *E coli* and Tolerance to Oxygen Toxicity

Several lines of evidence are available which support the superoxide theory of oxygen toxicity[32] and the physiologic role of superoxide dismutases. Thus, only strict anaerobes that are oxygen-sensitive, or organisms that do not utilize oxygen, were found to lack SOD.[32] SOD activity was reported in some obligately anaerobic bacteria.[33] However, recent studies[34] demonstrated a positive correlation between the levels of SOD in different anaerobes, and their degree of tolerance to oxygen. Therefore, the presence of SOD in some anaerobes is an indication of their vulnerability to the superoxide radical, which they may encounter during transient exposures to oxygen. The presence of SOD in these obligate anaerobes is an evolutionary response to ensure their continued survival.

Genetic losses of Mn-SOD and/or catalase in *E coli* resulted in oxygen-sensitive mutants that suffered structural damage and viability losses upon exposure to air.[29] In order to assess the relative importance of superoxide dismutase and catalase in providing protection against the toxicity of oxygen, we studied the effect of oxygen on the viability of a catalase-negative mutant when exposed to air in the presence and absence of puromycin (Figure 13-5). The data clearly show that both H_2O_2 and O_2^- are

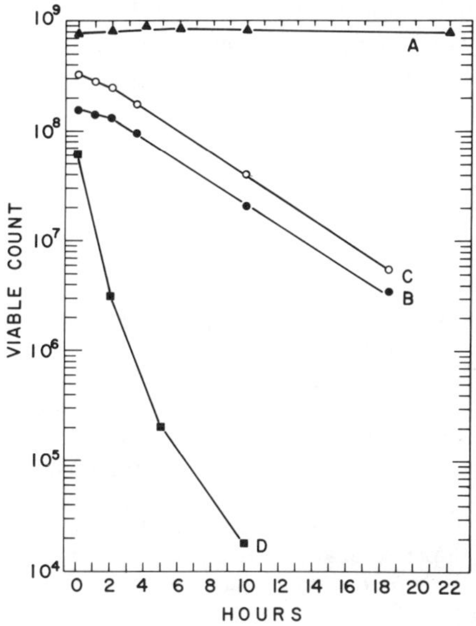

Figure 13-5 The lethal effects of oxygen on the parental strain *E coli* K12 and on the oxygen-intolerant mutant AN-1. Cells were grown anaerobically in TSY medium and then plated onto anaerobic TSY-agar. Control plates were kept anaerobically; the other plates were exposed to air for the indicated intervals before they were returned to the anaerobic chamber. The parental strain (line A) was resistant to the toxic effects of oxygen, whereas AN-1 (a catalase-negative mutant) (line B) was more sensitive to O_2 than the parental strain. The addition of catalase (150 μg per plate) (line C) increased the count of viable cells by a constant fraction but did not change their sensitivity to oxygen. When the induction of the Mn-SOD was prevented with puromycin (0.5 mg/ml) in liquid medium (line D), the intolerance of AN-1 to oxygen was increased. Reprinted from Hassan HM, Fridovich I. *Rev Infect Dis* 1:357–367, 1979, by permission of The University of Chicago Press.

important agents of oxygen toxicity. When the synthesis of catalase was restricted, the time required for 90% loss in viability was 11.3 hours (Figure 13-5, line B). However, when the synthesis of both superoxide dismutase and catalase was restricted, the corresponding decrease in viability required 1.6 hours (Figure 13-5, line D). These results indicate that the ability of the cells to induce SOD upon exposure to oxygen lowered the rate of killing by seven-fold over the rate observed with catalase alone.

Furthermore, increased levels of Mn-SOD in *E coli,* as induced by 1) oxygenation, 2) increasing the specific growth rate in a chemostat, or 3) growth in presence of redoxactive compounds, imparted increased resistance to the toxicity of hyperbaric oxygen or to the oxygen enchancement of streptonigrin lethality.[10,11,14,15]

Conclusions

The data strongly support the notion that O_2^- is an important agent of the toxicity of oxygen, paraquat, streptonigrin, and several redox active compounds, and that superoxide dismutase is the primary defense against the reactivities of this radical.

REFERENCES

1. Fridovich I: Superoxide radical and superoxide dismutase. *Accounts Biochem Res* 5:321–326, 1972.

2. Fridovich I: Superoxide dismutases. *Adv Enzymol* 41:35–97, 1974.

3. Fridovich I: Superoxide dismutases. *Annu Rev Biochem* 44:147–159, 1975.

4. Fridovich I: Oxygen radicals, hydrogen peroxide, and oxygen toxicity, in Pryor WA (ed): *Free Radicals in Biology,* vol I. New York, Academic Press Inc, 1976, pp 239–277.

5. Fridovich I: The biology of oxygen radicals. *Science* 201:875–880, 1978.

6. Fridovich I: Superoxide and superoxide dismutases, in Eichhorn GL, Marzilli DL (eds): *Advances in Inorganic Biochemistry.* New York, Elsevier North-Holland Inc, 1979, pp 67–90.

7. Britton L, Malinowski DP, Fridovich I: Superoxide dismutase and oxygen metabolism in *Streptococcus faecalis* and comparison with other organisms. *J Bacteriol* 134:229–236, 1978.

8. Gregory EM, Fridovich I: Induction of superoxide dismutase by molecular oxygen. *J Bacteriol* 114:543–548, 1973.

9. Gregory EM, Yost FJ Jr, Fridovich I: Superoxide dismutases of *Escherichia coli:* intracellular localization and functions. *J Bacteriol* 115:987–991, 1973.

10. Hassan HM, Fridovich I: Enzymatic defenses against the toxicity of oxygen and of streptonigrin in *Escherichia coli* K12. *J Bacteriol* 129:1574–1583, 1977.

11. Hassan HM, Fridovich I: Physiological function of superoxide dismutase in glucose-limited chemostat cultures of *Escherichia coli. J Bacteriol* 130:805–811, 1977.

12. Hassan HM, Fridovich I: Catabolite repression and the synthesis of superoxide dismutase in *Escherichia coli. Fed Proc* 36:715, 1977.

13. Hassan HM, Fridovich I: Regulation of superoxide dismutase synthesis in *Escherichia coli:* glucose effect. *J Bacteriol* 132:505–510, 1977.

14. Hassan HM, Fridovich I: Regulation of the synthesis of superoxide dismutase in *Escherichia coli:* induction by methyl viologen. *J Biol Chem* 252:7667–7672, 1977.

15. Hassan HM, Fridovich I: Superoxide radical and the oxygen enhancement of the toxicity of paraquat in *Escherichia coli. J Biol Chem* 253:8143–8148, 1978.

16. Hassan HM, Fridovich I: Exacerbation of oxygen toxicity by redox active compounds, in King TE, Mason HS, Morrison M (eds): *Oxidases and Related Redox Systems.* Baltimore, Md, University Park Press (In press).

17. Hassan HM, Fridovich I: Intracellular production of superoxide radical and of hydrogen peroxide by redox active compounds. *Arch Biochem Biophys* (In press).

18. Asada K, Kanematsu S, Takehashi M, et al: Superoxide dismutases in photosynthetic organisms. *Adv Exp Med Biol* 74:551–564, 1976.

19. Crapo JD, Tierney DL: Superoxide dismutase and pulmonary oxygen toxicity. *Am J Physiol* 226:1401–1407, 1974.

20. Gregory EM, Goscin SA, Fridovich I: Superoxide dismutase and oxygen toxicity in a eukaryote. *J Bacteriol* 117:456–460, 1974.

21. Keele BB Jr, McCord JM, Fridovich I: Superoxide dismutase from *Escherichia coli* B: a new manganese-containing enzyme. *J Biol Chem* 245:6176–6181, 1970.

22. Yost FJ Jr, Fridovich I: An iron-containing superoxide dismutase from *Escherichia coli. J Biol Chem* 248:4905–4908, 1973.

23. Dougherty HW, Sadaowski SJ, Baker EE: A new iron-containing superoxide dismutase from *Escherichia coli. J Biol Chem* 253:5220–5223, 1978.

24. Beauchamp C, Fridovich I: Superoxide dismutase: improved assays and an assay applicable to acrylamide gel. *Anal Biochem* 44:276–287, 1971.

25. Bus JS, Cagen SZ, Olgaard M, et al: A mechanism of paraquat toxicity in mice and rats. *Toxicol Appl Pharmacol* 35:501–513, 1976.

26. Montgomery MR: Interaction of paraquat with pulmonary microsomal enzyme systems. *Toxicol Appl Pharmacol* 37:106–107, 1976.

27. Fridovich I, Hassan HM: Paraquat and the exacerbation of oxygen toxicity. *Trends Biochem Sci* 4:113–115, 1979.

28. Hassan HM, Fridovich I: Paraquat and *Escherichia coli:* mechanism of production of extracellular superoxide radical. *J Biol Chem* (In press).

29. Hassan HM, Fridovich I: Superoxide, hydrogen peroxide, and oxygen tolerance of oxygen-sensitive mutants of *Escherichia coli. Rev Infect Dis* 1:357–367, 1979.

30. McCord JM, Beauchamp CO, Goscin S, et al: Superoxide and superoxide dismutase, in King TE, Mason HS, Morrison M (eds): *Oxidases and Related Redox Systems.* Baltimore, Md, University Park Press, 1973, pp 51–76.

31. Gregory EM, Fridovich I: Oxygen metabolism in *Lactobacillus plantarum. J Bacteriol* 117:166–169, 1974.

32. McCord JM, Keele BB Jr, Fridovich I: An enzyme-based theory of obligate anaerobiosis: the physiological function of superoxide dismutase. *Proc Natl Acad Sci USA* 68:1024–1027, 1971.

33. Hewitt J, Morris JG: Superoxide dismutase in some obligately anaerobic bacteria. *FEBS Lett* 50:315–318, 1975.

34. Tally FP, Goldin HR, Jacobus NV, et al: Superoxide dismutase in anaerobic bacteria of clinical significance. *Infect Immun* 16:20–25, 1977.

14 Superoxide Anion and the Erythrocyte Membrane

Robert E. Lynch, MD
G. Richard Lee, MD

The non-nucleated erythrocyte is a paradigm of specialization. It has divested itself of excess baggage, having either extruded or degraded its nucleus, mitochondria, and ribosomes. What remains is a plasma membrane which encloses a solution of mostly hemoglobin. In the evolutionary process that resulted in this marvel of simplicity one supposes that those organelles and enzymes that were discarded were no longer useful. Conversely, others may have been retained by virtue of their value to the erythrocyte, or at least to the organism in which the red cell circulated. Consequently, the retention of substantial concentrations of superoxide dismutase (SOD) suggests that it serves a useful purpose.

In procaryotes SOD appeared to be required for survival by those cells which reduced molecular oxygen.[1,2] Species that lived in the presence of O_2 without the protection of SOD were found not to reduce O_2. It was proposed that these organisms could not make O_2^- and therefore had no need

Supported by Grant #AM20207-03 from the Institute of Arthritis and Metabolic Diseases, National Institutes of Health.

for SOD. To explain these and other observations[3-8] it was suggested[9,10] that O_2^- was a byproduct of the reduction of molecular oxygen and that SOD had evolved to scavenge O_2^- and, thereby, prevent the potentially disastrous effects of its reactions in the cell.

The abundance of SOD in erythrocytes does not appear to fit comfortably into this scheme. Red cells, lacking organelles, scarcely reduce any O_2. The source of O_2^- is therefore not clear, even though the iron-oxygen bond in oxyhemoglobin (oxyHb) has been most recently represented as $HbFe^{3+}-O_2^-$.[11-15] In the majority of instances the dissociation of O_2 to form deoxyhemoglobin (deoxyHb) results in O_2 (rather than O_2^-) and $HbFe^{2+}$ (rather than $HbFe^{3+}$).

$$\underset{\text{(oxyHb)}}{HbFe^{3+}-O_2^-} \rightarrow O_2 + \underset{\text{(deoxyHb)}}{HbFe^{2+}}$$

Misra and Fridovich[16] suggested that the pathologic dissociation of hemoglobin produced O_2^- and $HbFe^{3+}$ (methemoglobin or metHb).

$$\underset{\text{(oxyHb)}}{HbFe^{3+}-O_2^-} \rightarrow \underset{\text{(metHb)}}{HbFe^{3+}} + O_2^-$$

However, this reaction was shown in the hemoglobin of the shark, which is notable for the propensity of its hemoglobin to autoxidize. Whether O_2^- is produced in mammals by the autoxidation of hemoglobin at a rate sufficient to account for the persistence of substantial amounts of SOD in mature red cells is not clear. Nor is it yet clear which component of the red cell reacts with O_2^- (or its progeny) to cause harm.

To determine which part of the cell O_2^- attacks, we first examined reactions of O_2^- with hemoglobin, and later, reactions with the plasma membrane. The reaction of O_2^- with oxyhemoglobin was shown to form metHb.[17,18]

$$O_2^- + 2H^+ + \underset{\text{(oxyHb)}}{HbFe^{3+}-O_2^-} \rightarrow H_2O_2 + O_2 + \underset{\text{(metHb)}}{HbFe^{3+}}$$

This conclusion was also reached by Winterbourn and associates[19,20] using an alternate means of producing O_2^-. The formation of metHb, which is virtually useless as a carrier of oxygen, is an event clearly disadvantageous to the host, but whether it occurs sufficiently to pose a threat to the erythrocyte's function is debatable. In studies of the copper-deficient pig, in which a pronounced decrease in the content of SOD in erythrocytes occurs,[21] no spontaneous methemoglobinemia was found even though the activity of SOD in the red cell had already declined (Table 14-1). In contrast, when O_2^- was generated photochemically in vitro in lysates of red cells from copper-deficient and from control pigs, the rate of metHb formation was clearly greater in lysates from animals in which the activity of SOD was reduced by dietary copper deficiency. The data suggested that although the SOD did prevent the formation of metHb by O_2^- in vitro, an analogous situation did not occur in vivo. This might be so if the methemoglobin reductases could compensate readily for the differing rates of metHb formation, if further degradation of metHb occurred rapidly preventing metHb accumulation, or if no flux of O_2^-, like the artificial photochemical flux in vitro, occurred in vivo.

Table 14-1
Methemoglobinemia and SOD Activity in Copper-Deficient and Control Pigs

Animals	SOD Activity*	% MetHb
Control (n = 4)	295 ± 23.2	1.18 ± 0.16
Copper-deficient (n = 4)	86 ± 26.8†	0.93 ± 0.37‡

*In U/ml red cells. Pigs were rendered copper-deficient on a condensed-milk diet[22] and littermate controls received an identical diet, with the exception of an oral $CuSO_4$ supplement. SOD activity was measured in the Tsuchihashi extracts of saline-washed red cells by the method of McCord and Fridovich.[23] The metHb concentrations were determined by the method of Hegesh et al.[24]
†$p \ll 0.001$.
‡$p > 0.3$.

To study the effects of O_2^- on the red cell membrane, vesicles were prepared from washed red cell ghosts.[25] Into the vesicles were sealed xanthine oxidase and ^{14}C-sucrose. Xanthine oxidase served as the source of O_2^- as it acted on its substrate, acetaldehyde, diffusing in from the medium. ^{14}C-sucrose was merely a convenient small compound, the leakage of which from the vesicles could easily be measured as an indication of disruption of the membrane. When O_2^- was generated enzymically inside these vesicles, lysis occurred, which was dependent on O_2^- as shown by the inhibitory effects of SOD. However, O_2^- was not itself judged to be the lytic agent because compounds that can scavenge $^1O_2$, but not catalase or scavengers of OH•, were also found to inhibit lysis.

These results contrasted with the observations which had been made in liposomes and in intact red cells.[26] When red cells in vitro were subjected to an external flux of O_2^-, lysis was inhibited by catalase and SOD as well as by scavengers of $^1O_2$. In red cell suspensions the lytic agent was postulated to be an obligatory product of the interaction of O_2^- with H_2O_2, possibly $^1O_2$.

$$H_2O_2 + O_2^- \rightarrow OH\bullet + O_2\,(^1\Delta g) + OH^-$$

In the experiments with vesicles, since catalase did not inhibit lysis but scavengers of $^1O_2$ did, reactants other than H_2O_2 were posited to be obligatory precursors of the lytic species. These putative precursors, lipid hydroperoxides, were proposed to have been formed spontaneously during the process of lysing the erythrocytes and washing away the cell sap, during which time the membrane was deprived of its intimate contact with certain cytoplasmic enzymes, notably glutathione peroxidase. In accord with this proposal, it was found that vesicles in which lipid peroxides were purposefully introduced by photoxidation with 4,5,6,7-tetrachloro-2′,4′,5′,7′-tetraiodofluorescein (Rose Bengal) were strongly sensitized to lysis by a subsequent flux of O_2^-. Although these and other experiments[27,28] supported the ability of O_2^- and/or its progeny to cause damage to cell membranes, the nature of that damage, which is presumed to be lipid peroxidation, remains unknown.

An unexpected finding in the experiments with resealed erythrocyte vesciles was the reduction of external cytochrome c when O_2^- was produced enzymatically within the vesicle. This suggested that O_2^- formed inside could escape into the medium. The ability of superoxide dismutase, either inside or outside the vesicles, to inhibit the reduction of external cytochrome c confirmed this interpretation. Since there is a "channel" in the red cell membrane through which stable anions exchange,[29-32] it seemed possible that O_2^- might share this same route. If it did, one of the derivatives of stilbene, 4-acetimido-4′-isothiocyano-2,2′-disulfonic acid stilbene (SITS) or 4,4′-diisothiocyano-2,2′-disulfonic acid stilbene (DIDS),[31] which inhibit the exchange of anions through the channel, should prevent the reduction of external cytochrome c by internally generated O_2^-. This was found to occur and the inhibition of the reduction of external cytochrome c by SITS was reversed by lysis of the vesicles with Triton X 100.[33]

If SITS did affect the reduction of external cytochrome c by blocking the egress of O_2^- from the vesicles through the anion channel, two predictions based on this interpretation could be tested. Effects of O_2^- on the vesicles should be augmented when SITS inhibited the egress of O_2^-, and the inhibition by external SOD should be overcome. These effects were evaluated in two situations, the auto-inactivation of xanthine oxidase turning over inside the vesicles, and the lysis of the vesicles. In both instances SITS overcame the inhibition by external SOD, and SITS alone accelerated the auto-inactivation of xanthine oxidase, all in accord with expectation. The effect of SITS on the lysis of vesicles was unpredictable, however, because it diminished the net flux of O_2^- by accelerating the auto-inactivation of xanthine oxidase, thereby inhibiting lysis, but at the same time SITS confined O_2^- to the vesicle's interior, thereby enhancing lysis. The effect observed, slight inhibition of lysis, suggested that the effect on the auto-inactivation of xanthine oxidase predominated.[33]

Whether the transit of O_2^- through the red cell membrane has any physiologic importance is unclear. One guess is that SOD in red cells may serve to scavenge O_2^- produced in plasma, by neutrophils for example, as they attempt to kill microorganisms. Such a function would parallel the action of carbonic anhydrase, which exists in the red cell and not in plasma, and whose substrate and product HCO_3^- must enter and exit through the anion channel in exchange for Cl^-. It should be noted that the products of the disproportionation reaction that SOD catalyses, H_2O_2 and O_2, are both conveniently handled by erythrocytes, the former by catalase and glutathione peroxidase, and the latter by hemoglobin.

Whether SOD serves in vivo to prevent damage to the red cell membrane is still not clear. If it did, one might predict that in animals in which the activity of SOD is reduced, there would be abnormalities in the red cells caused by injury to the cell membrane. In the copper-deficient pig the activity of SOD is markedly reduced.[21] In these animals a severe lack of ceruloplasmin occurs, which in turn causes abnormalities of iron transport.[22,34–36] Consequently, in spite of the availability of iron in hepatocytes (from which it is normally easily mobilized), the serum iron concentration declines, and red cells are produced by the bone marrow under conditions of limited iron supply. These red cells, like the erythrocytes of the iron-deficient animal, are small and contain a reduced concentration of hemoglobin.

To determine whether the copper-deficient, iron-replete animal bears stigmata of membrane injury we studied the red cells of swine deficient in iron but not in copper, deficient in copper but not in iron, and deficient in neither. If the entire defect in the copper-deficient pig were the lesion in the transport of iron, the red cells of copper-deficient animals and of iron-deficient animals should be indistinguishable. If, however, there were an additional defect in copper deficiency there might be other abnormalities resulting from that defect. As a simple means of searching for the lesion in the red cell membrane that might represent the effects of attack by superoxide of its progeny, osmotic fragility was studied. It was found that the red cells from copper-deficient and from iron-deficient animals were clearly different (Figure 14-1); as the anemia worsened the osmotic fragility of the copper-deficient animals increased whereas the osmotic fragility of the iron-deficient animals changed little. A defect not present in red cells in simple iron lack was present in erythrocytes in copper deficiency. If the defect was extrinsic to the cell, it was not readily reversible by incubation in normal plasma (control pig plasma) in vitro in 24 hours.

One way to explain the presence of osmotically fragile red cells is to attribute them to a senescent population of circulating erythrocytes. Since a defect in the production of red cells is one characteristic of the anemia of copper deficiency, the mean red cell age may increase and the resulting senescent population of circulating red cells would be expected to be osmotically fragile.[38] If this were correct the osmotic fragility of a cohort of young cells should not be increased in copper deficiency. To test this prediction the osmotic fragility of cohorts of 3- to 4-day-old red cells from copper-deficient, from iron-deficient, and from control pigs was evaluated, by injecting ^{59}Fe-lebeled plasma and measuring the osmotic fragility of the cohort of cells that had synthesized hemoglobin and entered the circulating blood in the time since the injection of isotope. The osmotic fragility of the cohort of young cells was determined by measuring the ^{59}Fe, rather than the hemoglobin, released by cells into the hypotonic medium. With this technique a 3- to 4-day-old red cell cohort from an iron-deficient pig was found to be osmotically resistant as compared to the cohorts from two control animals (Figure 14-2). The osmotic fragility of the cohort of red cells from an animal with early copper deficiency (VPRC 0.36) was like that of the control cells, but when copper deficiency was more severe (VPRC 0.20), the young cohort of copper-deficient cells was found to be osmotically fragile. Clearly the osmotic fragility of a cohort of young cells is already increased in copper deficiency. Senescence, therefore, cannot explain the abnormality.

One way for red cells to become osmotically fragile is to swell, but swelling cannot be the sole abnormality in copper deficiency, since swollen cells would be large whereas the red cells of the copper-deficient pig are small.[22] It seems more consistent with all the known effects of copper on the metabolism of iron to consider the red cells in copper deficiency to be small and underhemoglobinized due to the defects in iron transfer but to have, in addition, another lesion causing them to gain water and to acquire a reduced surface-to-volume ratio. If this interpretation were true,

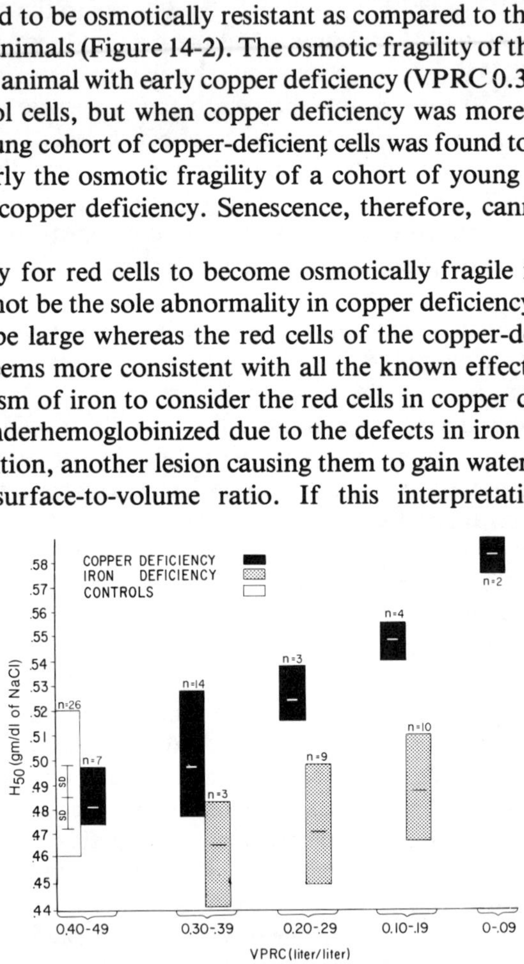

Figure 14-1 The relation of osmotic fragility[37] to severity of anemia in iron-deficient and copper-deficient pigs. Osmotic fragility on the ordinate is represented by the saline concentration at which 50% of the cells lysed (H_{50}). The concentration of erythrocytes is represented by the volume of packed red cells, determined by centrifugation at 2260 g for 30 min in "Wintrobe" tubes[37] (VPRC), on the abscissa. The range and standard deviation (SD) in littermate controls is shown in the far left-hand (open) bar. The vertical bars represent ranges, the horizontal line within each bar indicates the mean, and n is the number of determinations.

erythrocytes of the copper-deficient pig should exhibit a reduced critical hemolytic volume and an increased water content. This was found to be the case (Table 14-2).

Table 14-2
Water Content and Critical Hemolytic Volume in Copper-Deficient and Control Pig Red Cells

	Red Cell Water (gm/ml cells)*	Critical Hemolytic Volume†
Control		
No.	11	15
Mean ± SEM	0.714 ± 0.0071	1.47 ± 0.014
Copper-deficient		
No.	11	18
Mean ± SEM	0.748 ± 0.0053	1.35 ± 0.02
Significance *(p)*	< 0.005	< 0.001

*The red cell water content was determined by drying weighed 1.0-ml aliquots of plasma and of heparinized blood, adjusted to a VPRC of approximately 0.55 liter/liter, for 20 hours in an oven at 70 C. From the VPRC of the blood determined by radioiodinated albumin isotope dilution, and the weight lost by the cell suspension and by plasma, the red cell water in gm/ml of red cells could be calculated from the formula:

$$\frac{(\text{wet wt. blood} - \text{dry wt. blood}) - (1 - \text{VPRC})(\text{wet wt. plasma} - \text{dry wt. plasma})}{\text{VPRC}}$$

†The critical hemolytic volume[39] was determined by measuring the maximal volume, corrected for lysis, to which washed red cells swelled in hypotonic, phosphate-buffered saline solutions,[37] compared with the volume observed in 0.9% phosphate-buffered saline. The values represent the ratio of the maximal volume to the volume in 0.9% saline as determined by centrifugation in capillary tubes at 11,800 g for 5 min.

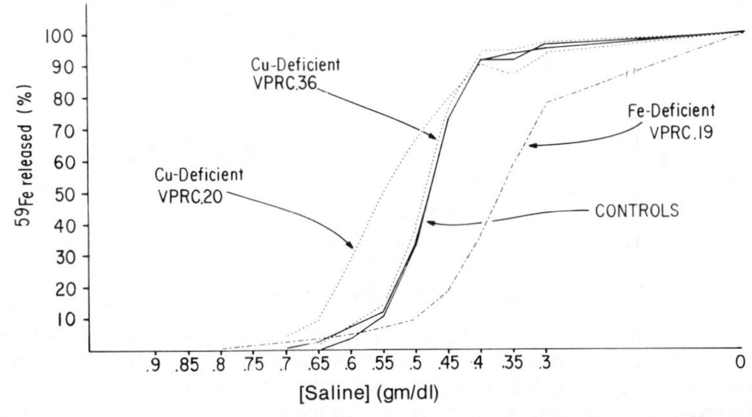

Figure 14-2 Osmotic fragility[37] curves of 3- to 4-day-old ^{59}Fe-labeled red cell cohorts in copper-deficient, iron-deficient, and control pigs. Note that the ordinate represents the percent^{59} Fe released into the supernatant, rather than percent hemoglobin released. Controls ————; iron deficient -•-•-•-; copper deficient -----.

The inference that tempted us was that the additional defect was the reduced activity of SOD in red cells. If this were so, the severity of the defect should increase as the SOD activity in red cells decreases. This did occur (Figure 14-3). With slight reductions in SOD activity no increase in osmotic fragility was detected, but as the activity declined further the osmotic fragility rose.

The data suggest that in the copper-deficient pig there is a defect in erythrocytes that is not attributable to the abnormalities in the movement of iron characteristic of copper deficiency. The defect has some characteristics of injury to the cell membrane, although the chemical nature of the injury is not known. A reasonable interpretation of these data is that they occur as the consequence of the decrease in the activity of SOD in red cells because membrane disruption[25,26,28,40] and lipid peroxidation[41–45] have been shown to be the consequence of exposure to a flux of O_2^-. It should be noted, however, that the loss of activity of some other constituent that contains, or depends on, copper could also account for these abnormalities.

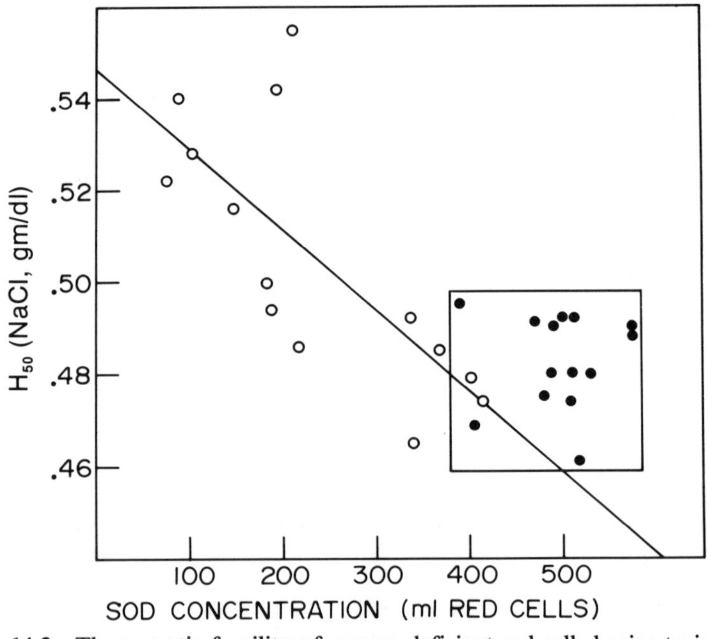

Figure 14-3 The osmotic fragility of copper-deficient red cells begins to increase after the SOD activity decreases. Osmotic fragility and SOD activity of red cells from control (•) and copper-deficient (O) animals were determined serially as copper deficiency developed. The points are the values for osmotic fragility and SOD activity in chloroform-ethanol extracts of saline-washed erythrocytes[23] for individual animals determined at the same time. Osmotic fragility is here represented by the saline concentration at which 50% of the cells lysed (H_{50}). The box encloses the values for the control animals and the line was generated by least-squares analysis of the values for the copper-deficient pigs.

REFERENCES

1. McCord JM, Keele BB Jr, Fridovich I: An enzyme-based theory of obligate anaerobiosis: the physiological function of superoxide dismutase. *Proc Natl Acad Sci USA* 68:1024–1027, 1971.

2. Gregory EM, Fridovich I: Oxygen metabolism in *Lactobacillus plantarum*. *J Bacteriol* 117:166–169, 1974.

3. Gregory EM, Fridovich I: Induction of superoxide dismutase by molecular oxygen. *J Bacteriol* 114:543–548, 1973.

4. Gregory EM, Fridovich I: Oxygen toxicity and the superoxide dismutase. *J Bacteriol* 114:1193–1197, 1973.

5. Hassan HM, Fridovich I: Enzymatic defenses against the toxicity of oxygen and of streptonigrin in *Escherichia coli*. *J Bacteriol* 129:1574–1583, 1977.

6. Hassan HM, Fridovich I: Physiological function of superoxide dismutase in glucose limited chemostat cultures of *Escherichia coli*. *J Bacteriol* 130:805–811, 1977.

7. Hassan HM, Fridovich I: Regulation of superoxide dismutase synthesis in *Escherichia coli:* glucose effect. *J Bacteriol* 132:505–510, 1977.

8. Hassan HM, Fridovich I: Regulation of the synthesis of superoxide dismutase in *Escherichia coli:* induction by methyl viologen. *J Biol Chem* 252:7667–7672, 1977.

9. Fridovich I: Biological aspects of superoxide radical and superoxide dismutases, in Hayaishi O, Asada K (eds): *Biochemical and Medical Aspects of Active Oxygen*. Baltimore, Md, University Park Press, 1977, pp 171–181.

10. Fridovich I: The biology of oxygen radicals. *Science* 201:875–880, 1978.

11. Peisach J, Blumberg WE, Wittenberg BA, et al: The electronic structure of protoheme/proteins. III. Configuration of the heme and its ligands. *J Biol Chem* 243:1871–1880, 1968.

12. Viale RO, Maggiora GM, Ingraham LL: Molecular orbital evidence for Weiss's oxyhaemoglobin structure. *Nature* 203:183–184, 1964.

13. Weiss JJ: Nature of the iron-oxygen bond in oxyhaemoglobin. *Nature* 202:83–84, 1964.

14. Wittenberg JB, Wittenberg BA, Peisach J, et al: On the state of the iron and the nature of the ligand in oxyhemoglobin. *Proc Natl Acad Sci USA* 67:1846–1853, 1970.

15. Yamamoto T, Palmer G, Gill D, et al: The valence and spin state of iron in oxyhemoblogin as inferred from resonance Raman spectroscopy. *J Biol Chem* 248:5211–5213, 1973.

16. Misra HP, Fridovich I: The generation of superoxide radical during the autoxidation of hemoglobin. *J Biol Chem 247:6960*–6962, 1972.

17. Lynch RE, Lee GR, Cartwright GE: Inhibition by superoxide dismutase of methemoglobin formation from oxyhemoglobin. *J Biol Chem* 251:1015–1019, 1976.

18. Lynch RE, Thomas JE, Lee GR: Inhibition of methemoglobin formation from purified oxyhemoglobin by superoxide dismutase. *Biochemistry* 16:4563–4567, 1977.

19. Sutton HC, Roberts PB, Winterbourn CC: The rate of reaction of superoxide radical ion with oxyhaemoglobin and methaemoglobin. *Biochem J* 155:503–510, 1976.

20. Winterbourn CC, McGrath BM, Carrell RW: Reactions involving superoxide and normal and unstable haemoglobins. *Biochem J* 155:493–502, 1976.

21. Williams DM, Lynch RE, Lee GR, et al: Superoxide dismutase activity in copper deficient swine. *Proc Soc Exp Biol Med* 149:534–536, 1975.

22. Lee GR, Nacht S, Lukens J, et al: Iron metabolism in copper deficient swine. *J Clin Invest* 47:2058–2069, 1968.

23. McCord JM, Fridovich I: Superoxide dismutase. An enzymic function for erythrocuprein (hemocuprein). *J Biol Chem* 224:6049–6055, 1969.

24. Hegesh E, Gruener N, Cohen S, et al: A sensitive micromethod for the determination of methemoglobin in blood. *Clin Chim Acta* 30:679–682, 1970.

25. Lynch RE, Fridovich I: Effects of superoxide on the erythrocyte membrane. *J Biol Chem* 253:1838–1845, 1978.

26. Kellogg EW III, Fridovich I: Liposome oxidation and erythrocyte lysis by enzymically generated superoxide and hydrogen peroxide. *J Biol Chem* 252:6721–6728, 1977.

27. Goldberg G, Stern A: The role of the superoxide anion as a toxic species in the erythrocyte. *Arch Biochem Biophys* 178:218–225, 1977.

28. Goldberg B, Stern A: Superoxide anion as a mediator of drug-induced oxidative hemolysis. *J Biol Chem* 251:6468–6470, 1976.

29. Cabantchik ZI, Rothstein A: Membrane proteins related to anion permeability of human red blood cells. II. Effects of proteoloytic enzymes on disulfonic stilbene sites of surface proteins. *J Membr Biol* 15:227–248, 1974.

30. Cabantchik ZI, Rothstein A: Membrane proteins related to anion permeability of human red blood cells. 1. Localization of disulfonic stilbene binding sites in proteins involved in permeation. *J Membr Biol* 15:207–226, 1974.

31. Cabantchik ZI, Rothstein A: The nature of the membrane sites controlling anion permeability of human red blood cells as determined by studies with disulfonic stilbene derivatives. *J Membr Biol* 10:311–330, 1972.

32. Tosteson DC: Halide transport in red blood cells. *Acta Physiol Scand* 46:19–41, 1959.

33. Lynch RE, Fridovich I: Permeation of the erythrocyte stroma by superoxide radical. *J Biol Chem* 253:4697–4699, 1978.

34. Ragan HA, Nacht S, Lee GR, et al: Effect of ceruloplasmin on plasma iron in copper deficient swine. *Am J Physiol* 217:1320–1323, 1969.

35. Roeser HP, Lee GR, Nacht S, et al: The role of ceruloplasmin in iron metabolism. *J Clin Invest* 49:2417, 1970.

36. Williams DM, Loukopoulos D, Lee GR, et al: Role of copper in mitochondrial iron metabolism. *Blood* 48:77–85, 1976.

37. Cartwright GE: *Diagnostic Laboratory Hematology,* 4th ed. New York, Grune & Stratton Inc, 1968.

38. Marks PA, Johnson AB: Relationship between the age of human erythrocytes and their osmotic resistance: a basis for separating young and old erythrocytes. *J Clin Invest* 37:1542–1548, 1958.

39. Seeman P, Sauks T, Argent W, et al: The effect of membrane-strain rate and of temperature on erythrocyte fragility and critical hemolytic volume. *Biochim Biophys Acta* 183:476–489, 1969.

40. Goldstein IM, Weissman G: Effects of the generation of superoxide anion on permeability of liposomes. *Biochem Biophys Res Commun* 75:604–609, 1977.

41. Kellogg EW III, Fridovich I: Superoxide, hydrogen peroxide, and singlet oxygen in lipid peroxidation by a xanthine oxidase system. *J Biol Chem* 250:8812–8817, 1975.

42. Pederson TC, Aust SD: The role of superoxide and singlet oxygen in lipid peroxidation promoted by xanthine oxidase. *Biochem Biophys Res Commun* 52:1071–1078, 1973.

43. Docampo R, Cruz FS, Boveris A, et al: Lipid peroxidation and the generation of free radicals, superoxide anion, and hydrogen peroxide in β lapachone-

treated trypanosoma cruzi epimastigotes. *Arch Biochem Biophys* 186:292–297, 1978.

44. Noguchi T, Nakano M: Effect of ferrous ions on microsomal phospholipid peroxidation and related light emission. *Biochim Biophys Acta* 368:446–455, 1974.

45. Petkau A, Chelack WS: Radioprotective effect of superoxide dismutase on model phospholipid membranes. *Biochim Biophys Acta* 433:445–456, 1976.

15 The Superoxide-Forming Enzyme from Human Neutrophils

Bernard M. Babior, MD

The neutrophil is a phagocytic cell whose primary function is the destruction of invading microorganisms. Attracted by chemotaxis into sites of infection, these cells are able to ingest and destroy the microorganisms they encounter there. The destruction of ingested bacteria is accomplished in part by a series of oxygen-dependent processes that employ a group of powerful oxidizing agents as microbicidal agents. The system responsible for the generation of these oxidants is the subject of this report.

The first evidence that oxygen might play a role in microbial killing was the discovery by Baldridge and Gerard[1] that neutrophils exposed to *Sarcina lutea* underwent a sharp increase in oxygen consumption. This increase in oxygen uptake was initially thought to provide energy for phagocytosis, but the discovery by Sbarra and Karnovsky[2] that phagocytosis proceeded as efficiently under nitrogen as under room air indicated that this was not the explanation. Shortly thereafter, Iyer et al[3]

Supported in part by a grant (AI-11872) from the National Institute of Allergy and Infectious Diseases, US Public Health Service.

found that phagocytes exposed to heat-killed *Escherichia coli* released H_2O_2 into the medium, and postulated that this compound, manufactured from the oxygen taken up by stimulated phagocytes, was used for the destruction of ingested microorganisms. Six years later Holmes et al[4] and Baehner and Nathan,[5] working independently, showed a defect in O_2 uptake and H_2O_2 production by neutrophils from patients with chronic granulomatous disease (CGD) an inherited condition characterized by an unusual susceptilibity to bacterial infections. This finding established the importance of oxygen-dependent microbicidal mechanisms in neutrophil function. A role for myeloperoxidase in oxygen-dependent microbicidal mechanisms was established by Klebanoff,[6] who showed that a system composed of H_2O_2, a halide ion, and myeloperoxidase (an enzyme abundant in neutrophils and certain mononuclear phagocytes) was able to kill many species of bacteria with remarkable efficiency. Finally, the observation was made that stimulated neutrophils released O_2^- into the medium,[7] and that this liberated O_2^- could react with endogenous peroxides to form powerful oxidizing radicals such as OH•,[8] which could supplement the H_2O_2-halide-myeloperoxidase mechanism as microbicidal agents.[9] Oxygen-dependent microbial killing by phagocytes is thus seen to be a complicated process involving many individual reactions.

The generation of the oxidizing agents used in the oxygen-dependent microbicidal mechanisms may be summarized as follows. The crucial reaction is the one-electron reduction of O_2 to O_2^- at the expense of a reduced pyridine nucleotide. Exactly which of the two pyridine nucleotides is the physiologic reducing agent has been a controversial matter for over 15 years, but for reasons that will become evident later and that are discussed more fully elsewhere,[10] I believe it to be NADPH. The O_2^--forming reaction is:

$$2O_2 + NADPH \rightarrow 2O_2^- + NADP^+$$

Most of the O_2^- formed in this reaction undergoes dismutation,[11] producing the H_2O_2 that is liberated by stimulated phagocytes:

$$2O_2^- + 2H^+ \rightarrow H_2O_2 + O_2$$

Studies of the stoichiometric relationships between oxygen uptake, O_2^- production, and H_2O_2 production have shown that all the H_2O_2 released by stimulated neutrophils is formed by the dismutation of O_2^-.[11] Finally, the $NADP^+$ produced in the O_2^--forming reaction is reconverted to NADPH by the reactions of the hexose-monophosphate shunt:

$$NADP^+ + \text{glucose-6-phosphate} \rightarrow NADPH + \text{6-phosphogluconic acid}$$
$$NADP^+ + \text{6-phosphogluconic acid} \rightarrow NADPH + \text{ribulose-5-P} + CO_2$$

Some $NADP^+$ is also produced by the glutathione-dependent detoxification of part of the H_2O_2 produced in this series of reactions[12]; this $NADP^+$ is also converted back to NADPH via the hexose-monophosphate shunt.

A question of great interest is the nature of the enzyme that catalyzes the O_2^--forming reaction of neutrophils. This question has been studied extensively over the past two decades, and a number of candidate oxidases have been proposed as the enzyme responsible for the production of the microbicidal oxidizing agents of the neutrophil. To decide among these candidates, the criterion has been informally established that the oxidase of interest will be the one whose properties in isolation most closely resemble those expressed in the intact cell.[13] In particular, cell-free preparations of this oxidase should reduce oxygen to O_2^-, and such preparations should not be active if obtained from cells that are not themselves able to produce O_2^-. In practice, the latter point means that O_2^- should be produced by cell-free preparations obtained from stimulated normal cells, but not by similar preparations from resting cells, or from stimulated cells obtained from patients with CGD.

Over the past several years, investigators in my laboratory have been studying an O_2^--forming oxidase whose properties strongly suggest that it is the enzyme that produces the microbicidal oxidants of neutrophils. This oxidase, a membrane-bound flavoprotein,[14,15] is found in homogenates of human neutrophils that have been stimulated with opsonized zymosan.[14,16] It can be released in soluble form by treatment with nonionic detergents under suitable conditions.[17] Studies have shown that the membrane-bound and the soluble oxidase display very similar behavior.

The activity in neutrophil homogenates is found in the $27{,}000 \times g$ pellet. Though this pellet contains several subcellular organelles, fractionation studies have shown that the activity is associated exclusively with material migrating with plasma membranes, and is separated cleanly from the lysosomes that are also found in this preparation.[15] Table 15-1 shows the level of activity of $27{,}000 \times g$ pellets from various neutrophil preparations. Substantial quantities of O_2^- are produced from NADPH by the oxidase from normal stimulated neutrophils, but very little is made by

Table 15-1
O_2^- Production by Particles from Various Neutrophil Preparations

Source of Particles	O_2^- Production (nmoles/min/mg)
Normal, activated	18.5
Normal, resting	0.5
CGD, activated	0.0

Subjects were: normal, or had CGD.
Cells were: resting, or zymosan-activated.

preparations from resting neutrophils or from stimulated neutrophils obtained from patients with CGD. Thus, this preparation fulfills the major criterion used to judge whether a given candidate enzyme is involved in the production of microbicidal oxidants by neutrophils.

The properties of this oxidase are listed in Table 15-2. As mentioned previously, the enzyme is a component of the plasma membrane. As electron donor, it can use either reduced pyridine nucleotide, but prefers NADPH, as indicated by the Michaelis constants for the two reductants.[16] The enzyme is active over a wide pH range, showing a broad peak in the region of neutrality. Finally, the enzyme appears to be an FAD-requiring flavoprotein, a conclusion based on the finding that O_2^- production is abolished by Triton X-100 but can be restored by the addition of FAD (or FMN, at substantially higher concentrations), provided the detergent is sufficiently dilute.[14]

Table 15-2
Properties of the Particulate O_2^--Forming Enzyme

1. Found in the plasma membrane fraction of activated neutrophils.
2. Catalyzes the following reaction:

$$2O_2 + NADPH \rightarrow 2O_2^- + NADP^+ + H^+$$

3. K_m values for reduced pyridine nucleotides:

NADPH:	0.02 mM
NADH:	1.0 mM

4. An FAD-requiring enzyme.

Sonication of O_2^--forming particles with detergent in the presence of 25% ethylene glycol released a substantial fraction of the O_2^--forming activity into solution. The solubilized activity did not sediment on centrifugation for 1 hour at 105,000 × g, and was able to pass through a filter that retained molecules larger than 300,000 daltons. No O_2^--forming activity was released by this procedure from particles obtained from resting neutrophils or from stimulated cells from patients with CGD (Table 15-3), indicating that the solubilized O_2^--forming activity, like that in the particles, was that of the enzyme responsible for microbicidal oxidant production.

Upon solubilization, the oxidase displayed a requirement for phospholipid that was not seen with the particulate preparation. Table 15-4 shows the activity of the oxidase in the presence of various phospholipids. There was a greater than two-fold increase in O_2^- production in the presence of either mixed phospholipids or phosphatidylethanolamine; phosphatidylcholine and phosphatidylserine, however, had no effect on the activity of the solubilized oxidase. These results show not only that phospholipid is required by the solubilized oxidase, but also that the requirement is specific by lipid class.

Table 15-3
O_2^- Production by Solubilized Enzyme Preparations from Various Neutrophil Preparations

Source of Preparation	O_2^- Production (nmoles/min/mg)
Normal, activated	18.1
Normal, resting	0.4
CGD, activated	0.0

Subjects were: normal, or had CGD.
Cells were: resting, or zymosan-activated.

Table 15-4
Effects of Specific Phospholipids on O_2^- Production by the Solubilized Enzyme

Phospholipid	O_2^- Production (nmoles/min/mg)
None	13.7
Crude phospholipid	39.9
Phosphatidylethanolamine	38.1
Phosphatidylserine	13.1
Phosphatidylcholine	13.3

In the presence of phospholipid, the properties of the solubilized oxidase were found to resemble closely those of the particulate enzyme. O_2^- production declines slowly with time, but is still reasonably rapid after 40 minutes (Figure 15-1). The reaction rate is strictly proportional to enzyme

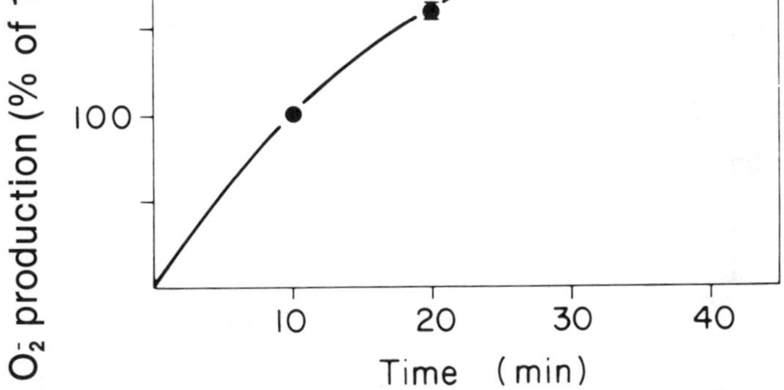

Figure 15-1 O_2^- production by the solubilized enzyme as a function of time.

concentration (Figure 15-2). The activity as a function of pH (Figure 15-3) is similar to that obtained with the particulate preparation. Finally, the Michaelis constants for NADPH and NADH are 33 μM and 930 μM respectively, values very close to those determined using the particulate preparation (see Table 15-2). These figures provide further evidence that the physiologic substrate is NADPH.

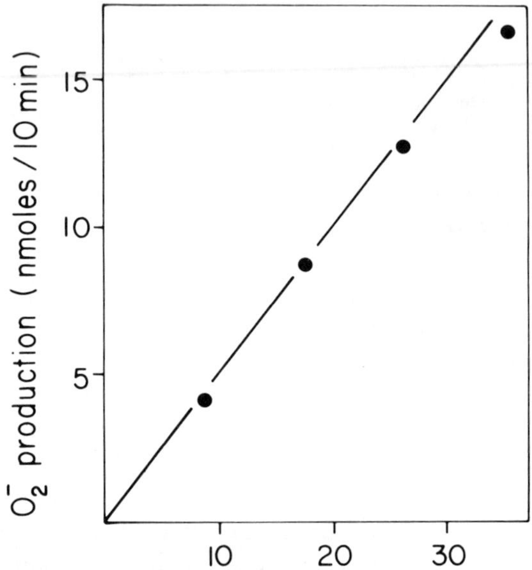

Figure 15-2 O_2^- production as a function of protein concentration.

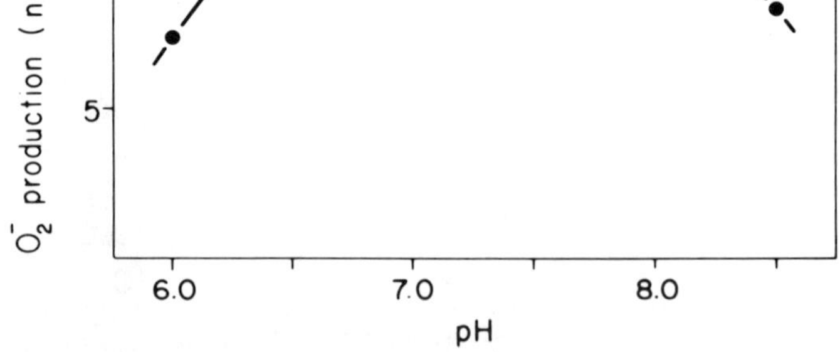

Figure 15-3 O_2^- production as a function of pH.

REFERENCES

1. Baldridge CW, Gerard RW: The extra respiration of phagocytosis. *Am J Physiol* 103:235–236, 1933.

2. Sbarra AJ, Karnovsky ML: The biochemical basis of phagocytosis. I. Metabolic changes during the ingestion of particles by polymorphonuclear leukocytes. *J Biol Chem* 234:1355–1362, 1959.

3. Iyer GYN, Islam MF, Quastel JH: Biochemical aspects of phagocytosis. *Nature* 192:535–541, 1961.

4. Holmes B, Page AR, Good RA: Studies of the metabolic activity of leukocytes from patients with a genetic abnormality of phagocytic function. *J Clin Invest* 46:1422–1432, 1967.

5. Baehner RL, Nathan DG: Leukocyte oxidase: defective activity in chronic granulomatous disease. *Science* 155:835–836, 1967.

6. Klebanoff SJ: A peroxidase-mediated antimicrobial mechanism. *J Exp Med* 126:1063–1078, 1967.

7. Babior BM, Kipnes RS, Curnutte JT: Biological defense mechanisms: the production by leukocytes of superoxide, a potential bactericidal agent. *J Clin Invest* 52:741–744, 1973.

8. Tauber AI, Babior BM: Evidence for hydroxyl radical production by human neutrophils. *J Clin Invest* 60:374–379, 1977.

9. Johnston RB Jr, Keele BB Jr, Misra HP, et al: The role of superoxide anion generation in phagocytic bactericidal activity: studies with normal and chronic granulomatous disease leukocytes. *J Clin Invest* 55:1357–1372, 1975.

10. Babior BM: Oxygen-dependent microbial killing by phagocytes. *N Engl J Med* 298:659–668, 721–725, 1978.

11. Root RK, Metcalf JA: H_2O_2 release from human granulocytes during phagocytosis: relationship to superoxide anion formation and cellular catabolism of H_2O_2: studies with normal and cytochalasin B treated cells. *J Clin Invest* 60:1266–1279, 1977.

12. Reed PW: Glutathione and the hexose monophosphate shunt in phagocytizing and hydrogen peroxide-treated rat leukocytes. *J Biol Chem* 244:2459–2464, 1969.

13. Badwey JA, Curnutte JT, Karnovsky ML: The enzyme of granulocytes that produces superoxide and peroxide: an elusive pimpernel. *N Engl J Med* 300:1157–1160, 1979.

14. Babior BM, Kipnes RS: Superoxide-forming enzyme from human neutrophils: evidence for a flavin requirement. *Blood* 50:517–524, 1977.

15. Dewald B, Baggiolini M, Curnutte JT, et al: Subcellular localization of the superoxide-forming enzyme in human neutrophils. *J Clin Invest* 63:21–29, 1979.

16. Barbior BM, Curnette JT, McMurrich BJ: The particulate superoxide-forming system from human neutrophils: properties of the system and further evidence supporting its participation in the respiratory burst. *J Clin Invest* 58:989–996, 1976.

17. Gabig TG, Kipnes RS, Babior BM: Solubilization of the O_2^--forming activity responsible for the respiratory burst in human neutrophils. *J Biol Chem* 253:6663–6665, 1978.

16 A Possible Role for Calmodulin in the Metabolic Activation of Leukocytes

Joe M. McCord, PhD
William F. Petrone, PhD
Milton J. Cormier, PhD
Harold P. Jones, PhD

Polymorphonuclear leukocytes (or neutrophils) may be isolated from peripheral blood in the so-called resting state, characterized by low oxygen consumption and a generally quiescent metabolic status. When these cells are exposed to any of a number of stimuli (opsonized bacteria,[1] immune complex,[2,3] complement component C5a,[3] endotoxins,[4] lymphokine,[4] calcium ionophore A23187,[5] and others) various metabolic activities may be triggered. These activities include chemotaxis, phagocytosis, greatly enhanced oxygen consumption manifested in the production and release of superoxide and hydrogen peroxide, and degranulation manifested in the intraphagosomal or extracellular release of granular enzymes. Together, these events result in the detection and elimination of invading microorganisms—a desirable end. These same events, however, are responsible for much of the tissue damage that results from an inflammatory episode, and for this reason the control of these events is of great clinical concern.

The exact biochemical mechanisms involved in the metabolic activation of neutrophils are poorly understood, despite the accumulation of a considerable body of data describing various features of the process. In this report we suggest a mechanism for the initiation of the respiratory burst in neutrophils, which is based on a number of observations taken from the literature plus some new observations that lend additional support to the hypothesis.

Our proposed mechanism centers on a regulatory protein known as calmodulin (also known as calcium-dependent regulator, activator, or modulator protein). In the presence of calcium, calmodulin has been shown to function as a common regulatory subunit for a variety of different enzymes. Among the enzymes activated by calmodulin are phosphodiesterase,[6] adenylate cyclase,[7] myosin light chain kinase,[8] protein kinase,[9] phosphorylase kinase,[10] Ca^{++}-Mg^{++} ATPase,[11] and NAD kinase.[12] The protein appears to offer a generalized mechanism for coupling calcium transients to subsequent biochemical events.

A unique feature of calmodulin is its inhibition by micromolar concentrations of drugs of the phenothiazine family.[13,14] The interaction of the drug with calmodulin at these concentrations is highly specific and calcium-dependent.[15] While the inhibitory effects of phenothiazines on a wide variety of enzyme activities and more complex biologic phenomena have been observed and reported for many years, only recently has it become apparent that most, if not all, of the drug effects are due to its highly specific affinity for calmodulin. This is, although inhibition of a phenomenon by a phenothiazine certainly does not constitute proof of calmodulin's involvement in the process, it may be considered a strong implication. This is particularly true if the phenomenon at hand is also known to involve a calcium transient.

Such is the case with the metabolic activation of neutrophils. Calcium transients, either as an influx from the extracellular milieu due to a sudden change in the permeability of the cell's plasma membrane to calcium or as a release of sequestered intracellular calcium, have been implicated in the processes of chemotaxis,[16] phagocytosis,[17] degranulation,[18] and the oxidative burst.[19] Likewise, the phenothiazines have been shown to inhibit chemotaxis,[20] phagocytosis,[21] the burst of oxygen consumption,[22] chemiluminescence,[23] and the killing of bacteria.[21] While these observations supported our hypothesis, a search of the literature turned up a problem. Of all the eucaryotic tissues and cell types examined for calmodulin content, only one was reportedly devoid of calmodulin activity—leukocytes.[24]

Do Leukocytes Contain Calmodulin?

Because of the development of an improved assay for calmodulin[25] since the negative report appeared, we decided to re-examine the question. Another possible factor in the negative finding may have been rapid loss of calmodulin in the sonicated leukocyte extract due to proteolysis by neutral proteases released from granules. While calmodulin is extremely heat-resistant, it is relatively sensitive to proteolysis.

Human blood was collected in EDTA-containing Vacutainer tubes, diluted with an equal volume of 0.9% saline, and layered onto Ficoll-Paque cushions. After centrifuging at 400 × g, the pellets were resuspended in erythrocyte lysing medium (0.155 M ammonium chloride, 10 mM $KHCO_3$, and 0.14 mM Na_2EDTA) and incubated at 37 C for five minutes. After centrifugation at 250 × g, the pellet was subjected to the lysing treatment four more times. The neutrophils were suspended to a concentration of 1.2×10^7 cells/ml and heated to 90 C for five minutes to destroy protease activity. The extract was then assayed for calmodulin by a modification of the bioluminescence assay devised by Matthews and Cormier.[25] Calmodulin was present, and accounted for 0.04% of the total cellular protein. This corresponds to about 2.5×10^6 molecules of calmodulin per neutrophil, or an intracellular concentration of approximately 5 μM.

Does Calmodulin Directly Activate the NADPH Oxidase?

Upon metabolic activation of the neutrophil, superoxide is produced by a membrane-bound NADPH oxidase.[26,27] NADPH is provided by the hexose-monophosphate shunt, which shows markedly enhanced activity upon activation of the cell. The requirement for calcium and inhibition by phenothiazines have been demonstrated for the initiation of the burst of oxygen consumption by intact neutrophils, thereby suggesting a role for calmodulin somewhere in the sequence of events involved in the overall activation process, but not necessarily suggesting a direct interaction with the oxidase. For example, calmodulin could activate the rate-limiting enzyme in the hexose-monophosphate shunt, causing a rapid rise in intracellular NADPH concentration, providing substrate to a permanently active oxidase. Alternatively, NAD kinase, which has been shown to be calmodulin-activited,[12] could suddenly increase the $NADP^+$ concentration in the cell, providing substrate for the shunt, which would in turn provide substrate for the oxidase. To address this question we examined the ability of phenothiazines to inhibit the NADPH oxidase in a cell-free system.

The only member of the phenothiazine family examined with regard to inhibition of oxidative metabolism in leukocytes has been promethazine,[22,23] with half-maximal inhibitions reported in the range of 50 to 100μM. Weiss and Levin[13] have reported the relative affinities of several members of the phenothiazine family for calmodulin, finding trifluoperazine (Stelazine) to be 30 or 40 times more effective in inhibiting the calmodulin-dependent phosphodiesterase activity than promethazine. Accordingly, we examined the ability of trifluoperazine to inhibit superoxide production by intact neutrophils. The results are seen in Figure 16-1. The cells were preincubated with trifluoperazine for one hour at the indicated concentrations, then activated with opsonized zymosan. Half-maximal inhibition was achieved at a trifluoperazine concentration of 1 μM, exactly the dissociation constant reported by Weiss and Levin[13] for the binding of trifluoperazine to free calmodulin.

A cell-free extract containing neutrophil NADPH oxidase was prepared as described by Gabig et al[27] from zymosan-activated cells. The 105,000 × g pellet containing plasma membrane, granules, and other organelles was resuspended and NADPH-dependent superoxide generation was monitored. As seen in Figure 16-2, this activity was inhibited by trifluoperazine with half-maximal inhibition at about 20 μM. The activity was likewise shown to be calcium-dependent by EGTA sensitivity. Hence, the oxidase activity per se demonstrates both criteria necessary for regulation by calmodulin.

At first glance there appears to be a discrepancy between Figures 16-1 and 16-2; more trifluoperazine is required to inhibit superoxide production by the cell-free extract than by intact cells. One possible explanation is as follows: in intact cells, trifluoperazine presumably effects its inhibition of

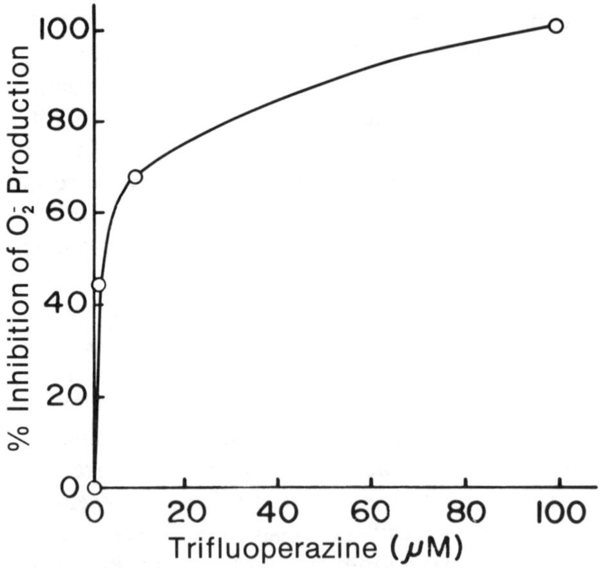

Figure 16-1 Trifluoperazine inhibition of superoxide production by activated intact neutrophils. Neutrophils were isolated from heparinized human blood by gravity sedimentation of red cells from dextran solution. The neutrophils were resuspended in Hank balanced salts solution to a density of 1.5 × 1o^7 cells/ml. Aliquots were incubated with various concentrations of trifluoperazine for one hour at 37 C. Assay mixtures contained 9 × 10^6 pretreated cells, 1 mg of opsonized zymosan, and 0.1 μM ferricytochrome c in 1.0 ml of Hank solution. Final trifluoperazine concentrations are shown on the figure. Trifluoperazine inhibition is relative to zymosan-dependent cytochrome c reduction in the absence of the drug.

activation by binding to free calmodulin; in the case of cell-free extract, the phenothiazine presumably must bind to the preformed active calmodulin-oxidase complex in order to reverse the association. That is, there may be a kinetic restriction on the attainment of thermodynamic equilibrium, such that the drug is actually binding to a different species in each case. This argument is supported by the data of Weiss and Levin.[13] They directly measured a binding constant between calmodulin and trifluoperazine of 1 μM. The amount of the drug required to cause half-maximal reversal of calmodulin activation of phosphodiesterase, however, was 10μM. These data closely correspond to our observations with NADPH oxidase.

The most direct approach for determining the role of calmodulin in the activation of the NADPH oxidase would be the activation of latent oxidase by the addition of purified calmodulin and calcium to a cell-free extract. Preliminary attempts in this vein have not met with success for reasons that are presently unclear. Active cell-free preparations may be inhibited with EGTA, but the readdition of calcium results in recovery of only a small fraction of the lost activity, at best. Thus, it may be difficult or impossible to form or reform the active oxidase complex once the cellular integrity has been destroyed. It is conceivable that yet another factor is required, beyond calcium, calmodulin, and oxidase. It is, in fact, necessary to invoke a higher level of control than the simplest hypothesis can provide in order to explain how a neutrophil can respond to a chemotactic stimulus without activating its NADPH oxidase, assuming that both processes are calmodulin-controlled.

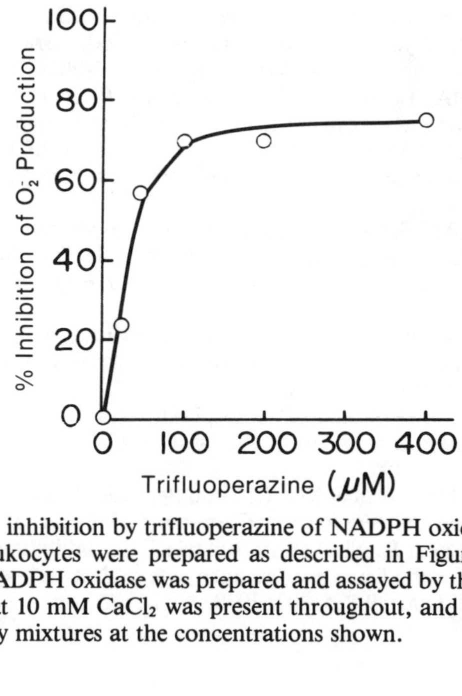

Figure 16-2 The inhibition by trifluoperazine of NADPH oxidase activity in cell-free extracts. Leukocytes were prepared as described in Figure 16-1. A cell-free preparation of NADPH oxidase was prepared and assayed by the method of Gabig et al,[27] except that 10 mM $CaCl_2$ was present throughout, and trifluoperazine was added to the assay mixtures at the concentrations shown.

In conclusion, compelling circumstantial evidence exists in support of a calmodulin-regulated NADPH oxidase in the neutrophil. Leukocytes do contain calmodulin. Activation of the oxidase requires calcium and is potently inhibited by phenothiazines. The relative efficacies of different phenothiazines are in excellent quantitative agreement with their binding affinities for calmodulin. Furthermore, the studies reported here provide strong support that the NADPH oxidase prepared according to Gabig et al[27] is indeed the source of superoxide production in intact neutrophils; both activities show a dependence on calcium and a sensitivity to inhibition by trifluoperazine.

REFERENCES

1. Curnutte JT, Barbior BM: Biological defense mechanism. The effect of bacteria and serum on superoxide production by granulocytes. *J Clin Invest* 53:1662–1672, 1974.
2. Johnston RB, Lehmeyer JE: Elaboration of toxic oxygen by-products by neutrophils in a model of immune complex disease. *J Clin Invest* 57:836–841, 1976.
3. Goldstein IM, Roos D, Kaplan H, et al: Complement and immunoglobulins stimulate superoxide production by human leukoyctes independently of phagocytosis. *J Clin Invest* 56:1155–1163, 1975.
4. McCord JM, Salin ML: Free radicals and inflammation: studies on superoxide-mediated NBT reduction by leukocytes, in Brewer GJ (ed): *Erythrocyte Structure and Function*. New York, Alan R Liss Inc, 1975, pp 731–752.
5. Schell-Frederick E: Stimulation of the oxidative metabolism of polymorphonuclear leukocytes by the Ca-ionophore A23187. *FEBS Lett* 48:37–40, 1974.
6. Cheung WY: Cyclic 3′,5′-nucleotide phosphodiesterase evidence for the properties of a protein activator. *J Biol Chem* 246:2859–2869, 1971.
7. Brostrom MA, Brostrom CO, Breckenridge B, et al: Calcium-dependent regulation of brain adenylate cyclase. *Adv Cyclic Nucleotide Res* 9:85–99, 1978.
8. Yagi K, Yazawa M, Kakiuchi S, et al: Identification of an activator protein for myosin light chain kinase as the Ca^{++} dependent modulator protein. *J Biol Chem* 253:1338–1340, 1978.
9. Waisman DM, Singh TJ, Wang JH: The modulator-dependent protein kinase—a multifunctional protein kinase activatable by the Ca^{++}-dependent modulator protein of the cyclic nucleotide system. *J Biol Chem* 253:3387–3390, 1978.
10. Cohen P, Burchell A, Foulkes J, et al: Identification of the Ca^{++}-dependent modulator protein as the growth subunit of rabbit skeletal muscle phosphoxylase kinase. *FEBS Lett* 92:287–293, 1978.
11. Gopinath RM, Vincenzi FF: Phosphodiesterase protein activator mimics red blood cell cytoplasmic activator of (Ca^{++}-Mg^{++}) ATPase. *Biochem Biophys Res Commun* 77:1203–1209, 1977.
12. Anderson JM, Cormier MJ: Calcium-dependent regulator of NAD kinase in higher plants. *Biochem Biophys Res Commun* 84:595–602, 1978.
13. Weiss B, Levin R: Mechanism for selectively inhibiting the activation of cyclic nucleotide phosphodiesterase and adenylate cyclase by antipsychotic agents. *Adv Cyclic Nucleotide Res* 9:285–303, 1978.

14. Levin R, Weiss B: Binding of trifluoperazine to the calcium-dependent activator of cyclic nucleotide phosphodiesterase. *Mol Pharmacol* 13:690–697, 1977.

15. Charbonneau H, Cormier MJ: Purification of plant calmodulin by fluphenazine-Sepharose affinity chromatography. *Biochem Biophys Res Commun* 90:1039–1047, 1979.

16. Boncek MM, Snyderman R: Calcium influx requirement for human neutrophil chemotaxis: inhibition by lanthanum chloride. *Science* 193:905–907, 1976.

17. Stossel TP: Quantitative studies of phagocytosis. Kinetic effects of cations and heat-labile opsonin. *J Cell Biol* 58:346–356, 1973.

18. Hoffstein S, Weissmann G: Microfilaments and microtubules in calcium ionophore-induced secretion of lysosomal enzymes from human polymorphonuclear leukocytes. *J Cell Biol* 78:769–781, 1978.

19. Simchowitz L, Spilberg I: Generation of superoxide radicals by human peripheral neutrophils activated by chemotactic factor. Evidence for the role of Ca. *J Lab Clin Med* 93:583–593, 1979.

20. Bjorksten B, Quie P: Inhibition of human chemotaxis *in vitro* by phenothiazines and related compounds. *Infect Immun* 14:948–950, 1976.

21. Ruutu T: Effect of phenothiazines and related compounds on phagocytosis and bacterial killing by human neutrophilic leukocytes. *Ann Med Exp Biol Fenniae* 50:24–36, 1972.

22. DeChatelet LR, Qualliotine-Mann D, Caldwell R, et al: Effects of promethazine HCl on human polymorphonuclear leukocytes. *Infect Immun* 7:403–408, 1973.

23. Trush MA, Van Dyke K: Effect of promethazine on human polymorphonuclear chemiluminescence. *Pharmacology* 16:314–320, 1978.

24. Smoake JA, Song S-Y, Cheung WY: Cyclic 3′ 5′-nucleotide phosphodiesterase. Distribution and developmental changes of the enzyme and its protein activator in mammalian tissues and cells. *Biochim Biophys Acta* 341:402–411, 1974.

25. Matthews JC, Cromier MJ: Rapid microassay for the calcium-dependent protein modulator of cyclic nucleotide phosphodiesterase. *Methods Enzymol* 57:107–112, 1978.

26. DeWald B, Baggiolini M, Curnutte J, et al: Subcelular localization of the superoxide-forming enzyme in human neutrophils. *J Clin Invest* 63:21–29, 1979.

27. Gabig TG, Kipnes RS, Babior BM: Solubilization of the O_2-forming activity responsible for the respiratory burst in human neutrophils. *J Biol Chem* 253:6663–6665, 1978.

17 How, Where, and Why Phagocytic Leukocytes Produce Superoxide and Peroxide

Manfred L. Karnovsky, PhD
John A. Badwey, PhD
Alfred I. Tauber, MD

During phagocytosis, or upon stimulation of the plasalemma by surface-active and other agents, phagocytic leukocytes produce substantial quantities of hydrogen peroxide and superoxide (O_2^-).[1-3] These products may subsequently interact to form hydroxyl radicals and possibly singlet oxygen.[1,3] Each of these substances has been implicated to varying degrees in the ability of phagocytic leukocytes to kill bacteria, yeast, mycoplasma, viruses, and tumor cells.[1,3-5]

Since this "respiratory burst" was first observed nearly a half-century ago by Gerard and Baldridge,[6] many of the biochemical events that accompany phagocytosis have been elucidated. Surprisingly, however, the nature of the oxidase(s) that initiate this burst remain controversial.[1,7] We shall briefly review some of the new data from our laboratory concerning the oxidase(s), and offer information on the cytochemical localization and the

Supported by US Public Health Service Grant No. AI-03260 from the National Institute of Allergy and Infectious Diseases.

possible regulatory mechanisms that may modulate the activity of the enzyme(s) during phagocytosis. Some differences between species (guinea pig and man) are intimated.

NADH Oxidase of Guinea Pig Granulocytes

This enzyme was first reported in quinea pig granulocytes in 1964 by Cagan and Karnovsky.[8] The principal properties of the protein are summarized in Table 17-1. The enzyme produces both H_2O_2 and O_2, and is present in guinea pig granulocytes in sufficient amounts to account for all of the net oxygen uptake observed during phagocytosis of a wide variety of particles (for a summary of values obtained by several workers, see reference 9).

Table 17-1
Summary of the Characteristics of Soluble NADH Oxidase from Guinea Pig Neutrophils

Property		Refs.
Specificity	$\frac{NADH}{NADPH}$ = 40; other electron donors ≤ 15% of activity with NADH	8,10
Stoichiometry	$NADH + H^+ + O_2 \longrightarrow H_2O_2 + NAD^+$ (via O_2^-); 1, 1, (0.03), 1, 1	8, 10
V_{max}	Resting cells: $9.7 \times 10^{-7} \pm 1.6 \times 10^{-7}$ nmoles O_2/min. Phagocytizing cells: 50% increase in extractable activity	10,11
K_m	0.4 mM (based on NADH consumption or $O_2^{-}\cdot$ production)	10
pH optimum	4.5 to 5.0	8
Molecular weight	310,000 ± 14,000 (Sepharose 6-B)	10
Flavins	80% FAD (excitation 450 nm, emission 535 nm)	2
Control	Polyvalent anions (ie, ATP, ADP, PP_i) depress activity. Divalent cations release inhibitions.	10
Cellular location	a) Homogenization in sucrose, 50% is membrane-bound	11
	b) Phagosomes (prepared in sucrose) contain 25% of total activity	12
	c) Isotonic KCl solubilizes all enzyme	11
	d) EM cytochemistry: resting cells are inactive; during phagocytosis—only phagosomes positive.	13

Since granulocytes at rest consume very little oxygen, the oxidase must be under strict metabolic control. We have recently observed that the NADH oxidase is potently inhibited by a variety of nucleotides (eg, ATP,

ADP, and AMP) and other polyvalent anions (eg, PP_i and citrate). Divalent cations can reverse this inhibition by nucleotides in a fashion that suggests the free (unchelated) forms of nucleotides are the major inhibitory species.[10]

ATP is a reasonable candidate to effect most of the inhibition of the enzyme in resting cells, because of the high concentration of total ATP (about 2.7 mM) in these cells.[14] It has been hypothesized that the initial event in the stimulation of the respiratory burst consists of a rapid shift of divalent cations from the environment and/or plasmalemma binding sites toward the cytoplasm.[15] Influx or redistribution of divalent cations could conceivably contribute to the activation of the oxidase by diminishing the levels of free (unchelated) ATP. This model is supported by the known requirement for extracellular Ca^{++} or Mg^{++} for O_2^- production,[14,15] and the ability of various ionophores to stimulate the oxidative metabolism of these cells.[15,16]

During phagocytosis, the oxidase functions in association with the phagosomal membrane (see Table 17-1, reference 13, and below). This association of the enzyme with membrane, and presumably lipids, may further enhance its activity, although these interactions have yet to be explored. Substantial amounts of NADH oxidase are also present in phagocytic monocytes from guinea pigs and mice, and in "resident" mouse peritoneal macrophages.[17]

The Oxidase of Human Granulocytes: An Unresolved Question

Both a membrane-bound NADPH oxidase,[18,19] and a "soluble" NADH oxidase[11,20] have been described in human neutrophils. Both activities are reportedly decreased in chronic granulomatous disease (CGD),[20,21] a genetic condition in which neutrophils fail to display the respiratory burst. Table 17-2 presents a comparison of the levels of these enzymes in human neutrophils reported in the literature, with the respiratory activity manifested by the whole cell. It is obvious that at the time of this writing neither enzyme, nor both enzymes acting together, can account for the whole-cell phenomenon.

One possible explanation for this is that "the oxidase," hitherto approached as a single entity, may in fact be a multicomponent system that is disrupted upon homogenization in such a fashion as to make reconstitution difficult. Evidence for such a system was first provided by Shinagawa et al,[28] who described an unusual b cytochrome in rabbit leukocytes. This cytochrome could not be reduced by either NADH or NADPH under aerobic conditions, but was reduced by these compounds under anaerobic conditions in a reaction that was insensitive to both KCN and antimycin A.[28] The reduced pigment was autoxidizable. A similar cytochrome has

Table 17-2
Superoxide Production and Oxygen Consumption of Phagocytosing Human Neutrophils
Comparison of the Magnitude of the Whole-Cell Phenomenon and Isolated Enzyme Systems

Preparation	Stimulus	$O_2^{-}\cdot$ Production (nmoles/min/10^7 cells)	O_2 Consumption (nmoles/min/10^7 cells)	References
Human neutrophils	Phorbol myristate acetate	≃ 78.0		22
		74.2 ± 7.4 (n = 6)		—*
	Opsonized Zymosan	≃ 47.0		23
			43.61 ± 5.11 (n = 6)	—†
	Opsonized *Staphylococcus aureus*	≃ 8.0	≃ 126	24
Particulate NADPH oxidase (membrane)	Opsonized zymosan	1.8 to 2.6‡§		25
				18
			0.55§‖	26
	Phorbol myristate acetate	≃ 3.8¶		19
Soluble NADH oxidase	Resting cells		0.35 to 0.5#	11

Note: The concentrations of NADPH and NADH in phagocytosing human neutrophils have recently been determined employing a very sensitive enzyme cycling procedure, and have been reported to be, respectively, 0.057 and 0.033 mM.[27]
*Curnutee and Karnosvky (unpublished data, 1979)
†Badwey and Karnovsky (unpublished data, 1979)
‡Assays were performed at 22 C, pH = 7.0, NADPH concentration of 0.1 mM.
§Calculations based on the assumption that 0.17 mg of particle protein are obtained from 10^7 cells.[18]
‖Assays were performed at 37 C, pH = 5.5, utilizing an NADPH concentration of 1.25 mM.
¶Assays were performed at 23 C, pH = 7.0, utilizing an NADPH concentration of 0.16 mM. This value was calculated on the assumption that 0.525 mg protein of the postnuclear supernatant (250 gm) is equivalent to 10^7 cells.
#Assays were performed at 37 C, pH = 7.0, NADH concentration of 0.16 mM.

been reported recently in human neutrophils,[29] and is reportedly missing in most,[30] but not all[31] cases of CGD.

We have recently been reinvestigating the soluble NADH oxidase of human neutrophils to determine if, in fact, it may function as a possible component of such a complex system, and if difficulties in our own and other laboratories in obtaining reproducible preparations of the latter might be explained by its mutlicomponent nature. Supernatants from extracts of human cells were found to reduce ferricyanide far more rapidly than molecular oxygen (Table 17-3). This activity was unaffected by dialysis and obliterated by boiling. The activity was directly proportional to the concentration of protein (Figure 17-1), and there was no evidence of kinetically relevant dissociation or association of the enzyme in the concentration range employed. The quinones juglone and gamma-benzoquinonebutyrate also displayed considerable activity as electron acceptors with these preparations (Table 17-3). With ferricyanide as the electron acceptor, NADH was more active than NADPH as the electron donor (Table 17-4). The kinetics of ferricyanide reduction are compared in Figure 17-2 A and B. The enzyme exhibited approximately the same K_m (about 40μM) toward NADH and NADPH. However, the maximum velocity was approximately ten-fold larger for NADH.

We are currently evaluating the possibility that this diaphorase may, in fact, function as an early component of a respiratory chain in these cells. It

Table 17-3
Specificity of Human NADH-$Fe(CN)_6$ Reductase for Electron Acceptors

Electron Acceptor	Concentration (mM)	NADH Oxidation (nmoles/min/10^7 cell equivs)*
$Fe(CN)_6$	1.0	30.00
Gamma-benzoquinonebutyric acid	0.2	7.24
	0.4	9.48
Juglone	0.1	3.36
	0.2	5.29
2,6-Dichlorophenol-indophenol	0.1	3.58
Cytochrome c†	0.1	1.88
	0.2	1.71
Menadione	0.1	0.35
	0.2	0.54
0_2		0.48

The reaction mixture, 1.0 ml, contained 0.16 mM NADH and 50 mM Hepes (pH = 7.0), plus a suitable amount of enzyme, which was added last to initiate the reaction. Temperatures were maintained at 37 C. Values represent the means of at least three closely agreeing determinations.
*"10^7 cell equivs" refers to the 27,000-g supernatant fraction obtained from 10^7 cells.
†35% of this reaction was inhibited by 60μg of superoxide dismutase.

Table 17-4
Specificity of Ferricyanide Reductase for Electron Donors

Substrate	Relative Oxidase Activity* (%)
NADH	100
NADPH	12
Purine, hypoxanthine, acetaldehyde, L-leucine, D-alanine, glycine, α-DL-glycerophosphate, benzylamine, histamine, L-lactate, pyruvate, D-glucose, D-galactose	0

The $Fe(CN)_6$ reductase activities of the cytosol fraction of human neutrophils were determined for a variety of different compounds known to serve as electron donors for a variety of reactions. The reaction mixture, 1.0 ml, contained 1.0 mM $K_3Fe(CN)_6$, 50 mM Hepes (pH = 7.0), 1.0 mM substrate, and a suitable amount of enzyme, which was added last to initiate the reaction. Temperatures were maintained at 37 C.
*Given as the percentage of activity compared to that observed for the substrate NADH.

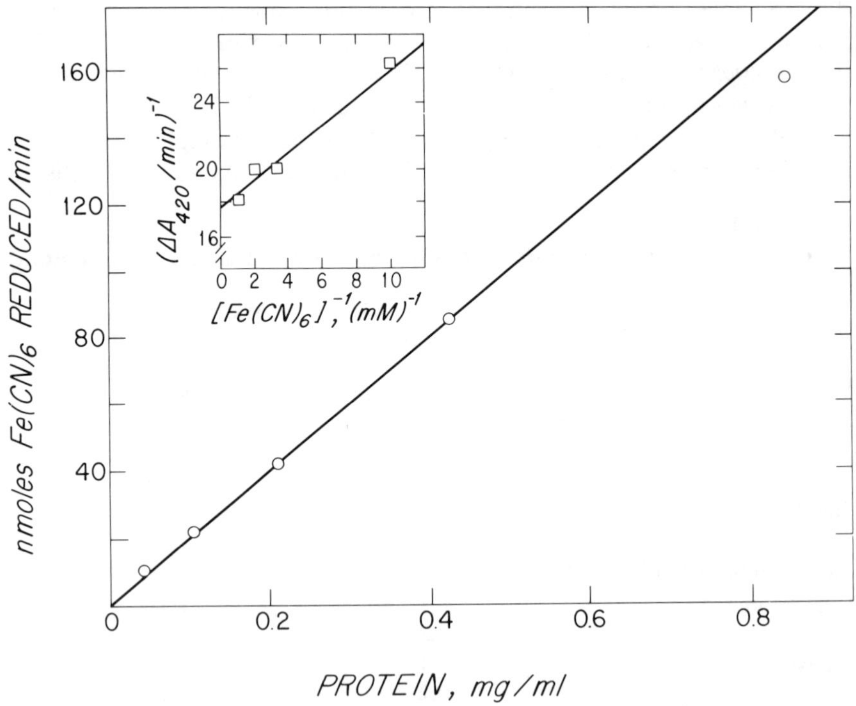

Figure 17-1 Human granulocyte NADH-$Fe(CN)_6$ reductase: effect of protein concentration on activity. The reaction mixture, 1.0 ml, contained 0.3 mM $K_3Fe(CN)_6$, 0.16 mM NADH, 50 mM Hepes (pH = 7.0), and the concentrations of protein indicated. Temperatures were maintained at 37 C. The enzyme preparation utilized in these studies consisted of a 27,000-g supernatant fraction obtained after homogenization of the cells in alkaline KCl as described by Baehner et al.[11] The insert shows the effect of the concentration of ferricyanide on activity.

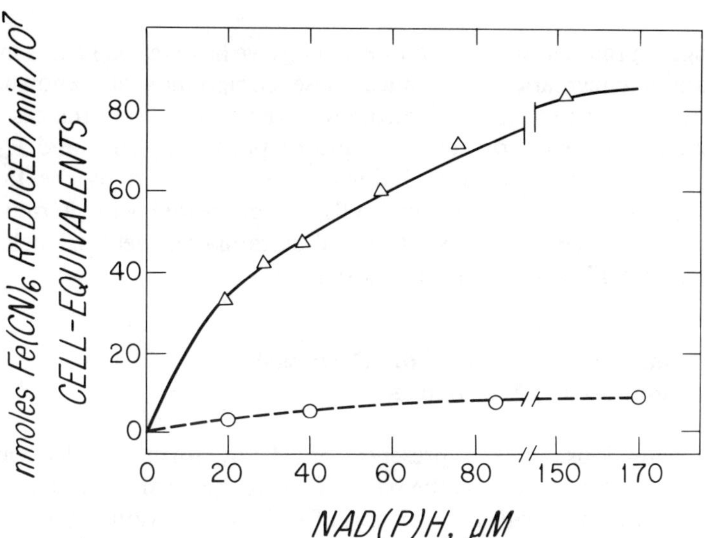

A

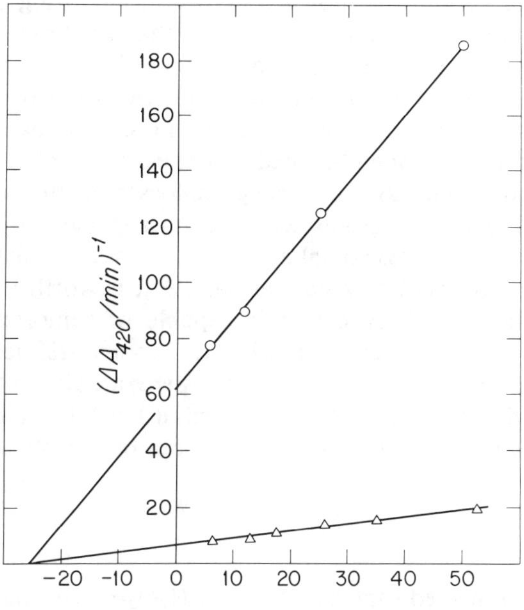

B

Figure 17-2 Saturation kinetics of ferricyanide reductase of human granulocytes. The reaction mixture of 1.0 ml, contained 0.3 mM $K_3Fe(CN)_6$ and 50 mM Hepes (pH = 7.0). Temperature was maintained at 37 C. The enzyme preparation utilized was the 27,000-g supernatant fraction after homogenization of the cells in alkaline isotonic KCl.[11] Initial velocities were measured by monitoring the reduction of ferricyanide with both NADH (Δ-Δ) and NADPH (O-O) serving as the electron donor and **A** plotted in terms of the total activity for comparative purposes. The data of **A** were used to construct the double-reciprocal plots shown in **B.**

is intriguing that use of the quinone analogs we have tested results in rather substantial enzymatic rates, since these compounds are known to be capable both of reducing cytochromes of the b-type, as in the case of the mitochondrial respiratory chain,[32] and of producing $O_4^{-}\cdot$ and H_2O_2 by autoxidation.[33,34] These speculations are strengthened by the fact that stimulated cells, but not resting cells, when treated with a radioactive nonpenetrating reagent (^{35}S-diazotized sulfanilate), yield a labeled lipid product—possibly a derivative of a semiquinone.

Ctyochemical Localization of the Oxidase(s) in Granulocytes and Macrophages

Cerous ions, in the presence of H_2O_2, form an electron-dense precipitate. This reaction is the basis of a cytochemical technique that was developed for the localization of sites of H_2O_2 production in phagocytes.[35] When resting guinea pig granulocytes were examined by this technique, precipitation of cerium was not detected. Phagocytozing cells, however, had extensive deposits of reaction product that was present only on the phagosomal membrane.[13] These data are in marked contrast to the situation in human and mouse granulocytes, in which reaction product was detected both in the phagosome and on eht plasmalemma.[13,35]

Both normal resident and elicited peritoneal macrophages of mice gave staining patterns similar to guinea pig granulocytes by this technique.[13] On the other hand, immunologically activated macrophages displayed reaction product both on the phagosomal and on the plasmalemmal surface when stimulated with phorbol myristate acetate.[13] It is worth noting that it is this last mentioned cell type that is capable of tumoricidal activity,[36] presumably by virtue of cell-cell contact. In some defined systems, the phenomenon occurs through the agency of peroxide alone[4,37]; in others, in conjunction with peroxidase.[38] The potential importance of the plasmalemma enzyme system in this context has been previously indicated.[13]

Conclusions

We have indicated that the O_2^{-}- and H_2O_2-producing oxidase(s) for reduced pyridine nucleotides differ in various phagocytic leukocytes from various species. The point is made that although granulocytes have been studied extensively in this context, macrophages in various states (of activation) must also be considered, because this enzymatic activity has a role in their functions. The differences referred to above pertain particularly to the preferred substrate (NADH or NADPH), and the cellular location of the enzyme(s). A key problem is the question of how the enzymatic activity is unmasked when the cells are stimulated, to manifest the respiratory burst.

In the case of the guinea pig granulocyte the enzyme responsible is obtained in soluble form, is cytochemically manifested on the phagosomal membrane, and can be modulated by polyanions and divalent cations (eg, ATP and Mg^{++}). In human granulocytes, the extracted (soluble) enzymatic activity is less reproducibly linked to oxygen as electron acceptor, and more readily studied with ferricyanide or certain quinones as acceptor. Peroxide-producing enzyme activity is cytochemically demonstrable in both plasmalemmal and phagosomal locations. The cytochemical picture with the other cell types (macrophages, and granulocytes of other species) studied resembles one or other of the above situations. The relation of the extracted oxidase of granulocytes (preferred substrate NADH) to that found in an isolated membrane fraction (preferred substrate NADPH) by Babior and his colleagues[1] still requires clarification.[7]

REFERENCES

1. Babior BM: Oxygen-dependent microbicidal killing by phagocytes. *N Engl J Med* 298:659–668, 1978.

2. Karnovsky ML: Chronic granulomatous disease—pieces of a cellular and molecular puzzle. *Fed Proc Am Soc Exp Biol* 32:1527–1533, 1973.

3. Klebanoff SJ: Anitmicrobial mechanisms in neutrophilic polymorphonuclear leukocytes. *Semin Hematol* 12:117–142, 1975.

4. Nathan CF, Silverstein SC, Brukner LH, et al: Extracellular cytolysis by activated macrophages and granulocytes. II. Hydrogen peroxide as a mediator of cytotoxicity. *J Exp Med* 149:100–113, 1979.

5. Rosen H, Klebanoff SJ: Bactericidal activity of a superoxide anion-generating system. A model for the polymorphonuclear leukocyte. *J Exp Med* 149:27–39, 1979.

6. Gerard RW, Baldridge CW: The extra respiration of phagocytosis. *Am J Physiol* 103:235–236, 1933.

7. Badwey JA, Curnutte JT, Karnovsky ML: The enzyme of granulocytes that produces superoxide and peroxide: an elusive pimpernel. *N Engl J Med* 300:1157–1160, 1979.

8. Cagan RH, Karnovsky ML: Enzymatic basis of the respiratory stimulation during phagocytosis. *Nature* 204:255–257, 1964.

9. Cheson BD, Curnutte JT, Barbior BM: The oxidative killing mechanisms of the neutrophil. *Prog Clin Immunol* 3:1–65, 1977.

10. Badwey JA, Karnovsky ML: Production of superoxide and hydrogen peroxide by an NADH-oxidase in guinea pig polymorphonuclear leukocytes: modulation by nucleotides and divalent cations. *J Biol Chem* (In press)

11. Baehner R, Gilman N, Karnovsky ML: Respiration and glucose oxidation in human and guinea pig leukocytes: comparative studies. *J Clin Invest* 49:692–700, 1970.

12. Stossel TP, Pollard TD, Mason R, et al: Isolation and properties of phagocytic vesicles from polymorphonuclear leukocytes. *J Clin Invest* 50:1745–1757, 1971.

13. Karnovsky ML, Badwey J, Briggs R, et al: Respiratory pathways of macrophages—cellular localization and relation to cidal activity, in van Furth (ed): *Mononuclear Phagocytes*. London, Blackwell Scientific Publications Ltd, 1980.

14. Cohen HJ, Chovaniec ME: Superoxide production by digitonin-stimulated guinea pig granulocytes: the effects of N-ethyl maleimide, divalent cations, and glycolytic and mitochondrial inhibitors on the activation of the superoxide generating system. *J Clin Invest* 61:1088–1096, 1978.

15. Romeo D, Zabucchi G, Miani N, et al: Ion movement across leukocyte plasma membrane and excitation of their metabolism. *Nature* 253:542–544.

16. Schell-Frederick E: Stimulation of the oxidative metabolism of polymorphonuclear leukocytes by the calcium ionophore A23187. *FEBS Lett* 48:37–40, 1974.

17. Simmons SR, Karnovsky ML: Iodinating ability of various leukocytes and their bactericidal activity. *J Exp Med* 138:44–63, 1973.

18. Babior BM, Curnutte JT, McMurrick BJ: The particulate superoxide-forming system from human neutrophils: properties of the system and further evidence supporting its participation in the respiratory burst. *J Clin Invest* 58:989–996, 1976.

19. Dewald D, Baggiolini M, Curnutte J, et al: Subcellular localization of the superoxide-forming enzyme in human neutrophils. *J Clin Invest* 63:21–29, 1979.

20. Baehner RL, Karnovsky ML: Deficiency of reduced nicotinamide adenine dinucleotide oxidase in chronic granulomatous disease. *Science* 162:1277–1279, 1968.

21. Curnutte JT, Kipnes RS, Babior BM: Defect in pyridine nucleotide dependent superoxide production by a particulate fraction from the granulocytes of patients with chronic granulomatous disease. *N Engl J Med* 293:628–632, 1975.

22. Newberg PE, Cohen HJ, Rothchild SB, et al: Prenatal diagnosis of chronic granulomatous disease. *N Engl J Med* 300:178–181, 1979.

23. Rosen H, Klebanoff SJ: Chemiluminescence and superoxide production by myeloperoxide-deficient leukocytes. *J Clin Invest* 58:50–60, 1976.

24. Root RK, Metcalf JA: H_2O_2 Release from granulocytes during phagocytosis. Relationship to superoxide anion formation and cellular catabolism of H_2O_2: studies with normal and cytochalasin B-treated cells. *J Clin Invest* 60:1266–1279, 1977.

25. Gabig TG, Kipnes RS, Babior BM: Solubilization of the O_2^--forming activity responsible for the respiratory burst in human neutrophils. *J Biol Chem* 253:6663–6665, 1978.

26. McPhail LC, DeChatelet LR, Shirley PS: Further characterization of NADPH oxidase activity of human polymorphonuclear leukocytes. *J Clin Invest* 58:774–780, 1976.

27. Aellig A, Maillard M, Phavorin A, et al: The energy metabolism of the leukocyte. IX. Changes in the concentrations of the coenzymes NAD, NADH, NADP and NADPH in polymorphonuclear leukocytes during phagocytosis of *Staphylococcus albus* and due to the action of phospholipase c. *Enzyme* 22:207–212, 1977.

28. Shinagawa Y, Tanaka C, Teraoka A, et al: A new cytochrome in neutrophilic granules of rabbit leukocyte. *J Biochem* 59:622–624, 1966.

29. Segal AW, Jones OTG: Novel cytochrome b system in phagocytic vacuoles of human granulocytes. *Nature* 276:515–517, 1978.

30. Segal AW, Jones OTG, Webster D, et al: Absence of a newly described cytochrome b from neutrophils of patients with chronic granulomatous disease. *Lancet* 2:446–449, 1978.

31. Borregaard N, Johansen KK, Taudorff E, et al: Cytochrome b is present in neutrophils from patients with chronic granulomatous disease. *Lancet* 2:949, 1979.

32. Crane FL: Hydroquinone dehydrogenases. *Annu Rev Biochem* 46:439–469, 1977.

33. McCord JM, Fridovich I: The utility of superoxide dismutase in studying free radical reactions. II. The mechanism of the mediation of cytochrome c reduction by a variety of electron carriers. *J Biol Chem* 245:1374–1377, 1970.

34. Misra HP, Fridovich I: The univalent reduction of oxygen by reduced flavins and quinones. *J Biol Chem* 247:188–192, 1972.

35. Briggs RT, Drath DB, Karnovsky ML, et al: Localization of NADH oxidase on the surface of human polymorphonuclear leukocytes by a new cytochemical method. *J Cell Biol* 67:566–586, 1975.

36. Hibbs JB, Chapman HA, Weinberg JB: The macrophage as an antineoplastic surveillance cell: biological perspectives. *J Reticuloendothel Soc* 24:549–570, 1978.

37. Nathan CF, Brukner LH, Silverstein SC, et al: Extracellular cytolysis by activated macrophages and granulocytes. I. Pharmacologic triggering of effector cells and the release of hydrogen peroxide. *J Exp Med* 149:84–99, 1979.

38. Clark RA, Klebanoff SJ: Neutrophil-mediated tumor cell cytotoxicity: role of the peroxidase system. *J Exp Med* 141:1442–1447, 1975.

18 Neutrophils as Secretory Organs

Gerald Weissman, MD
Helen M. Korchak, PhD
H. Daniel Perez, MD
James E. Smolen, PhD
Ira M. Goldstein, MD
Sylvia T. Hoffstein, PhD

Endocytosis and Exocytosis in the Granulocyte

The relationship between phagocytosis and inflammation was established nearly one hundred years ago when Metchnikoff appreciated that phagocytic cells contained cytases (now called lysosomal enzymes) and that these cytases were responsible not only for digestion of material entering the phagocyte, but might provoke inflammation. However, Metchnikoff believed that the cytases were elaborated chiefly from dying cells.

Metchnikoff, recapitulating his studies of twenty years,[1] suggested that "the leukocytes having arrived at the spot where the intruders are found seize them after the manner of the amebae and within their bodies subject them to intracellular digestion. This digestion takes place in the vacuoles in which usually is a weakly acid fluid which contains digestive ferments,

Aided by grants (AM-11949, HL-19072, HL-19721, GM-23211, AM-18531, AM-00463) from the National Institutes of Health, the National Foundation—March of Dimes, The National Science Foundation (78-13221), and the Arthritis Foundation.

cytases, of these a very considerable number are now recognized. The cytases must also be grouped with the soluble ferments which are not thrown off by the phagocytes so long as these remain intact. Immediately these cells are injured however, they allow a part of their cytases to escape." We now know that the cells need not, in fact, be injured in order for cytases and other materials to be liberated.

Figure 18-1 illustrates that at least three categories of inflammatory material are released from phagocytes when these are exposed to a phagocytic stimulus. When a white cell encounters opsonized particles, such as zymosan or immune complexes, it is keyed by surface receptors that encounter these ligands to invaginate a protein of its plasma membrane. The phagocytic vacuole or phagosome is pinched off and drawn toward the cell interior. Flowing from the organelle-rich cytocenter, lysosomes merge with the forming phagosome and if the particle:cell ratio is great, lysosomal enzymes (most notably the neutral proteases found in the azurophil granules) escape to the surrounding tissues (1). There they can attack appropriate substrates in the connective tissue and interact with circulating cascades of inflammation.[2] The surface response of the cell to phagocytosis is also accompanied by a burst of oxidative metabolism launched by the generation of toxic oxygen products (2), chiefly superoxide anion ($O_2^{-}\cdot$), but also hydrogen peroxide (H_2O_2) (useful in the killing of bacteria), as well as singlet oxygen ($^1O_2^{*}$) and hydroxyl radical ($OH\cdot$).[3] Concomitantly, plasma membrane fatty acids such as arachidonate, are transformed (3) by phospholipases into hydroperoxides, endoperoxides, thromboxanes (Txs), and stable prostaglandins (PGs).[4,5]

These three classes of inflammatory mediators are, however, prevented from acting unopposed. Ceruloplasmin scavenges $O_2^{-}\cdot$. Antiproteases such as α_1-antitrypsin (α_1-AT) or α_2-macroglobulins (α_2-M) in-

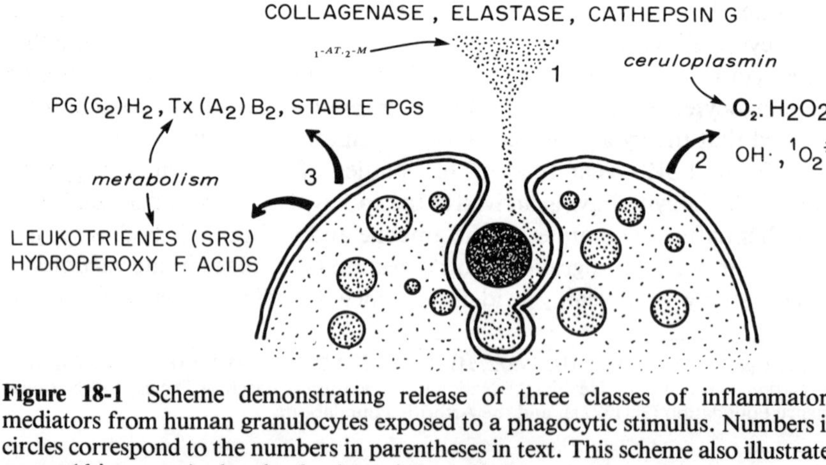

Figure 18-1 Scheme demonstrating release of three classes of inflammatory mediators from human granulocytes exposed to a phagocytic stimulus. Numbers in circles correspond to the numbers in parentheses in text. This scheme also illustrates means (thin arrows) whereby the three classes of inflammatory mediators released by granulocytes are prevented from acting unopposed. SRS: slow-reactive substances.

hibit secreted lysosomal hydrolases. Rapid transformation of biologically pro-inflammatory products of arachidonate (PGG_2, PGH_2, and Tx A_2) results in inert metabolites or anti-inflammatory stable prostaglandins (PGI_2, PGE_1, and PGE_2). The latter inhibit secretion of lysosomal hydrolases, presumably by interacting with membrane receptors to boost cAMP.

We will now discuss the morphologic correlates of this process. When human polymorphonuclear leukocytes (PMNs) are studied by scanning electron microscopy, one observes rounded cells still tethered to substrate by means of filopodia, after the cells' encounter with zymosan appropriately opsonized with fresh serum.[6] Transmission electron photomicrographs show that when these particles engage the surface of the polymorphonuclear leukocyte by means of appropriate receptors, there is a sub-plasmalemmal condensation of microfilaments. Following capture, and invagination of the vacuole, particles are ingested and discharge their osmiophilic contents into the as-yet-incompletely closed phagocytic vacuole.[6,7] Thus, the critical event in the secretion of Metchinikoff's cytases to the surrounding milieu is fusion of the membrane of the phagocytic vacuole with that of the lysosome. Indeed, it has become possible to observe fusion by means of the freeze-fracture technique, which demonstrates that the ectoplasmic face of the lysosomal membrane comes into contact with, and joins, the protoplasmic face of the plasma membrane.[8] It is this true, secretory event of inflammation that we will now analyze.

When a white cell encounters such a particle, consequent to fusion of the lysosome with the as-yet-incomplete phagocytic vacuole, a portion of its lysosomal enzymes such as β-glucuronidase and the neutral proteases are secreted into the medium in a concentration- and time-dependent fashion. Because the plasma membrane is not impaired, cytoplasmic enzymes such as lactate dehydrogenase do not escape.[9] The bulk of lysosomal enzymes under these circumstances do not escape into the external milieu. Most of the vacuoles close efficiently and the bulk of lysosomal hydrolases are designed to degrade the invader inside the phagocytic vacuoles. When, however, the cells are treated with cytochalasin B, which prevents the appropriate interaction of sub-plasmalemmal actin-binding protein with cytoplasmic actin,[10] the normal purse-string closure of the phagocytic vacuole is impaired and lysosomal hydrolases can be quantitatively recovered in the external milieu. Thus cytochalasin B turns the leukocyte from a phagocytic cell into a kind of secretory organ of inflammation, in which materials designed to remain in the interior of cells escape instead to the external environment.[11] Moreover, we have demonstrated by cytochemical ultrastructural techniques the secretion of lysosomal myeloperoxidase from azurophil granules after cytochalasin B-treated cells have engaged zymosan particles. The enzyme is secreted entirely to the outside because the cell cannot close the stoma formed by the invaginated protein of the plasma membrane.

Stimulus-Secretion Coupling in the Granulocyte

Now, opsonized particulate stimuli are not the only surface ligands that, when recognized by appropriate receptors on the PMN, call forth the secretory response. We have summarized elsewhere[2,7–9,11,12] many of the various surface stimuli that provoke the release of inflammatory mediators from white cells. Phagocytosis leads to release of the contents of azurophil granules, such as the neutral proteases and myeloperoxidase; release of materials from specific granules, such as lysozyme and lactoferrin; and toxic oxygen derivatives. The cells respond similarly to immune complexes, or to surrogate immune complexes produced by simply heat-aggregating IgG. The contents of both types of granules are discharged by such chemotactic stimuli as the peptide FMLP (*m*-formylmethionylleucylphenylalamine) and the complement component C5a, but only when the cell has been treated with cytochalasin B. Even without cytochalasin, both types of granules discharge in response to the ionophore A23187 (in the presence of calcium), if the concentration of ionophore is high enough.[7] In contrast, concanavalin A,[13] phorbol myristate acetate,[14] and low concentrations of the ionophore provoke only secretion of the contents of specific granules and generation of superoxide anions. Finally, complement components C3b and C5a (in the absence of cytochalasin B) simply cause the cell to generate superoxide anion at the cell surface.[15,16]

The release of these products from the stimulated cell may be described as stimulus-secretion coupling in human PMNs. Figure 18-2 summarizes previously published evidence[13,14,17–19] that stimulus-secretion coupling in human PMNs involves a kind of secretory code. The binding of such ligands as concanavalin A, FMCP, and Fc portions of immunoglobulins to specific surface receptors initiates a complex series of events in the plasma membrane. These may be characterized as

1. Enhanced membrane fluidity
2. A change in membrane potential ($\Delta\psi$)
3. Loss of calcium from the plasma membrane subjacent to the area of ligand-receptor interaction
4. Coupling of receptors (β-adrenergic, PGE_1) to membrane-bound adenylate cyclase.

These events, which can be demonstrated within five to ten seconds after presentation of an appropriate stimulus[17,20] are followed by the generation of superoxide anion ($O_2^{-}\cdot$) effected by a surface oxidase, and the liberation of arachidonic acid from membrane phospholipids. The arachidonate thus formed can then be transformed to biologically active products via the lipoxygenase or cyclooxygenase pathways. Directly beneath the forming phagocytic vacuole (or areas where ligand-coated sur-

Ligand Binding to Surface Receptors

↓

Enhanced Membrane Fluidity	Change in Surface Potential	Loss of Membrane Ca^{++}	Receptor-Cyclase Coupling

↓

Ionophore A23187 →

Generation of $O_2^{-}\cdot$, H_2O_2, $OH\cdot$, $^1O_2^*$
Transformation of arachidonate via lipoxygenase, cyclooxygenase, $O_2^{-}\cdot$, etc
Condensation of sub-plasmalemmal microfilaments
Microtubule assembly
Fusion of lysosomes with the phagocytic vacuole or plasma membrane
Secretion of lysosomal enzymes

Figure 18-2 The secretory code of the neutrophil. SRS: slow-reactive substances; NSAIA: nonsteroidal anti-inflammatory agents.

faces engage the phagocyte), sub-plasmalemmal filaments form a dense, contractile web, and cytoplasmic microtubules are assembled. Only then can one observe fusion of azurophile and/or specific granules with the forming phagocytic vacuole (or the invaginated plasma membrane of cytochalasin B-treated cells). Secretion of lysosomal constituents into the vacuole or regurgitation to external medium is the final step in secretion.

Early Events in Membrane Stimulation

Let us begin with ligand binding to surface receptors. Among the better characterized receptors of the PMN for external stimuli are the Fc receptors at the surface of the plasma membrane.[21] These are engaged by aggregated, rather than by native, immunoglobulins. Indeed, in cytochalasin B-treated PMNs, we demonstrated IgG-mediated release of lysosomal β-glucuronidase, linear with respect to concentration of aggregated immunoglobulin and with time of incubation. Surface engagement of Fc receptors led to secretion of lysosomal hydrolases, as aggregated (but not native) immunoglobulin contacted receptors on the surface of the cell.[22]

These events could be brought about if aggregated, but not native, immunoglobulins engaged hydrophobic portions of Fc receptors, or if they were capable of interacting with naked lipid membranes. To test the hypothesis that these surrogate immune complexes would interact with biologic membranes only when the aggregates displayed hydrophobic sites in the Fc portions of the Ig molecules, we prepared liposomes, artificial lipid bilayers, first described by Bangham and associates in 1964.[23] They compared the interaction of liposomes both with native and with aggregated IgG to determine if these might interact with the hydrophobic zones of lipid bilayers. In studies reported by Weissmann et al,[24] it was found that aggregated immunoglobulins interacted more avidly with liposomes, whether one measured anion or glucose release from liposomes. Therefore, aggregated immunoglobulins were more effective than native IgG in perturbing the bilayers of artificial liposomes. The perturbing capacity appeared to be class- and subclass-specific, with various IgGs being more effective than IgM, IgA_1 being more effective than IgA_2, and the Fc portion of the molecule more effective than Fabs. Moreover, when sucrose density gradients were performed on the aggregates, it was found that intermediate complexes were more effective than either native immunoglobulins or large completely sedimentable precipitates, a finding in keeping with the Fc receptor activity of the aggregated immunoglobulins.

These experiments suggest the possibility that the Fc fragments of aggregated IgG might insert into lipid bilayers. To test this hypothesis, we have made use of spin-labeled liposomes in which the cholestane analog 4HCAT (4-[3-β-hydroxycholest-5-enamido]-tempo) was inserted into the bilayers.[25] The unpaired electron on the nitroxide spin label therefore reported mobility of the lipids 15 Å into the bilayer. Spectra obtained with this label indicated that interaction of liposomes with aggregated immunoglobulin caused enhanced mobility in the hydrophobic region. The order parameter, tau, was calculated. It was found that aggregated immunoglobulin produced enhanced mobility of the bilayer (manifested as decreases in tau), whereas native immunoglobulin had no such effect: the usual interpretation of such changes is that membrane fluidity has been enhanced. In contrast, when we incorporated into liposomes the reporter molecule 3-doxylcholestate (3-DC), which gave information as to the hydrophobic portions of the bilayer, no change was noted. These data indicate that ligand-receptor interactions might result when immunoglobulins in the antigen-reactive configuration engage either hydrophobic portions of protein receptors or the naked lipid bilayers themselves.

Decrease in rotational correlation times observed in Ig-stimulated membranes resemble those reported for the more fluid surface membranes of PMN from mice and humans with the Chediak-Higashi syndrome.[26]

Thus, ligand-receptor interactions at the membrane seem to be requisite for the signaling of the "stranger at the gate": the early consequence of which is recognized as enhanced membrane fluidity. Whether concanavalin A (which binds to sugar receptors) or immune complexes (which bind to the Fc receptors) are employed, a surface signal is generated. Surface responses in secretory cells are usually accompanied by changes in membrane potential. To quantify this in the leukocyte, which is too small for the insertion of microelectrodes, we have used the lipophilic triphenymethylphosphonium cation, $TPMP^+$.[17] The distribution of this lipophilic cation is determined by the electrochemical gradient across cell membranes. Changes in the ratio of $TPMP^+_{in}$ to $TPMP^+_{out}$ therefore reflect changes in the membrane potential, which can be calculated from the Nernst equation. The distribution of TPMP has previously been used to monitor the surface potential of bacteria, lymphocytes, and thyroid cells.

We have studied the time course of $TPMP^+$ uptake by the granulocyte as measured by the technique of Schuldiner and Kaback.[27] After approximately 75 minutes, in unstimulated cells, cell-associated $TPMP^+$ had reached a plateau. Gramicidin caused a large depolarization of the cells, whether added at time 0 or after $TPMP^+$ equilibration, indicating that $\Delta\psi$ was due to gradients of monovalent cations. The protonophore CCCP (carbonycianide-*m*-chlorophenylhydrazone) caused only a small depolarization, indicating that $TPMP^+$ uptake was not primarily due to mitochondria. Using the intracellular volume of human PMNs of 3.94 μl/mg of protein determined experimentally, it was possible to calculate a resting potential of human PMNs of -26 mV.

The effects of concanavalin A and immune complexes on the surface potential of human PMNs (measured as $TPMP^+$ uptake) were next studied. We have examined both control cells, which can take up these particulates, and cytochalasin B-treated cells, which cannot. When purified granulocytes were exposed to concanavalin A, they responded within ten seconds by a sharp hyperpolarization, followed by depolarization, followed by long slow hyperpolarization. When endocytosis of concanavalin A was blocked by cytochalasin B, similar results were seen; if anything, the surface potential changes were observed even earlier. In experiments[19] in which the cells were exposed to immune complexes of bovine serum albumin/anti-bovine serum albumin (BSA/anti-BSA), to the chemotactic peptide FMLP, and to latex beads, a similar hyperpolarization response was observed. In contrast, cells exposed to ionophore A23187 did not exhibit a hyperpolarization response.

Loss of Membrane Calcium

Because calcium has been held necessary for stimulus-secretion coupling in granulocytes, we repeated the above experiments both in the usual Krebs Ringer's medium, which contained 1.3 mmoles of calcium, and in media from which calcium had been removed by means of magnesium EGTA. If anything, the surface potential changes were even more rapid in the absence of extracellular calcium. This was unexpected, since other events in phagocytosis, such as superoxide anion production and enzyme secretion, are enhanced by extracellular calcium. It was therefore of interest to determine by ultrastructural and cytochemical techniques whether mobilization of cellular loci of calcium was another early event in phagocytosis.[19,28] Cations were precipitated in situ by fixing cells in aqueous solutions of osmium hydroxide and calcium pyroantimonate. Although pyroantimonate precipitation is particularly suitable for the ultrastructural localization of calcium, it can precipitate other cations. Therefore, in our studies, pyroantimonate precipitates were considered to contain calcium only if they could be reduced or eliminated by treatment with EGTA *en grid* or *en bloc*. Cells were exposed to zymosan particles and permitted to ingest them; they were then immediately fixed in glutaraldehyde and stained by the pyroantimonate method. We noted that whereas the plasma membrane stained heavily for calcium, the membranes of the phagocytic vacuoles were entirely devoid of the precipitates. When we observed granulocytes in the process of ingesting zymosan particles, those areas of the plasma membrane of the leukocyte not in proximity to the zymosan particles still retained calcium at the surface. However, at sites of the leukocyte membrane that were at the region of contact with the zymosan particles, calcium had been lost.

Perhaps an even more dramatic demonstration was observed in electron photomicrographs taken of granulocytes that were permitted to adhere to Sepharose beads to which the ligand concanavalin A had been covalently bound. Aherence of this sort is also inhibited by α-methyl-D-mannoside. The plasma membrane not in contact with the Sepharose beads still contained calcium, as did the granules and the nucleus. In contrast, calcium was absent from the broad area of contact between the ligand-coated bead and the receptor-containing surface of the white cell. The contention that this method localizes calcium and not other cellular cations was validated by x-ray microanalysis. The only spectral lines obtained by microanalysis of stained cells, in addition to those in the supporting medium in which the cells were processed, were those of the added antimonate and of endogenous calcium.[19,28] Significantly, neither magnesium nor sodium is precipitated by this method.

One of the consequences of ligand-receptor interaction in cells that possess membrane adenylate cylcase is the prompt coupling of engaged receptors with the cyclase.[29] Coupling has been attributed both to methylation of phosphatidylethanolamine and to enhanced membrane fluidity (decreased viscosity). We are therefore able to reinterpret our earlier observations and the inhibitory effects of cyclic AMP on enzyme secretion from the granulocyte. We have found that cytochalasin B-treated cells exposed to β-adrenergic agents or PGE_1 increased their baseline cAMP from roughly 10 to 20 pmoles/10^7 cells.[30] This would represent engagement of receptors. When such cells were exposed to opsonized zymosan particles, they accumulated up to 40 pmoles; this can be interpreted as enhanced coupling due to fluidity increase induced by the particles. It was at this level that enzyme secretion was significantly inhibited.[30] We have repeated these experiments with more purified granulocytes (Hypaque/Ficoll) and have studied cAMP responses to a variety of stimuli.[20] We have found that stimuli such as immune complexes, chemotactic factor (FMLP or C5a), or opsonized zymosan provoke prompt doubling of baseline cAMP levels (receptor engagement) within 5 to 10 seconds, but enzyme secretion is stimulated as expected. Similarly, when PGE_1 or PGI_2 are added, the cells also double their baseline cAMP. However, when FMLP, opsonized particles, or immune complexes are now added (to Pg-treated cells), levels of cAMP rise four- to six-fold and expected enzyme secretion is diminished. These data present strong evidence that receptor-cyclase coupling is operative in the granulocyte, and that it results from increases in membrane fluidity provoked by stimulating ligands.

The Sequencing of Secretion

Having shown early events in secretion, we went on to try to determine the sequence in which these events were associated with the later release of superoxide anion, prostaglandins, thromboxanes, and finally, lysosomal enzyme secretion.

When human PMNs are exposed to a surface stimulus, they generate superoxide anion that spontaneously dismutases to H_2O_2 in aqueous solution. The dismutation of $O_2^{-}\cdot$ can be markedly enhanced by superoxide dismutase (SOD). When this takes place in the phagocytic vacuole, H_2O_2, together with myeloperoxidase and a halide, constitute the potent microbicidal system of Klebanoff.[31] H_2O_2 apparently acts to generate a series of linked reactions that culminate in the respiratory burst of the hexose-monophosphate shunt.

Since Briggs et al,[32] and Goldstein et al[33] have presented evidence that the superoxide-generating system is located at the plasmalemma, we have timed the onset of superoxide generation in response to the same ligands that were used to study surface membrane potential (concanavalin A and BSA/anit-BSA). Using the continuous recording technique in which ferricytochrome c reduction is monitored in suspensions of leukocytes,[19,34] it was found that after a lag period of 30 to 40 seconds both these stimuli caused the cell to respond with a burst of superoxide generation. When, however, we compared the onset of superoxide generation to the membrane potential changes, we found that the membrane potential changes (initial hyperpolarization measured as uptake of $TPMP^+$) took place as early as 5 to 10 seconds after stimulation of granulocytes.[19] As previously indicated, superoxide generation displayed a distinct lag, confirming results obtained by Cohen and Chovaniec.[34] Thus, surface membrane potential changes are the first detectable response of the PMN to surface stimulation and are followed, only after a significant lag period, by superoxide generation. Increases in cAMP were never observed before changes in $\Delta\psi$, but always anteceded $O_2^{-}\cdot$ generation.[20]

Oxygen-Derived Free Radicals, Prostaglandins, and Thromboxanes

Free radicals have been implicated both in the transformation of the endoperoxides formed from arachidonic acid released from membrane phospholipids, and in the formation of hydroperoxides via the lipoxygenase pathway. Zurier and Sayadoff[35] have presented evidence that PMNs release the stable prostaglandins (PGs). We have gone on to demonstrate[4,5] that human PMNs also generate thromboxanes (Txs) upon appropriate stimulation by pathways shown in Figure 18-3. By means of radioimmunoassay of Tx B_2 we have demonstrated that this pathway is operative in the human PMN. Human PMNs exposed to opsonized zymosan particles released prodigious amounts of Tx B_2 in a time- and concentration-dependent fashion. The earliest appearance of these products could be detected by one to two minutes after stimulation. Phagocytosis was not required, since we have demonstrated that cytochalasin B-treated PMNs respond to surface stimulation as readily as untreated cells.[4] Thin layer radiochromatography[5] of the products of prelabeled arachidonic acid showed a peak in the thromboxane region that was missing when such cells had been treated with indomethacin, indicating that the products were generated via the cyclooxygenase pathway.

It was necessary to demonstrate not only that the inert metabolite Tx B^2 was generated, but also that the cells formed the biologically active Tx A_2, quantified by release of serotonin from indomethacin-treated

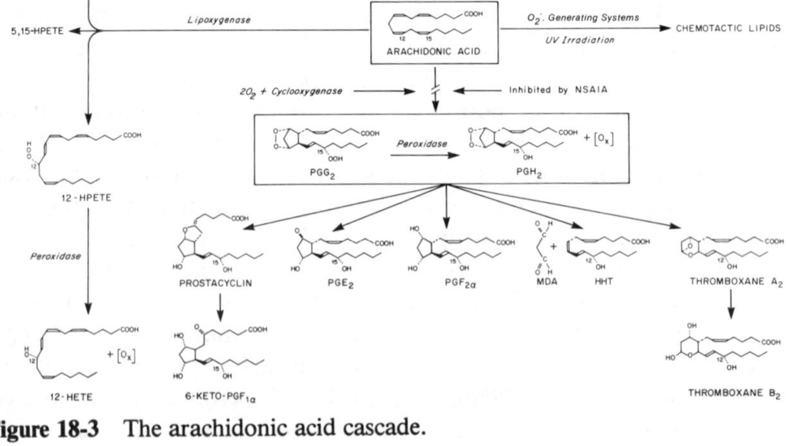

Figure 18-3 The arachidonic acid cascade.

platelets. These were exposed to supernatants of PMNs stimulated by opsonized zymosan. The decay of serotonin-releasing activity, 30 seconds, was in keeping with the biologic half-life of purified Tx A_2.

However, the relationship between human PMNs and prostaglandins is more complex than might be implied by the simple demonstration that stimulated cells release thromboxanes and prostaglandins. Indeed, we have discussed above that when stable prostaglandins are presented to PMNs, their inhibitory capacity is: $PGI_2 > PGE_1 > PGE_2$. Thus, if anything, these products of the cyclooxygenase pathway would appear to exert anti-inflammatory, antisecretory effects in the granulocyte and they do so apparently by raising cAMP four- to six-fold over basal levels. Moreover, inhibition of secretion, as reported in detail elsewhere,[2,6,30,36] can readily be achieved by adding other stimulators of cAMP.

We have already demonstrated that stimulated PMNs produce superoxide anion. We have recently done experiments[18] to show that superoxide-generating systems can convert membrane arachidonic acid, by nonenzymatic means, to a chemotactic lipid, presumably by forming oxidation products such as hydroxy-acids or a mixture of positional isomers of arachidonic acid hydroperoxides (middle right of Figure 18-3). Purified arachidonic acid was exposed to a superoxide-generating system, consisting of xanthine oxidase and acetaldehyde under aerobic conditions. The lipids were extracted in chloroform methanol and evaporated under nitrogen. The residue was studied by two means, first by thin-layer chromatography and second for its capacity to provoke chemotaxis of PMNs in a modified Boyden chamber. Arachidonic acid, exposed to a superoxide-generating system for varying periods of time, generated a product that was chemotactic for PMNs. That migration, measured by the leading-front method, represented true chemotaxis and was verified by appropriate "checkerboard" analyses. Generation of chemotactic product by the superoxide-generating system could be inhibited by SOD. Other studies indicated that the generation of this product could also be inhibited by scavengers of singlet oxygen and hydroxyl radical. It was possible to isolate the chemotactic product generated by the action of xanthine oxidase and acetaldehyde on thin-layer chromatography in which the precursor, arachidonate, had been radiolabeled. One peak of chemotactic activity was regularly isolated. In this system radiolabeled HETE (kindly provided by Dr Edward Goetzl) migrated at a position that permitted resolution of this newly described chemotactic factor.

Having the labeled product available made it possible to quantitate the dose-response curve for this new derivative of arachidonic acid, and it was found that as few as 5 ng of the material were capable of provoking chemotaxis in human PMNs. This compares very well with the amounts required for chemotaxis of the most potent chemotactic product of the complement system C5a. Moreover, this new product is approximately one to two orders of magnitude more potent than previously described chemotactic products of arachidonic acid such as HETE and HHT. These experiments indicate that when superoxide is generated in the vicinity of free arachidonic acid, products are generated that can amplify the inflammatory response.

The Final Secretory Event

Thus we have demonstrated that the surface stimulation of PMNs leads to a series of reactions that, even in the absence of lysosomal enzyme secretion, are capable of liberating into the external milieu toxic products generated from oxygen as well as biologically potent metabolites of membrane fatty acids. These events are associated with rearrangements of the cytoskeletal system of the cell.[7,37] When a PMN is exposed to a Sepharose bead to which concanavalin A has been covalently linked, the cell surface in contact with the bead recognizes this large nonphagocytosable particle as if it were a series of small particles that it is about to ingest. The first ultrastructural response, resembling that to zymosan particles, is a condensation of sub-plasmalemmal microfilaments, and appropriate studies have identified these as actin and myosin interacting with sub-plasmalemmal actin-binding protein. Associated with the contraction of these microfilaments is a massive assembly of microtubules.[13] Appropriate quantitative stereology has shown that concanavalin A, at doses that provoke all the changes we have been discussing, is capable of provoking massive assembly of cytoplasmic microtubules, both in the centriolar region and in the periphery of the cell. This assembly of microtubules is accompanied by secretion of enzymes ordinarily contained within granules. Concanavalin A (which provokes both surface stimulation of the cell and its attendant cytoskeletal changes) causes a concentration- and time-dependent release of the specific granule marker enzyme, lysozyme. Again, secretion could be appropriately inhibited if ligand-receptor interactions were blocked with α-methyl-D-mannoside.

We now come to the penultimate point. Lysosomal enzyme secretion involves the fusion of specific and/or azurophil granules with the plasma membrane. Recent experiments[18,19] have permitted us to time the onset of lysosomal enzyme secretion in response to surface stimulation. Our previous attempts to measure onset of lysosomal degranulation had indicated that enzymes are released within 1 minute, but continuous assay techniques for lysosomal enzymes had not been available. We have been able to demonstrate that lysosomal enzyme release can also be timed with a resolution of approximately 6 seconds.[18] A suspension of purified human leukocytes containing 1×10^8 to 2×10^8 cells was placed in the upper chamber of a modified-flow dialysis cell at 37 C. The cells in the upper chamber were separated from the lower chamber by a 1-μ millipore filter. The lower chamber was continuously eluted by fresh buffer from a reservoir and the flow rate was regulated by a peristaltic pump; the eluant was finally directed to a fraction collector wherein samples were collected every 6 seconds. Cells in the upper chamber were preincubated for 2 minutes with cytochalasin B. At this point a bolus containing both the desired stimulus and ^{14}C-inulin was added to the upper chamber. Inulin, the extracellular space marker, immediately began to cross the filter and was eluted. Lysosomal enzymes were then similarly eluted. Determinations of radioactivity and of enzyme activity in the collected fractions permitted us to quantify a lag period between presentation of the stimulus and the onset of enzyme secretion.

When a bolus of ^{14}C-inulin and BSA/anti-BSA immune complexes were added at time 0, inulin was driven through the apparatus after approximately 30 seconds, and an additional 25 seconds elapsed before lysozyme, an enzyme found predominantly in specific granules, was detected. This 25-second difference was considered the lag period for release of this enzyme. A similar profile with a lag of 40 seconds was observed for lysozyme released in response to concanavalin A. β-Glucuronidase, an enzyme found predominantly in azurophil granules, was similarly released from human granulocytes. Discharge of this enzyme took place after a lag period of approximately 30 seconds in response to the BSA/anti-BSA immune complex and to serum-treated zymosan. Appropriate control experiments indicated that these lag periods were intrinsic to each of the stimuli used and did not depend on the concentration or the final magnitude of enzyme release. Moreover, further control experiments indicated that the lag period was not due to adsorption of the enzyme to the cells or the filter, since in cells that were pre-exposed to stimuli, with enzymes present in the external milieu from the onset of the experiment, no lag period could be detected.[18]

Finally, we were able to compare the early changes in membrane potential of the PMN to the time required to release granule contents. When we compared the initial membrane potential changes, measured as $TPMP^+$ uptake, with the lag periods of superoxide anion production and lysosomal enzyme secretion, the lag periods clearly showed that membrane potential changes significantly anteceded the final event in stimulus-secretion coupling.

Endogenous Inhibitors of Inflammation

What turns off these secretory events and regulates them so that we do not undergo uncontrolled inflammation? We have recently identified a potent circulating scavenger of oxygen-derived free radicals: ceruloplasmin.[38]

Ceruloplasmin appears to act extracellularly in the way that SOD acts intracellularly. This copper-containing protein, an acute-phase protein, inhibited the reduction of cytochrome c in a dose-dependent fashion when xanthine oxidase and hypoxanthine were used as a superoxide-generating system. The protein had no effect on the assay system, nor did it interfere with the production of uric acid. Additionally, ceruloplasmin inhibited reduction of nitroblue tetrazolium (NBT) by the same superoxide-generating system, especially at concentrations of ceruloplasmin found in the blood as an acute-phase reactant to inflammatory stimuli.

Moreover, when we determined whether the superoxide-scavenging activity of this plasma protein corresponded to its oxidase activity, we found that it exhibited both activities. Finally, heating experiments indicated that whereas erythrocyte SOD was heat-stable, ceruloplasmin was heat-labile, both with respect to oxidase activity and to superoxide-scavenging activity. These experiments excluded the possibility that contaminating SOD from erythrocytes was responsible for the observed activity. Indeed, it was possible to show that purified ceruloplasmin could be demonstrated on polyacrylamide gel electrophoresis by virtue of its capacity to inhibit the reduction of NBT, and to enhance the photo-oxidation of orthodianisidine, a positive stain. Similar electrophoretic studies excluded the contamination of ceruloplasmin by red cell SOD.

Finally, to exclude the possibility that ceruloplasmin could be directly interacting with any of the indicator systems, we trapped NBT in liposomes and generated superoxide anion outside these lipid bilayers. Ceruloplasmin, which cannot penetrate the liposome, was added to this system. It inhibited the reduction of NBT within liposomes. As we have previously shown,[39] superoxide anion can traverse the membrane of a liposome in order to provoke biochemical reactions inside a bilayer, when it has been generated on the outside. Ceruloplasmin, at appropriate concentrations, clearly inhibited the intraliposomal reduction of NBT. This suggests a model of how this acute-phase reactant might inhibit the toxic effects of superoxide anion in cells and tissues that are infiltrated by phagocytic cells.

Thus, the predictions of Metchnikoff have been substantially validated. A phagocytic cell encountering an appropriate stimulus acts to release its cytases. However, the cell does not have to be dead in order to do so. Surface stimulation, or more usually phagocytosis, leads to the release by these inflammatory cells of various potent biologic products that can go on to harm cells and connective tissue. However, regulatory mechanisms are built into the inflammatory response (see Figure 18-1). Thus the rapid metabolism or short half-life of the prostaglandin intermediates assures tissues that they will not be exposed to these products too long. Plasma protease inhibitors such as α-antitrypsin or α_2-macroglobulin defend the tissues against escape of neutral proteases and finally, anti-oxidants or free-radical scavengers, such as ceruloplasmin, are generated as acute-phase reactants to dampen the response of the host to products secreted by granulocytes.

REFERENCES

1. Metchnikoff E: *Immunity in Infective Diseases.* New York, Johnson Reprint Corporation, 1905, p 70.

2. Smolen JE, Weissmann G: The granulocyte: metabolic properties and mechanisms of lysosomal enzyme release, in Havemann K, Janoff A (eds): *Neutral Proteases of Human Polymorphonuclear Leukocytes.* Baltimore, Md, Urban & Schwarzenberg Inc, 1978, pp 56–76.

3. Babior BM, Kipnes RS, Curnutte JT: Biological defense mechanisms: the production by leukocytes of superoxide, a potential bactericidal agent. *J Clin Invest* 52:741–744, 1973.

4. Goldstein IM, Malmsten CL, Samuelsson B, et al: Prostaglandins, thromboxanes, and polymorphonuclear leukocytes: mediation and modulation of inflammation. *Inflammation* 2:309–317, 1977.

5. Goldstein IM, Malmsten CL, Kindahl H, et al: Thromboxane generation by hyman peripheral blood polymorphonuclear leukocytes. *J Exp Med* 148:787–792, 1978.

6. Weissmann G, Zurier RB, Hoffstein S: Leukocyte proteases and the immunologic release of lysosomal enzymes. *Am J Pathol* 68:539–560, 1972.

7. Hoffstein S, Weissmann G: Microfilaments and microtubules in calcium ionophore-induced secretion of lysosomal enzymes from human PMNs. *J Cell Biol* 78:769–781, 1978.

8. Weissmann G, Smolen JE, Hoffstein S: Polymorphonuclear leukocytes as secretory organs of inflammation. *J Invest Dermatol* 71:95–99, 1978.

9. Weissmann G, Zurier RB, Spieler PJ, et al: Mechanisms of lysosomal enzyme release from leucocytes exposed to immune complexes and other particles. *J Exp Med* 134:149s–165s, 1971.

10. Hartwig JH, Stossel TP: Interactions of actin, myosin and an actin-binding protein of rabbit pulmonary macrophages. III. Effects of cytochalasin B. *J Cell Biol* 68:602–619, 1978.

11. Zurier RB, Hoffstein S, Weissmann G: Cytochalasin B: effect on lysosomal enzyme release from human leucocytes. *Proc Natl Acad Sci USA* 70:844–848, 1973.

12. Weissmann G: Lysosomes and rheumatoid joint inflammation. *Arthritis Rheum* 20:S193–S204, 1977.

13. Hoffstein S, Soberman R, Goldstein I, et al: Concanavalin A induces microtubule assembly and specific granule discharge in human polymorphonuclear leukocytes. *J Cell Biol* 68:781–786, 1976.

14. Goldstein IM, Hoffstein ST, Weissmann G: Mechanisms of lysosomal enzyme release from human polymorphonuclear leukocytes. Effects of phorbol myristate acetate. *J Cell Biol* 66:647–652, 1975.

15. Goldstein IM, Feit F, Weissmann G: Enhancement of nitroblue tetrazolium dye reduction by leukocytes exposed to a component of complement in the absence of phagocytosis. *J Immunol* 114:516–518, 1975.

16. Goldstein IM, Kaplan HB, Radin A, et al: Independent effects of IgG and complement upon human polymorphonuclear leukocyte function. *J Immunol* 117:1282–1287, 1976.

17. Korchak HM, Weissmann G: Changes in membrane potential of human granulocytes antecede the metabolic responses to surface stimulation. *Proc Natl Acad Sci USA* 75:3818–3822, 1978.

18. Weissmann G, Korchak HM, Perez HD, et al: Leukocytes as secretory organs of inflammation. *Adv Inflamm Res* 1:95–112, 1979.

19. Weissmann G, Hoffstein S, Korchak H, et al: The earliest membrane responses to phagocytosis: membrane potential changes and Ca^{++} loss in human granulocytes. *Trans Assoc Am Physicians* 91:90–104, 1978.

20. Smolen JF, Korchak HM, Weissmann G: The first minute of granulocyte stimulation: changes in cyclic nucleotide levels are not the first signal. *Clin Res* 27:307A, 1979.

21. Messner RP, Selinek J: Receptors for human gamma globulin on human neutrophils. *J Clin Invest* 49:2165–2173, 1970.

22. Goldstein IM, Roos D, Weissman G, et al: Influence of corticosteroids on human polymorphonuclear leukocyte function in vitro. Reduction of lysosomal enzyme release and superoxide production. *Inflammation* 1:305–315, 1976.

23. Bangham AD, Standish MM, Weissmann G: The action of steroids and streptolysin S on the permeability of phospholipid structures to cations. *J Mol Biol* 13:253–259, 1965.

24. Weissmann G, Brand A, Franklin EC: Interaction of immunoglobulins with liposomes. *J Clin Invest* 53:536–543, 1974.

25. Schieren H, Weissmann G, Seligman M, et al: Interactions of immunoglobulins with liposomes: an ESR study demonstrating protection by cortisol. *Biochem Biophys Res Commun* 82:1160–1167, 1978.

26. Haak RA, Ingraham LM, Baehner RL, et al: Membrane fluidity in human and mouse Chediak-Higashi leukocytes. *J Clin Invest* 64:138–144, 1979.

27. Schuldiner S, Kaback HR: Membrane potential and active transport in membrane vesicles from *Escherichia coli. Biochemistry* 14:5451–5461, 1975.

28. Hoffstein S: Ultrastructural demonstration of calcium loss from local regions of the plasma membrane of surface-stimulated human granulocytes. *J Immunol* 123:1395–1402, 1979.

29. Hirata F, Strittmatter WJ, Axelrod J: β-Adrenergic receptor agonists increase phospholipid methylation, membrane fluidity, and β-adrenergic receptor-adenylate cyclase coupling. *Proc Natl Acad Sci USA* 76:368–372, 1979.

30. Zurier RB, Weissmann G, Hoffstein S, et al: Mechanisms of lysosomal enzyme release from human leukocytes. II. Effects of cAMP and cGMP, autonomic agonists, and agents which affect microtubule function. *J Clin Invest* 53:297–309, 1974.

31. Klebanoff SJ: Antimicrobial systems of the polymorphonuclear leukocyte, in Bellanti JA, Dayton DH (eds): in *The Phagocytic Cell in Host Resistance.* New York, Raven Press, 1975, pp 45–56.

32. Briggs RT, Drath DB, Karnovsky ML, et al: Localization of NADH oxidase on the surface of the human polymorphonuclear leukocytes by a new cytochemical method. *J Cell Biol* 17:566–568, 1975.

33. Goldstein IM, Cerqueira M, Linda S, et al: Evidence that the superoxide-generating system of human leukocytes is associated with the cell surface. *J Clin Invest* 59:249–254, 1977.

34. Cohen HS, Chovaniec ME: Superoxide generation by digitonin-treated guinea pig granulocytes. A basis for a continuous assay for monitoring superoxide production and for the study of the activation of the generating system. *J Clin Invest* 61:1081–1087, 1978.

35. Zurier RB, Sayadoff DM: Release of prostaglandins from human polymorphonuclear leukocytes. *Inflammation* 1:93–101, 1975.

36. Weissmann G, Goldstein I, Hoffstein S: et al: Yin/Yang modulation of lysosomal enzyme release from polymorphonuclear leukocytes by cyclic nucleotides. *Ann NY Acad Sci* 256:222–232, 1975.

37. Hoffstein S, Goldstein IM, Weissmann G: Role of microtubule assembly in lysosomal enzyme secretion from human polymorphonuclear leukocytes: a re-evaluation. *J Cell Biol* 73:242–256, 1977.

38. Goldstein IM, Kaplan HB, Edelson HS, et al: Ceruloplasmin: a scavenger of superoxide anion radicals. *J Biol Chem* 254:4040–4045, 1979.

39. Goldstein IM, Weissmann G: Effects of the generation of superoxide anion on permeability of liposomes. *Biochem Biophys Res Commun* 75:604–609, 1977.

SECTION IV
Membranes in Cell-Cell Interactions

19 The Molecular Basis of Cell-Cell Recognition in a Simple Invertebrate System

Werner Burkart, PhD
Max M. Burger, Md, PhD

Morphogenetic processes in tissue formation are probably based on a multitude of mechanisms, such as migration, recognition, fixation, induction and cell death. Cell movement, whether random, guided by physical barriers, by low-specificity interactions with extracellular material, or by chemotaxis, will bring groups of cells into direct contact with selected subpopulations. Once direct cell-cell contact is achieved, several different types of intercellular information-exchange processes may play a role. Matching surface molecules could directly glue cells together (aggregation factors) or, in much lower concentrations, trigger events inside the cells, eg, shut down the migration machinery, which would result in a halt of cell movement. Then, in yet another step, formation of extracellular matrix components, trapping the cells, would stabilize the proper cell groups.

This work was supported by the Swiss National Foundation, Grant No. 3.721-076.

Exchange of cytoplasmic material through junctions and in vesicle form are additional possibilities for exchange of information between adjacent cells (Figure 19-1). Cell-cell recognition may occur via the interaction of specific surface molecules (A in Figure 19-1); or information signals may be passed between cells via vesicular exchange (B_1 in Figure 19-1) or via a gap or other junction (B_2 in Figure 19-1). Such connections are likely to have a short life span and can be considered as mutual probing, frequently seen at the ruffling edges of moving cells contacting neighbor cells. Both types of specific interactions can lead either to secondary, stabilizing linkages (eg, intercytoplasmic, intermembranous, or extracellular linkages, as in the cell doublet to the right in Figure 19-1), or to dissociation of the interacting cells.

For the formation of organs in higher animals a certain redundancy of such mechanisms may provide for more safety as well as better plasticity in reacting to distortions and errors. Cells of the slime mold *Dictyostelium,* for example, are attracted toward a center by diffusible chemotactic agents. After reaching the center, cells start to adhere to each other through sugar-specific macromolecules newly appearing on the cell surface.[2] Here, then, both chemotaxis and surface recognition at the target are used during morphogenesis, although one process would have been sufficient on purely theoretical grounds. Bearing in mind that such chemotactic agents, surface receptors, and aggregation factors need only be displayed during a short period in development, the drain on cellular resources imposed by such a redundancy seems not to be substantial.

Although most of the mechanisms mentioned above were found in systems ranging from organisms at the borderline between unicellular and

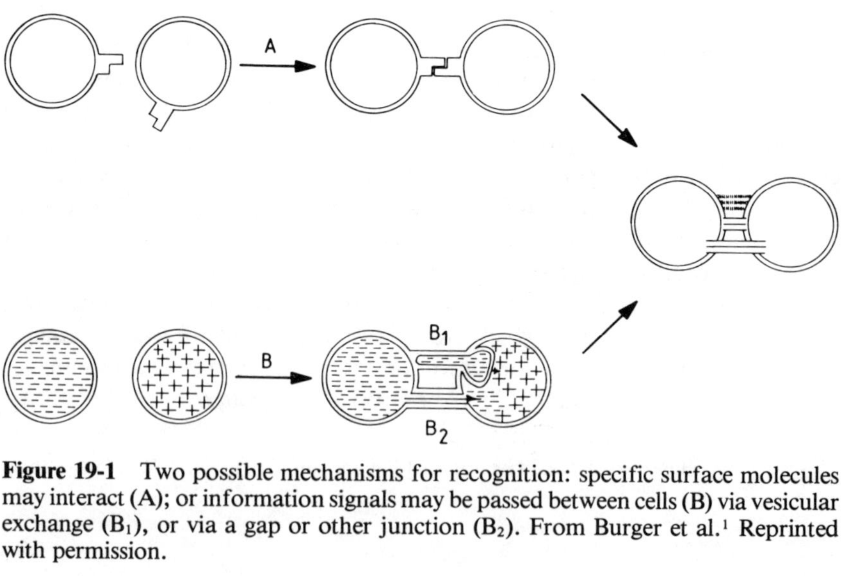

Figure 19-1 Two possible mechanisms for recognition: specific surface molecules may interact (A); or information signals may be passed between cells (B) via vesicular exchange (B_1), or via a gap or other junction (B_2). From Burger et al.[1] Reprinted with permission.

multicellular organization on the one hand, to the nervous system of higher vertebrates[3] on the other, the importance and abundance of these mechanisms at different levels of phylogenesis and ontogenesis cannot be assessed at this time. Furthermore, the outcome of such experiments is very strongly influenced by the type of assay used and the time allowed for reconstitution of surface components destroyed or altered during the enzymatic dissociation. Often, in aggregation of single-cell suspensions consisting of two different cell populations (ie, liver and heart) nonspecific adhesion is dominant in the early stages, and only later a more ordered arrangement is achieved.[4]

Keeping these restrictions in mind, critical studies of cell-cell interactions in vitro, as artificial as they may be, will nevertheless enable us to approach a multitude of open questions in the important fields of morphogenesis and cell differentiation.

A Two-Component Cell Recognition System

Sponges are considered to be the most primitive of the multicellular animals and are therefore not always included among the metazoa, but classified as parazoa. Primarily because they form loose federations of primitive but specialized cells, sponges are a useful model system for studying specific sorting of cell types into aggregates and the eventual buildup of the latter into differentiated tissues.

Cell-cell recognition processes in sponges were studied as early as 1907 by Wilson[5] at the Marine Biological Laboratory in Woods Hole. He found that sponge cells, dissociated by squeezing tissue fragments through a finemesh cloth, were able to attach to the dish, to reaggregate after migration, and to eventually reform small functional sponges by internal sorting-out processes leading to the specialized cell sheets and chambers making up a sponge. If cells from two different species are mixed, they will not collect in mixed aggregates, but form separate species-specific aggregates.[5-7]

Seventy years ago sponges had some crucial advantages as experimental material: their dissociation without the use of enzymes yields viable, undamaged cells; furthermore, cells from different species carry their inherent label, ie, pigments of different colors; and they had yet other advantages, not the least being the simplicity of their anatomy and number of cell types. Today, the refined biochemical tools available enable us to study recognition processes on cells as fragile as nerve cells. Nevertheless, sponge cell recognition is still one of the best models for embryonal cell-cell interactions, since some of the macromolecules involved were purified,[8,9] and hence amenable to chemical and physical characterization. Also, at today's very limited state of knowledge it is more promising to probe theories and concepts on simple organisms.

The *Microciona* aggregation factor (MAF) *Microciona prolifera* or *M parthena* cells, dissociated into Ca^{++}- and Mg^{++}-free seawater were shown by Humphreys[10] to release a high molecular weight "factor" into the supernatant. Such chemically dissociated cells devoid of MAF do not aggregate in cold seawater unless MAF is re-added. The *Microciona* factor was further purified and shown to be a 21×10^6-dalton proteoglycan complex,[8,11] consisting of about 30% protein and 70% carbohydrate, with high amounts of uronic acid and hexosamine. In the electron microscope, the molecule has a sunburst configuration with a central ring 800 Å in diameter and about 15 radiating arms, each about 1100 Å in length. Both ring and arm structures contain roughly equal proportions of protein and carbohydrate, plus hundreds of calcium binding sites.

In the early days of sponge aggregation and before the factors were well-characterized, several of them were examined by Humphreys[10] and by Moscona,[12] and were found to be species-specific, in that they promoted the reaggregation of the homospecific cells only. Although species specificity has been widely demonstrated, it is by no means universal. Cross-reactivity was found in several cases. For example, aggregation factor from *Haliclona occulata* is capable of aggregating *M prolifera* cells, but the minimal dose required is an order of magnitude higher.[13] The MAFs from *M prolifera* and *M parthena* are at present the only such aggregation factors that have been biochemically characterized, to some extent. Exactly how MAF can bring about specific aggregation, as well as the interaction of MAF with the cell surface, has yet to be elucidated.

The baseplate Hypotonic treatment of *M prolifera* cells, devoid of surface-exposed aggregation factor, or Triton X-100 extraction of plasma membranes isolated from such cells, yields a supernatant containing an aggregation-inhibiting activity, called "baseplate."[14,15] Hypotonically treated cells lose their capability to respond to aggregation factor. Baseplate can rebind to shocked cells, making them again responsive to aggregation factor.[16] Baseplate appears to be a peripheral rather than an integral protein, since it can be removed from the cell surface by a mild hypotonic treatment, and since it cannot be sedimented at 105,000 g for 90 minutes. It is heat-sensitive (60 C for 10 minutes) and nondialyzable. Unlike aggregation factor, it is stable to EDTA, to a pH range of 3 to 12, as well as to freezing, thawing, and lyophilization, suggesting that it is not simply a subunit of aggregation factor. A partial purification has been recently reported.[15,16] Since the baseplate did not promote, but rather inhibited aggregation of dissociated cells, and because it neither precipitated aggregation factor nor promoted aggregation of factor-coated beads, it must be a functionally monovalent macromolecule (see Figure 19-2).

Cell factor and factor-factor specificity It was of interest to use the two components, MAF and baseplate, to reconstitute the sponge recognition system in an acellular system. Originally we coupled crude baseplate of

MAF to Sepharose beads.[14] This cell-free system showed all the responses predicted from our working model. Beads covered with MAF could be aggregated in the presence of calcium; free baseplate inhibited this aggregation. Baseplate-covered beads needed MAF and calcium to form aggregates. Glutaraldehyde-fixed cells also retain species-specific binding and aggregation patterns. Müller et al[17] confirmed this concept in a living system, by transferring the baseplate released from *Suberites domuncula* by hypotonic treatment to baseplate-depleted cells from the sponge *Geodia cydonium,* which then formed aggregates in the presence of species-specific *Suberites* aggregation factor.

Figure 19-2 shows a very simplistic model of MAF-mediated cell recognition in *M prolifera.* Surprisingly, binding of ^{125}I-labeled MAF to cells showed no calcium dependence,[18,19] although factor is released by soaking cells in Ca^{++}- and Mg^{++}-free seawater. This apparent contradiction could be explained by the fact that cells possess several layers of MAF on their surface that, with the exception of the innermost layers, are held together by MAF-MAF interactions that are clearly calcium-dependent. Aggregation, which is brought about by the same MAF-MAF interactions, therefore needs high amounts of calcium ($\geqslant$ 10 mM).

Calcium is also involved at other levels in maintaining function and integrity of the MAF molecule. Monomeric MAF has to be kept in solutions containing at least 1 mM calcium to preserve its activity (Figure 19-3). Lack of calcium destroys the aggregation activity of MAF irreversibly. Finally, EDTA treatment will dissociate MAF into subunits in an extremely slow

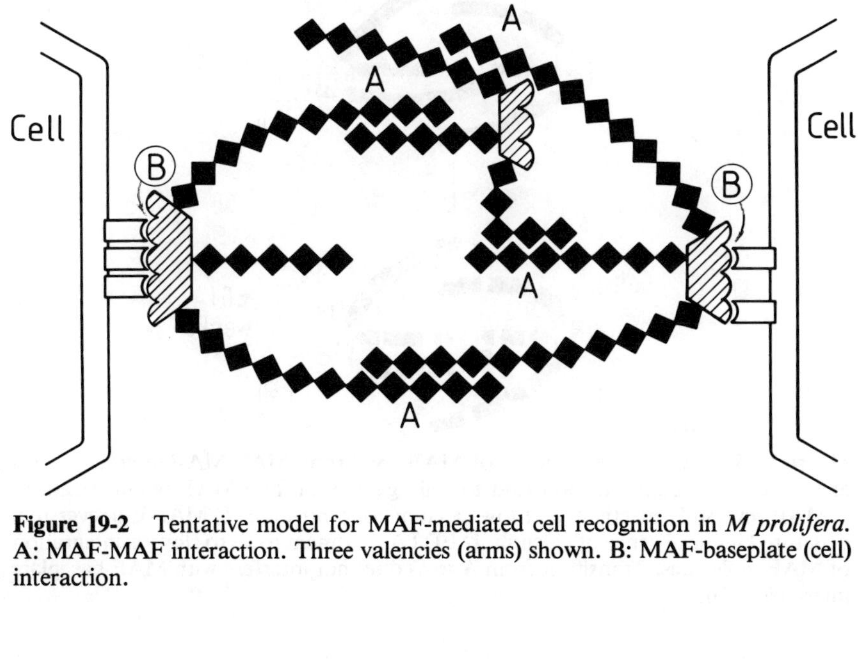

Figure 19-2 Tentative model for MAF-mediated cell recognition in *M prolifera.* A: MAF-MAF interaction. Three valencies (arms) shown. B: MAF-baseplate (cell) interaction.

process (weeks). However, MAF or parts of it still bind species-specifically to cells after inactivation by heat, EDTA, or urea. Since MAF polymerizes in the presence of high calcium concentrations, polyvalency is assumed for MAF-MAF interactions. However, monovalency is assumed for the highly species-specific MAF-baseplate recognition, since binding of MAF to cells at low calcium concentrations does not yet result in cell aggregation.

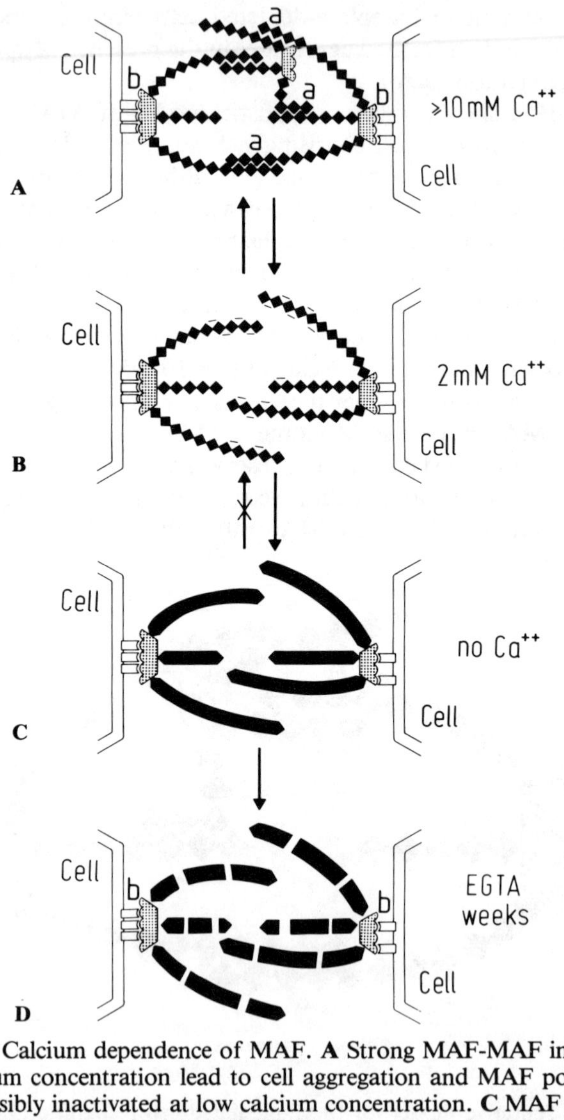

Figure 19-3 Calcium dependence of MAF. **A** Strong MAF-MAF interactions (a) at high calcium concentration lead to cell aggregation and MAF polymerization. **B** MAF reversibly inactivated at low calcium concentration. **C** MAF is irreversibly denatured in the absence of calcium. **D** EDTA treatment leads to slow disintegration of MAF molecules. Transition from **A** to **D** does not interfere with MAF-baseplate interaction (b).

The characteristics of these interactions can be summarized:

MAF-MAF interaction	— calcium-dependent (⩾ 10 mM)
	— several sites/MAF complex
	— ? some selectivity
MAF-baseplate (cell) interaction	— calcium-independent
	— one site/baseplate
	— highly species-specific

Species specificity at the level of MAF-cell interactions would explain cell-cell recognition or sorting-out, with cells having free baseplate exposed on their surface. This is certainly the case for cells that have no MAF. But even mechanically dissociated cells, which have a lot of MAF complexes on their surface that could cover the major portion of the baseplate molecules, may still have some baseplate sites available at exposed surfaces, as in pseudopodes. In live cells a certain degree of slow reshuffling of MAF complexes after primary adhesion may lead to a situation where some MAF complexes eventually reach the baseplate on neighboring cells directly, thereby stabilizing species-specific interactions better than nonspecific interactions. Such mechanisms should be investigated further.

On the other hand, the recognition between mechanically dissociated cells that have quite a lot of MAF complexes may also occur to some not-yet-determined degree, based on weak and semi-specific MAF-MAF interactions as well. The dependence of this interaction on high calcium concentrations makes it difficult to assume a very specific lock-and-key mechanism at this level, although this possibility cannot be ruled out. More probably, forces involved in this type of interaction are the kind observed in aqueous two-phase polymer systems. Macromolecules like polyethylene glycol or dextran, which have evenly spaced dipoles, charges, or sterically effective groups, tend to sort out and to enrich themselves in distinct phases. Slight differences in the spacing of the glucuronic acids present in different aggregation factors could result in a calcium-dependent aqueous polymer system displaying these kinds of affinities between cells. Although a single interaction in such a system can only be partially selective at best, and by no means specific, the display of a multitude of such factors on the surface of adjacent cells can result in fairly big energy differences between homologous and heterologous cell-cell interactions, as applies to any poor affinity link with a high valency.[20] Calculations made by Edwards[21] showed that such mechanisms could explain sorting in cellular systems.

So far, experimental evidence for a possible role of this kind of interaction in sponge cell sorting is circumstantial and rests on the following findings:

1. Binding of MAF to mechanically dissociated cells, whose surfaces are still covered with MAF, displays lowered species

specificity. The amount bound to heterologous cells *(Halichondria)* rises to 30% to 50% of the amount bound to homotypic cells, compared to less than 10% in factor-depleted cells (MAF-baseplate interaction).

2. Some polycations like P-120 or polylysine, with evenly spaced charges, will interfere with MAF-MAF interactions at very low concentrations (measured as number of charges/ml and compared with calcium concentrations), possibly suggesting large areas of interactions between MAF molecules. Figures 19-2 and 19-3 depict a model based on such a theory.

At the present time it remains to be seen whether factor-baseplate interaction is responsible for the correct instantaneous aggregation of species-specific cells, and to what degree during secondary sorting-out of randomly mixed aggregates less specific interactions, eg, two-phase demixing of surface polymers, are involved as well.

Possible Roles of MAF in Mating, Metamorphosis, and Tissue Formation

Several functions for the species-specific aggregation have been discussed earlier.[3] Since MAF is a proteoglycan complex, it may simply contribute to the intercellular ground substance. Since sperm is released into the seawater and has to find the egg in another sponge at another location, the species-specific aggregation factor may serve in preventing interspecies mating. Such a function is difficult to test, first because eggs and sperm are not confined to specialized organs but arise diffusely in time and place, and also because sperm cells seem to be phagocytosed by carrier cells before they interact with the egg in the recipient sponge.

There exists an often-forgotten life stage in most sponges: free-swimming larvae arising through sexual reproduction (Figure 19-4). Except for the problematic fact that *M prolifera* larvae are only produced during two to three weeks at the end of June, they offer the opportunity to study release, settlement, and a fast-proceeding metamorphosis in very tiny organisms easy to keep in culture. Larvae have none of the organs seen in the adult animal, but they enclose within their flagellated surface epithelium almost all adult cell types, randomly arranged and ready to rearrange into organs. They have a diameter of about 300 μ; they are formed in the large intercellular matrix (mesohyl) from fertilized eggs and leave the sponge through canals.

How fast settling and spreading of the larvae going through metamorphosis proceed can be seen in Figure 19-5. The fast morphogenesis observed after settlement leads to a functional sponglet, displaying canals, flagellated

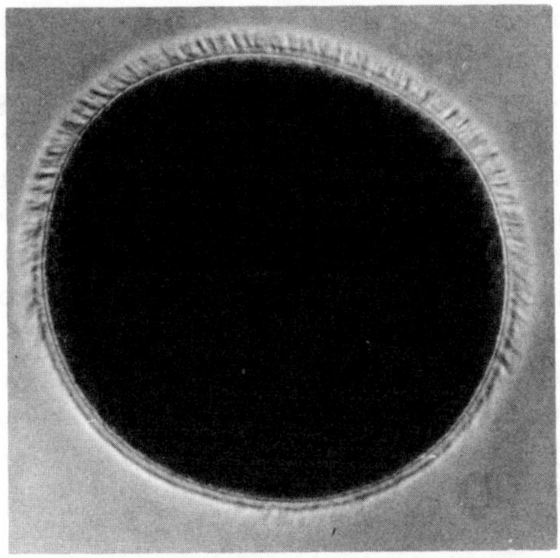

Figure 19-4 Free-swimming *M prolifera* larva. Bar = 100 μ.

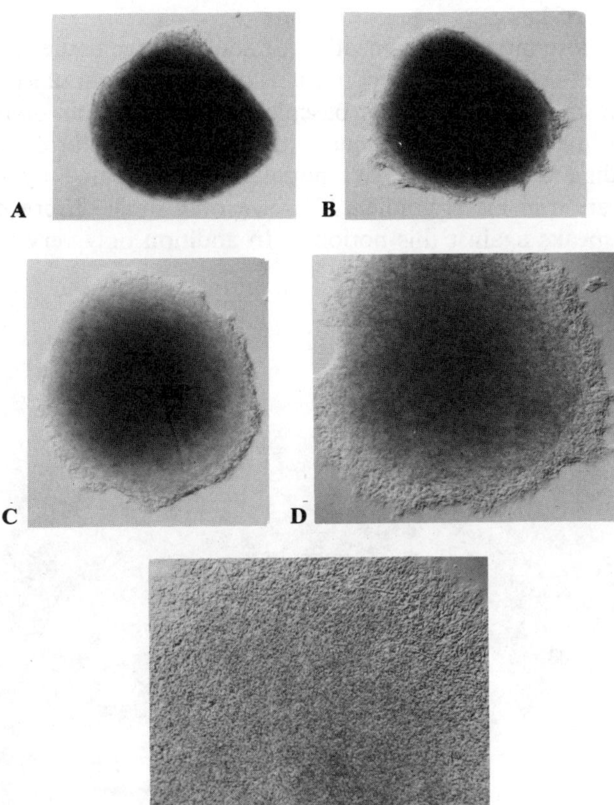

Figure 19-5 Spreading of *M prolifera* larva. **A** Spreading after less than one hour. **B–E** Pictures taken at one-hour intervals. Bar = 200 μ.

chambers, and an osculum (Figure 19-6). Major rearrangements of cells and a large increase of surface area make this system very interesting for the study of morphogenetic processes. In this context, the long-disputed reversal of layers, the migration of the flagellated cells (that form the surface of the swimming larva) to the interior to become the flagellated chambers of the merging sponglet, awaits further investigation. This key issue dates back to the last century; it has never been resolved, and based on its outcome sponges will have to be classified either as metazoa or as parazoa, ie, between protozoa and mesozoa.

Fusion of larvae is a process often observed in the laboratory. Although the advantage of this mechanism for survival is not clear, it offers another possibility to look at self versus nonself interactions, since only larvae from nearly related strains[23] and at the same stage of development seem to be able to fuse. Figure 19-6 shows a rare example of noncoalescence of two adjacent *M prolifera* larvae developing at the same time. So far, our culture experiments do not show any strong effect of MAF on settlement and fusion of larvae.

The loose internal organization of adult sponge tissue leaves most of the different cell types in direct mutual contact throughout the life span of the organism. Could the factor-baseplate system act as an omnipresent internal sorting mechanism to maintain this loose structural organization? Our finding that different cell populations, separated by 1-g velocity sedimentation in Ficoll gradients,[23] show only small differences in MAF binding speaks against this notion.[24] In addition only very few cells not

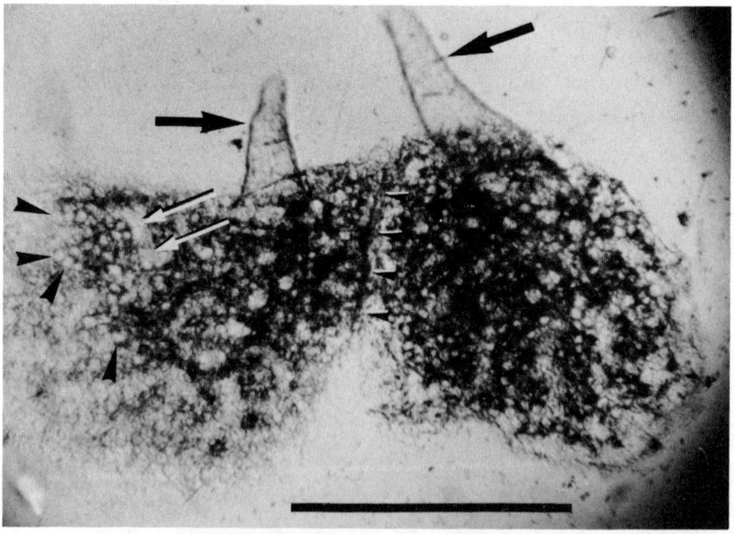

Figure 19-6 Noncoalescence between two *M prolifera* larvae. Although a large area of contact exists (▽▽), no fusion events take place. ► choanocyte chambers; ►canal; ►osculum. Bar = 500 μ.

belonging to any distinct cell type are excluded from the clumps formed in MAF-dependent reaggregation. The behavior of the different cell types in aqueous two-phase systems (polethylene glycol and dextran) is also quite similar. Despite these arguments, the possibility remains that embryogenesis or cell sorting in the adult colonial organism are factor-dependent phenomena, not based on the strong forces observed in primary reaggregation of cells by MAF. Cell sorting and tissue formation in vivo could involve weaker, qualitatively different interactions with factor, or possibly the modulation of the number or distribution of factor molecules on the surfaces of distinct sponge cell types. Different rates of synthesis, release, and breakdown of MAF molecules in different cell types also create sources and sinks of MAF matrix in the large intercellular space (mesohyl), which could account for the stabilization of the sponge tissue. These notions are backed by Leith,[25] who found some quantitative differences in the aggregative behavior of *Microciona* cell types partially purified in a sucrose step-gradient.

To test such hypotheses, very clean preparations of the distinct cell types have to be produced beforehand. It has also to be kept in mind that the dissociation procedure is selective with regard at least to some of the cell types. Only a fraction of the cells is released from the sponge fragments during dissociation, and cell types organized in sheets or chambers probably suffer great losses when squeezed through the 25-μ-pore mesh. Thus very rarely are any of the epidermal sheet cells (pinacocytes) isolated after the classic dissociation procedures using cheesecloth. Although refined 1-g velocity sedimentation techniques lead to successful enrichments,[24,26] the wide size fluctuations of discrete cell types calls for additional purification steps. In the case of one of the cell types (gray cells), we were able to purify them using the inherent fluorescence of the gray cell granules as a sorting criterion in a fluorescence-activated cell sorter.[27] Currently we are using enriched fractions to investigate species-specific sorting and tissue reformation in reconstitution experiments. The fact that in such experiments MAF-coated Sepharose beads are enclosed and accepted in forming *Microciona* aggregates, but rejected as nonself from *Mycale* aggregates, speaks for a role of MAF not only in primary aggregation events but also in tissue formation (Figure 19-7).

Summary

Although sponges are relatively simple organisms, they provide a very useful experimental system for studying specific cellular recognition of the kind occurring during tissue and organ formation in higher animals. In our present state of ignorance about molecular mechanisms for cell-cell recognition in development, the sponge system offers considerable advantages, ie, the availability of cells of specific types in great quantities without resorting to harsh disruptive treatments that might alter or obscure surface components involved in recognition.

The primary mechanism for species-specific sponge cell recognition consists of a minimum of two macromolecular components present on all the cells, namely the aggregation factor and the baseplate. This mechanism can be studied in detail using fixed cells or a completely cell-free system.

It is most unlikely that such cell-surface molecule recognition processes are sufficient to explain the entire morphogenetic process. As pointed out earlier[24] the latter can be formally divided into three components: cell or tissue movement, recognition, and stabilization of the new cell distribution. But probably not even the recognition process as such is restricted to molecular recognition by cell-surface molecules. Electrical coupling between sponge cells is not observed between cells that are missing the major portion of their aggregation molecules (MAF), although brought into close apposition. As soon as the species-specific recognition factor (MAF) is added again, coupling is induced.[28] Specific cell-surface recognition molecules may therefore provide the necessary basis on which secondary processes such as coupling and vesicular exchange between cells (see Figure 19-1) can be induced, which will permit the fine tuning for topologic recognition, leading in a yet later process to stabilization of the cell arrangement in a tissue. Such a two-step mechanism, illustrated in more detail in Figure 19-8,

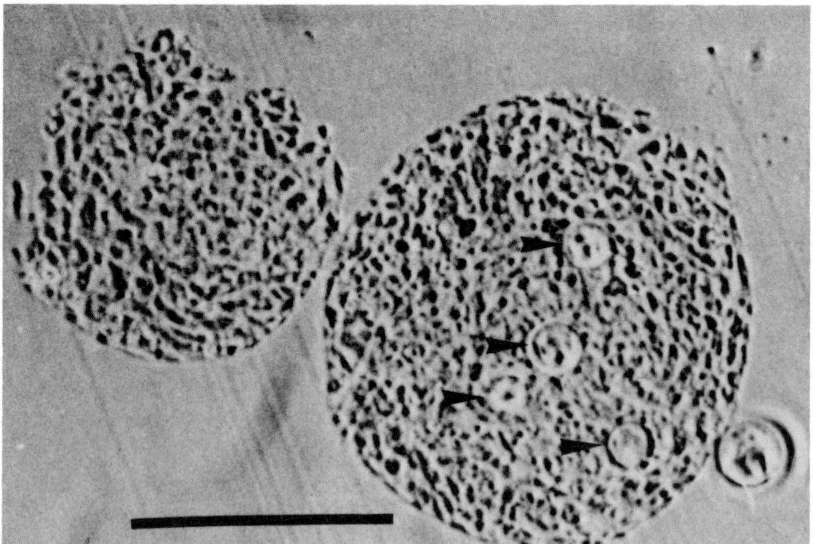

Figure 19-7 Dumbbell aggregate formed through sorting-out of mixed *Mycale/Microciona* aggregate: *Mycale* cells in left aggregate, *Microciona* cells in right aggregate. Note that only the *Microciona* aggregate incorporated MAF-coated Sepharose beads (►). Bar = 100 μ.

can be extended as a working hypothesis to recognition processes between cells and tissues during organ morphogenesis in higher animals.

The sponge cell recognition system has so far been used as a model only for developmental morphogenetic cell-cell recognition, since only species-specific recognition was investigated. It may turn out to offer itself for studies on real development (larvae) as well as in organ morphogenesis.

We would like to acknowledge the continued help and interest of other colleagues presently or formerly associated with the sponge project, which is carried out mainly at the Marine Biological Laboratory in Woods Hole (Drs J Jumblatt, W. Kuhns, F. Mir, and T. Simpson). In addition we would like to express special thanks to Dr Jody Corey, Ms Ruth Imhof, and Ms Verena Schlup for expert technical assistance.

REFERENCES

1. Burger MM, Turner RS, Kuhns WJ, et al: A possible model for cell-cell recognition via surface molecules. *Philos Trans R Soc Lond [Biol]* 271:379–393, 1975.

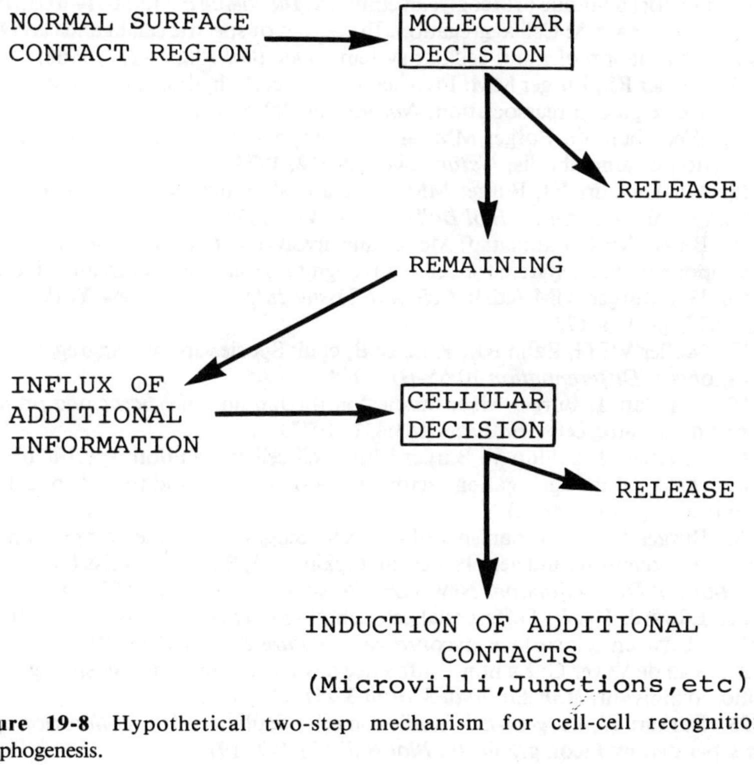

Figure 19-8 Hypothetical two-step mechanism for cell-cell recognition in morphogenesis.

2. Barondes SH, Rosen SD: Cellular recognition in slime molds. Evidence for its mediation by cell surface species-specific lectins and complimentary oligosaccharides, in Bradshaw RP, Frazier WA, Merrell RC et al (eds): *Surface Membrane Receptors.* New York, Plenum Publishing Corporation, 1976, pp 39–55.

3. Glaser L, Santala R, Gottlieb AI, et al: Cell-cell recognition in the embryonal nervous system, in Lash JW, Burger MM (eds): *Cell and Tissue Interactions.* New York, Raven Press, 1977, pp 197–208.

4. Steinberg MS: Does differential adhesion govern self-assembly processes? Equilibrium configuration and the emergence of a hierarchy among populations of embryonic cells. *J Exp Zool* 173:395, 1970.

5. Wilson HV: On some phenomena of coalescence and regeneration in sponges. *J Exp Zool* 5:245–258, 1907.

6. Wilson HV: On the behavior of the dissociated cells in hydroids, alcyonania and asterials. *J Exp Zool* 11:281–338, 1911.

7. Galtsoff PS: Regeneration after dissociation (an experimental study on sponges). *J Exp Zool* 42:183–251, 1925.

8. Henkart PS, Humphreys S, Humphreys T: Characterization of sponge aggregation factor: a unique proteoglycan complex. *Biochemistry* 12:3045–3050, 1973.

9. Humphreys S, Humphreys T, Sano J: Organization and polysaccharides of sponge aggregation factor. *J Supramol Struc* 7:330–351, 1977.

10. Humphreys T: Chemical and *in vitro* reconstruction of sponge cell adhesion. I. Isolation and functional demonstration of the components involved. *Dev Biol* 8:27–47, 1963.

11. Cauldwell C, Henkart P, Humphreys T: Physical properties of sponge aggregation factor: a unique proteoglycan complex. *Biochemistry* 12:3051–3055, 1973.

12. Moscona AA: Cell aggregation. Properties of specific cell ligands and their role in the formation of multi-cellular systems. *Dev Biol* 18:250–277, 1968.

13. Turner RS, Burger MM: Involvement of a carbohydrate group in the active site for surface guided reassociation. *Nature* 244:509–510, 1973.

14. Weinbaum G, Burger MM: A two component system for surface guided reassociation of animal cells. *Nature* 244:510–512, 1973.

15. Mir-Lechaire FJ, Burger MM: Isolation of an inhibitor of sponge cell aggregation from membranes. *Biol Bull* 155:456–457, 1978.

16. Burger MM, Jumblatt J: Membrane involvement in cell-cell interactions: a two-component model system for cellular recognition that does not require live cells, in Lash JW, Burger MM (eds): *Cell and Tissue Interactions.* New York, Raven Press, 1977, pp 155–172.

17. Müller WEG, Zahn RK, Kuselec B, et al: Species-specific aggregation factor in sponges. *Differentiation* 10:55–60, 1978.

18. Jumblatt J, Burger MM: Studies on the binding of *Microciona prolifera* factor to dissociated cells. *Biol Bull* 153:431, 1977.

19. Jumblatt J, Schlup V, Burger MM: Cell-cell recognition: specific binding of *Microciona* sponge aggregation factor to homotypic cells and the role of calcium ions. *Biochemistry* (In press)

20. Burger MM: Mechanisms of cell-cell recognition: some comparisons between lower organisms and vertebrates, in Karkinen M, Saxén L, Weiss L (eds): *Cell Interactions in Differentiation.* New York, Academic Press Inc, 1977, pp 357–376.

21. Edwards PAW: Differential cell adhesion may result from nonspecific interactions between cell surface glycoproteins. *Nature* 271:248–249, 1978.

22. Van de Vyver G: La non-confluence intra-spécifique chez les spongiaires et la notion d'individu. *Ann Embryol Morph* 3:251–262, 1970.

23. Burkart W, Burger MM: Studies on cell populations from *Microciona prolifera* separated by Ficoll gradients. *Biol Bull* 153:417, 1977.

24. Burger MM, Burkart W, Weinbaum G, et al: Cell-cell recognition: molecular aspects. Recognition and its relation to morphogenetic processes in general, in Curtis ASG (ed): *Cell-Cell Recognition*. Cambridge, Cambridge University Press, 1978, pp 1–24.

25. Leith A: Role of aggregation factor and cell type in sponge cell adhesion. *Biol Bull* 156:212–223, 1979.

26. Burkart W, Simpson TL, Burger MM: Cell surface properties of larva and adult cell types in the marine sponge *Microciona prolifera. Biol Bull* 155:430, 1978.

27. Burkart W, Jumblatt J, Simpson TL, et al: Macromolecules which mediate cell-cell recognition in *Microciona prolifera,* in Levi C, Bowy-Esnault N (eds): *Colloques Internationaux du Centre National de la Recherche Scientifique,* No. 291. Paris, Editions du CNRS, 1980, pp 239–246.

28. Loewenstein WR: On the genesis of cellular communication. *Dev Biol* 15:503–520, 1967.

20 Cell-to-Cell Communication in the Early Amphibian Embryo

Anne E. Warner, PhD

It has been recognized for some years that the cells within embryos of most species are able to communicate with each other through a direct pathway that does not include the extracellular space. The existence of this channel for communication has been detected using electrophysiologic techniques in which one intracellular microelectrode is used to inject current into one cell, while a second microelectrode records from its neighbors. Injection of a rectangular current pulse into one cell produces a potential change not only in the polarized cell, but also in cells at some distance: the cells are said to be electrically coupled to each other. The idea that such direct channels may be involved in the transfer of developmentally important information from one cell to the next was first expounded by Potter et al,[1] in their original description of this phenomenon in the squid embryo. Although this possibility has since been raised by many authors, it has proved difficult to obtain clear evidence either for or against the view

This work was supported by grants from the Medical Research Council.

that the direct cell-to-cell pathway plays a functional role during embryogenesis.[2]

This chapter will briefly review two approaches to this problem. First, the question of the permeability of the channel between embryonic cells will be considered, since information on the size of molecules afforded passage from one cell to the next is likely to be important when considering possible models for cellular interactions during development. Second, a rapid and reversible method for abolishing the direct pathway will be described; this may prove useful in elucidating whether direct cell-to-cell communication is essential for normal development.

Evidence to suggest that a wide range of substances may be transferred from one differentiated cell to another has accumulated from experiments on the *Chironomus* salivary gland. Using fluorescently tagged probes, Simpson et al[3] have demonstrated that molecules with molecular weight up to about 2000 can be transferred from one cell to the next through the direct pathway. Study of the transfer of radioactively labeled naturally occurring molecules between cells in tissue culture, sets a similar size limit to the channel.[4] Table 20-1 lists the substances so far examined for the ability to move from one embryonic cell to the next. Although the range of molecules tested is not as wide as that examined for transfer between communicating differentiated cells, the findings suggest that the channel between embryonic cells is more restrictive than that between adult cells. The upper limit for transfer lies in the region of molecular weight 450.

If this conclusion is true, then it has implications for models that may be set up to explain embryonic interactions. A number of these models[12,13] rely on the passive diffusion of molecules from one cell to the next for the transfer of developmentally relevant information. The findings given in Table 20-1 imply that any passively diffusing molecule must be relatively small; furthermore, consideration of the time available for developmental fields and gradients to be set up[12,14] suggests that small ions may be the only appropriate candidates. This conclusion does not exclude two other possibilities: that the permeability of the channel may be transiently increased at times when it is necessary to pass information; or, that some larger molecule may be transferred by a specific, fast transport mechanism.

An alternative approach to the question of whether direct cell-to-cell communication plays a functional role during embryogenesis is to examine the effects of abolishing the direct transfer pathway during development. It has been shown that the permeability of the channel between cells of the *Chironomus* salivary gland is reduced when the level of ionized intracellular calcium is raised above its resting level.[15] The disadvantage of this method is that only single cells can be uncoupled from their neighbors. It would clearly be useful to have a method that allows all cells in the embryo to become electrically isolated from each other both rapidly and reversibly. Recently it has been shown that the permeability of the cell-to-cell pathway in the early amphibian embryo is highly sensitive to alterations in in-

Table 20-1
Permeability of Communication Channels between Embryonic Cells

Molecule	Preparation	Transfer	Source
Fluorescein	*Xenopus* blastula	No	Slack and Palmer[5]
	Asterias 32-cell	No	Tupper and Saunders[6]
	Fundulus isolated pairs of blastomeres	No	Bennett[7]
	Fundulus isolated pairs of blastomeres	Yes	Bennett et al[8]
Lucifer yellow	*Fundulus* isolated pairs of blastomeres	Yes	Bennett et al[8]
Lead EDTA	*Xenopus* blastula	No	Turin[9]
Cobalt chloride	*Xenopus* blastula	No	Turin[10]
Sodium gold tetrachloride ($AuCl_4^-$)	*Xenopus* blastula	Yes	Turin[9]
Potassium argentocyanate [$Ag(CN)_2^-$]	*Xenopus* blastula	Yes	Turin[10]
EDTA	*Xenopus,* early cleavage to blastula	No	Baker and Warner[11]
Thymidine monophosphate	*Xenopus,* early cleavage to blastula	No	A.E. Warner, unpublished data, 1977

In all cases the cells were electrically coupled to each other at the time of the transfer test.

tracellular, but not extracellular pH.[16,17] One of these experiments is illustrated in Figure 20-1. Figure 20-1A shows the arrangement of the microelectrodes diagrammatically. One cell contains the current-passing electrode to inject hyperpolarizing rectangular current pulses, and a microelectrode to record the resting potential and the voltage deflection produced by the injection of current. The record from this electrode is shown on a slow time speed in Figure 20-1B as E_1. A second voltage-recording electrode inserted into the adjacent cell measures the membrane potential, E_2, and the voltage deflection produced by current flow through the intercellular junction during the injection of current. This record is shown as E_2 in Figure 20-1B. A fourth, pH-sensitive microelectrode is inserted into the same cell to determine the changes in intracellular pH (pH_i). The voltage electrode E_2 serves as a reference electrode for pH measurements. The record of intracellular pH is shown at the bottom of Figure 20-1B.

With all four electrodes inserted and the record stable, the resting potential at E_1 was -42 mV and at E_2 -35 mV. The electrotonic potentials produced by injection of current are shown as downward deflections from the resting potential. The coupling ratio, defined as the height of the electrotonic potential recorded at E_2 (V_2) divided by the electrotonic potential at E_1 (V_1), was 0.64. The resting intracellular pH was 7.8. During the period indicated by the horizontal bar the bathing solution was changed from an ACES-buffered solution (at pH 6.5) to one buffered with bicarbonate and equilibrated with 100% CO_2. This produced a rapid fall in intracellular pH, accompanied by a drop in the resting potential of both cells. The final pH_i value was 6.5; by this time the electrotonic potential on E_1 had risen substantially and no electrotonic potential could be picked up in the direct neighbor: the coupling ratio had fallen to 0, showing that the cells had uncoupled from each other. With the return to ACES-buffered saline the in-

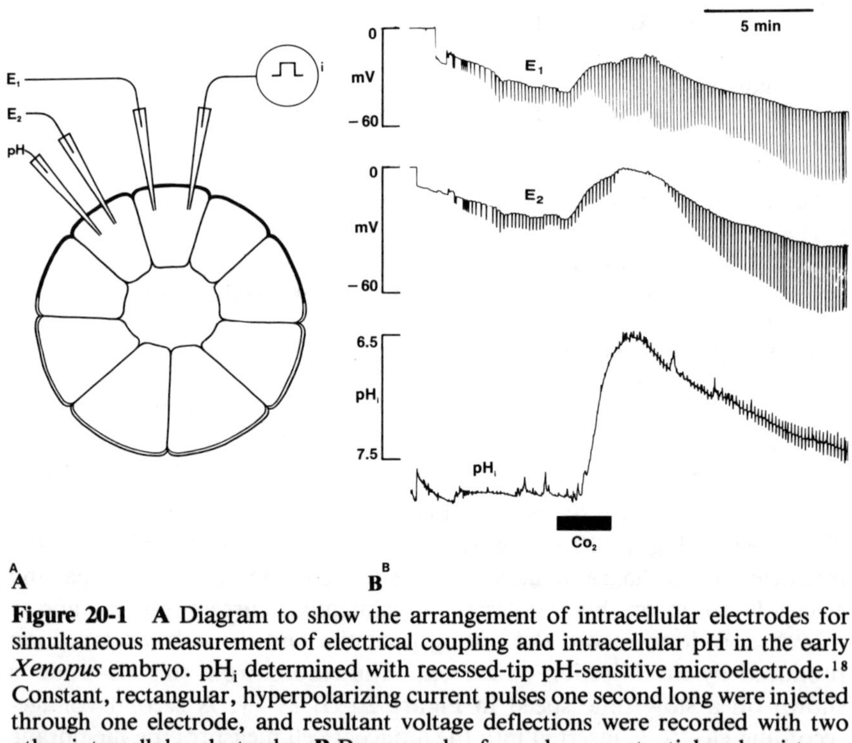

A B

Figure 20-1 **A** Diagram to show the arrangement of intracellular electrodes for simultaneous measurement of electrical coupling and intracellular pH in the early *Xenopus* embryo. pH_i determined with recessed-tip pH-sensitive microelectrode.[18] Constant, rectangular, hyperpolarizing current pulses one second long were injected through one electrode, and resultant voltage deflections were recorded with two other intracellular electrodes. **B** Pen records of membrane potential and resistance recorded in cell one (E_1), cell two (E_2), and intracellular pH (pH_i) of 32-cell *Xenopus laevis* embryo. During the time indicated by the bar, bathing solution changed from ACES-buffered Holtfreter solution (60mM NaCl, 1mM KCl, 3mM $CaCl_2$, 2mM ACES, pH 6.5) to HCO_3^--buffered, CO_2-saturated Holtfreter solution (20mM NaCl, 40mM $NaHCO_3$ 1mM KCl, 3mM $CaCl_2$, 100% CO_2, pH 6.4). Note fall in pH_i and disappearance of electrotonic potential from E_2 trace during CO_2 exposure. Reproduced from Turin and Warner[17] with permission.

tracellular pH gradually returned to its resting level and coupling between the cells returned concomitantly.

Figure 20-2 plots the relationship between the coupling ratio, measured at intervals during the recovery from CO_2 treatment, and the intracellular pH. At each interval the relationship may be considered to be in a steady state.[17] The cells start to recouple at pH_i 6.8 and are completely recoupled at pH_i 7.0. The true junctional conductance/pH_i relationship cannot be extracted from this data, because the measurements reflect alterations in the permeability of both the surface and junctional membranes, and because of the complicated paths available for current flow within the three-dimensional, electrically coupled embryo. However, calculation shows that the true relationship must be less steep than that illustrated and must lie to the right, ie, at more alkaline pH values.

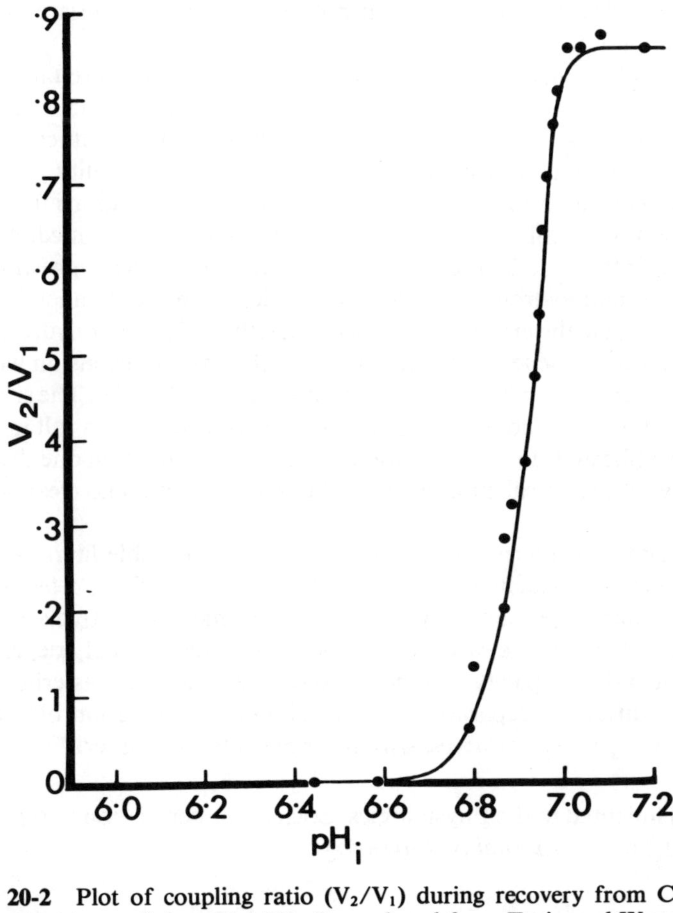

Figure 20-2 Plot of coupling ratio (V_2/V_1) during recovery from CO_2 treatment against intracellular pH (pH_i). Reproduced from Turin and Warner[17] with permission.

These experiments show, for the first time, that an embryonic system can be rapidly and reversibly uncoupled. They hold out the hope that it may now be possible to determine the role played by direct communication between cells during development.

The finding that the channel between communicating cells is sensitive to alterations in pH_i is of interest from another point of view. When the intracellular calcium concentration rises, it is likely that there will be a concomitant fall in pH_i.[19] The possibility therefore arises that the reduction in permeability of the direct channel produced by an increase in intracellular calcium is the consequence of a secondary fall in pH_i. Alternatively it could be that the effects observed when the pH_i is lowered are actually achieved by a secondary release of calcium from intracellular sequestering stores. A further possibility is that both calcium and hydrogen ions have direct effects on the channel permeability, and secondary release from sequestering sites modifies the direct response. It will not be easy to differentiate between these possibilities.

Table 20-2 collates results drawn from the rather scattered literature, to see whether any overall picture emerges. The table shows that in all cases tested, an increase in the intracellular concentration of either calcium or hydrogen ions causes a reduction in channel permeability (first two columns). This ability is shared by some other divalent cations (third column). The fourth column shows that the degree of calcium-mediated uncoupling is influenced by the intracellular pH, in a manner suggesting that calcium ions and hydrogen ions act synergistically on the channel. The last two columns give the effect of an increase in either calcium or hydrogen ion concentration on the levels of the other. On the two occasions on which it has been measured, an increase in calcium causes a fall in pH_i. The situation with regard to a secondary release of calcium as a result of a fall in pH_i is clearly complicated. In two preparations there is no effect, in one there is a secondary release of calcium, while in the other two cases no clean answer emerges.

The simplest interpretation of the presently available literature is to suppose that both calcium and hydrogen ions have a direct effect on the cell-to-cell channel permeability. The degree to which these effects are complicated by secondary release from sequestering sites probably depends on the relative buffer capacity and the speed of release and sequestering of the two ions in different preparations. This view implies that results obtained in one preparation will not necessarily be applicable to another.

I am indebted to the Physiological Society for permission to reproduce material from the *Journal of Physiology*.

Table 20-2
Effects of Intracellular Calcium (Ca_i) and pH (pH_i) on Electrical Coupling

Preparation	Ca_i on Coupling	pH_i on Coupling	Other Ions on Coupling	pH_i on Ca_i Uncoupling	pH_i on Ca_i	Ca_i on pH_i	Source
Chironomus salivary gland	Yes	...	Yes (Mg^{++} Sr^{++}, Ba^{++}, Mn^{++})	...	...	...	Oliveira-Castro and Loewenstein[20]
	Yes	Yes	...	...	Yes/No (10 out of 25 positive)		Rose and Rick[21]
Squid axon	...	...	...	...	No or reduced	...	Baker and Hönerjäger[22]
Barnacle muscle	...	...	...	...	Yes	...	Lea and Ashley[23]
Mammalian pancreatic acini	Yes (via ACh)	Yes	...	Yes	...	...	Iwatsuki and Petersen[24]
Heart muscle	Yes	...	Yes (Sr^{++})	...	...	...	De Mello[25]
Snail neurone	...	...	...	...	...		Meech and Thomas[19]
Xenopus blastomeres	...	Yes	...	...	...	...	Turin and Warner[16]
	...	...	...	...	No or small increase*	...	Rink et al[26]
Fundulus blastomere	...	Yes	...	...	No	...	Bennett et al[27]
Isolated gap junctions (structure)	Yes (at pH 7.4)	...	...	...	...	...	Peracchia[28]
	...	Yes (in absence of Ca^{++})	...	...	...	...	Perrachia and Peracchia[29]

Intracellular pH determined with pH-sensitive microelectrodes.
Ionized calcium concentration determined with the Ca-sensitive photo-protein aequorin, except for where a Ca-sensitive microelectrode was used.

REFERENCES

1. Potter DD, Furshpan EJ, Lennox ED: Connections between cells of the developing squid as revealed by electrophysiological methods. *Proc Natl Acad Sci USA* 55:328–336, 1966.

2. Wolpert L: Gap junctions: channels for communication in development, in Feldman J, Gilula NB, Pitts JD (eds): *Intercellular Junctions and Synapses.* London, Chapman & Hall Ltd, 1978.

3. Simpson I, Rose B, Loewenstein WR: Size limit of molecules permeating the junctional membrane channels. *Science* 1955:294–296, 1977.

4. Pitts JD: Direct communication between animal cells, in Brinkley BR, Porter KR (eds): *International Cell Biology.* New York, Rockefeller University Press, 1977, pp 43–49.

5. Slack C, Palmer JF: The permeability of intercellular junctions in the early embryo of *Xenopus laevis* studied with a fluorescent tracer. *Exp Cell Res* 55:416–419, 1969.

6. Tupper JT, Saunders JW: Intercellular permeability in the early *Asterias* embryo. *Dev Biol* 27:546–554, 1972.

7. Bennett MVL: Permeability and structure of electrotonic junctions and intercellular movement of tracers, in Kater SD, Nicholson C (eds): *Intracellular Staining Techniques in Neurobiology.* London, Chapman & Hall Ltd, 1973, pp 115–142.

8. Bennett MVL, Spira ME, Spray DC: Permeability of gap junctions between embryonic cells of *Fundulus:* a reevaluation. *Dev Biol* 65:114–125, 1978.

9. Turin L: New probes for studying intercellular permeability. *J Physiol* 269:6–7, 1977.

10. Turin L: *The Physiology of Intercellular Communication in the Early Amphibian Embryo.* PhD Thesis, University of London, 1978.

11. Baker PF, Warner AE: Intracellular calcium and cell cleavage in early embryos of *Xenopus laevis. J Cell Biol* 5:51–77, 1972.

12. Wolpert L: Positional information and the spatial pattern of cellular differentiation. *J Theor Biol* 25:1–47, 1969.

13. Goodwin BC, Cohen MH: A phase shift model for the spatial and temporal organization of developing systems. *J Theor Biol* 25:49–107, 1969.

14. Warner AE: Consequences of cell to cell signals in the early amphibian embryo, in Nichols JG (ed): *Life Science Research Report. The Role of Intercellular Signals: Navigation, Encounter, Outcome.* Weinheim, Verlag Chemie GMBH, 1979, pp 179–200.

15. Rose B, Loewenstein WR: Permeability of a cell junction and the local cytoplasmic free ionized calcium concentration. *J Membr Biol* 28:87–119, 1976.

16. Turin L, Warner AE: Carbon dioxide reversibly abolishes ionic communication between cells of the early amphibian embryo. *Nature* 270:56–57, 1977.

17. Turin L, Warner AE: Intracellular pH in *Xenopus* embryos: its effect on current flow between blastomeres. *J Physiol* 300:489–504.

18. Thomas RC: Intracellular pH of snail neurones measured with a new pH-sensitive micro-electrode. *J Physiol* 238:159–180, 1974.

19. Meech RW, Thomas RC: The effect of calcium injections on the intracellular sodium and pH of snail neurones. *J Physiol* 265:867–879, 1977.

20. Oliveira-Castro G, Loewenstein WR: Junctional membrane permeability: effect of divalent cations. *J Membr Biol* 5:51–77, 1971.

21. Rose B, Rick R: Intracellular pH, intracellular free Ca and junctional cell-cell coupling. *J Membr Biol* 44:377–415, 1978.

22. Baker PF, Hönerjäger P: Influence of carbon dioxide on the level of ionized calcium in the squid axon. *Nature* 273:160, 1978.

23. Lea TJ, Ashley CC: Increase in free Ca^{2+} in muscle after exposure to CO_2. *Nature* 275:236, 1978.

24. Iwatsuki N, Petersen OH: Pancreatic acinar cells: the effect of CO_2, NH_4Cl and acetylcholine on intercellular communication. *J Physiol* 291:317–326, 1979.

25. De Mello WC: Effect of injection of intracellular calcium and strontium on cell communication in heart. *J Physiol* 250:231–245, 1975.

26. Rink TJ, Tsien RY, Warner AE: Free calcium in *Xenopus* embryos measured with ion-sensitive micro-electrodes. *Nature* 283:658–660, 1980.

27. Bennett MVL, Brown JE, Harris AC, et al: Electrotonic junctions between *Fundulus* blastomeres: reversible block by low intracellular pH. *Biol Bull* 155:428, 1978.

28. Peracchia C: Calcium effects on gap junction structure and coupling. *Nature* 271:669–671, 1978.

29. Peracchia C, Peracchia LL: Orthogonal and rhombic arrays in gap junctions exposed to low extracellular pH. *J Cell Biol* 79:217a, 1978.

21 Ligatin from Neural Retina—A Potential Baseplate for Retinal Cell Adhesion

Richard B. Marchase, PhD
Emma R. Jakoi, PhD

In his hypothesis of neuronal specificity, Sperry[1,2] attributed the selectivity that characterizes the circuitry of the brain to interactions during development between cell-surface molecules on contacting cells. In the visual system of lower vertebrates, for example, axons of retinal ganglion cells form a topographically ordered projection in their synapses with neurons of the optic tectum.[3] Sperry proposed that the axonal tips of each retinal ganglion cell possessed a unique battery of cytochemical markers, as did their potential termini in the tectum. It was suggested that interactions between these molecules result in differential affinities between the individual axons and the various tectal loci. Sperry reasoned that each axon would adhere preferentially to, and finally synapse with, an area of tectum possessing a particular set of surface molecules, while bypassing sites with less appropriate displays.

Funds for this work were provided by US Public Health Service program project grant 9 PO1 GM 23911-06 and by research grant 1 RO1 EY 02845-01.

Our approach to testing the hypothesis of neuronal specificity and investigating its biochemical basis has utilized an assay capable of directly measuring the adhesion of dissociated chick neural retina cells to limited areas of tectal surface.[4,5] With this assay, cell bodies from dorsal retina were shown to adhere preferentially to ventral half-tecta, while cell bodies from ventral retina preferred dorsal half-tecta, thus mimicking the innervation pattern of the tectum by the retina.[3] While there is no assurance that the molecular mechanisms responsible for this in vitro adhesive specificity are in any way related to those responsible for the in vivo synaptic specificity, our strategy is to attempt to elucidate the biochemical basis for the simpler and more accessible in vitro adhesive selectivity, with the goal of ultimately testing these results for relevance to the more complicated in vivo situation.

A Baseplate Protein for Intercellular Adhesion among Retinal Cells

While molecules forming adhesive bonds between contacting retinal and tectal cells could directly interact with the lipid bilayers of their respective plasma membranes, a second possibility is that a protein baseplate serves to anchor them to the cell surface. In the system in which the biochemical basis for adhesion is best understood, sponges, the latter possibility seems to be the case (see Burkart and Burger, Chapter 19). In neural retina, two previous lines of evidence have suggested the presence of a baseplate. First, supernatants from cultured retina enhance the reaggregation of trypsinized retinal cells,[6,7] and contain components that bind back to them.[6] The tissue-specific nature of these effects is consistent with the notion of a baseplate. Second, Merrel et al[8] have partially characterized a preparation from retinal plasma membranes that inhibits the reassociation of single retinal cells. This inhibition could be due to the presence in their extract of baseplates that compete with cell-associated baseplates for molecules necessary for the formation of adhesive bonds. The inhibitory molecule(s) isolated by Merrell et al[8] has a molecular weight, determined by ultrafiltration, of approximately 10,000 daltons, and is sensitive to inactivation by trypsin. However, attempts to further characterize this preparation have been hampered by the finding that, after polyacrylamide gel electrophoresis, no stainable protein corresponding to the biologic activity was found.

Ligatin from Suckling Rat Ileum

In newborn mammals during the suckling period, ileal epithelial cells are characterized by a highly invaginated interconnecting system of plasma membrane beneath the microvilli, the endocytic complex.[9] The luminal surfaces of these membranes are covered in part by an ordered array of 7.5-nm β-N-acetylhexosaminidase molecules joined together with a center-to-center spacing of ~ 14.5 nm[10] (Figure 21-1). Rows of these particles can be released from the membranes by treatment with 20 mM Ca^{++} and retain their unidimensional periodicity (Figure 21-2).

Following their release from the membranes, dialysis against EGTA results in the depolymerization of these rows of particles, and molecular sieve chromatography resolves two components (Figure 21-3). The excluded fractions contain all of the β-N-acetylhexosaminidase activity and, by electron microscopy, particles identical to those seen on the membranes are observed[10] (E.R. Jakoi and J.D. Robertson, unpublished data, 1980). The included protein, named ligatin, has an apparent monomeric molecular weight of 10,000 daltons and, upon readdition of Ca^{++}, will repolymerize to form filaments 3 nm in diameter. Interestingly, when up to 60 mg of ileal ligatin are subjected to polyacrylamide gel electrophoresis, no bands are

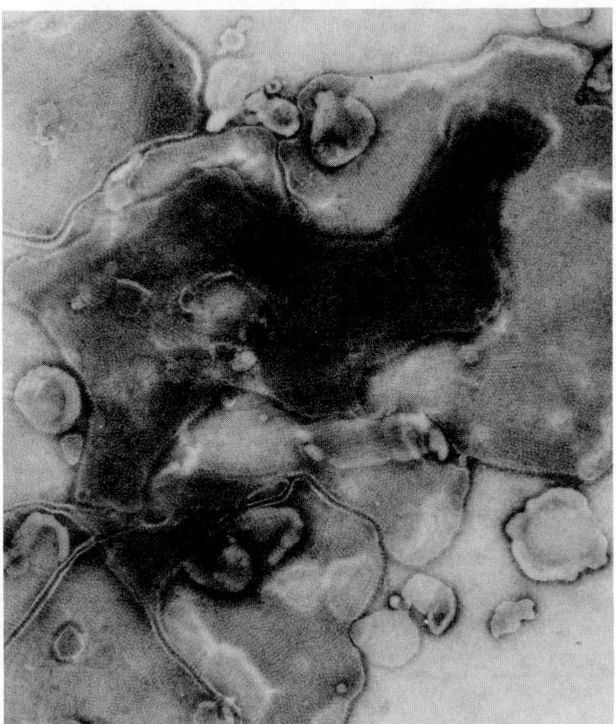

Figure 21-1 Negatively stained preparation (1% PTA with bacitracin) of the isolated endocytic membranes. The membrane surfaces are covered by a particulate array. × 37,000.

visualized by sodium anazolene (Coomassie Blue) or periodic acid-Schiff stain (E.R. Jakoi and J.D. Robertson, unpublished data, 1980). Other morphologic and biochemical observations[10] (E.R. Jakoi and J.D. Robertson, unpublished data, 1980) support the conclusion that the ordered arrays found on the plasma membranes of the endocytic complex are due to aligned filaments of ligatin decorated by β-N-acetylhexosaminidase molecules.

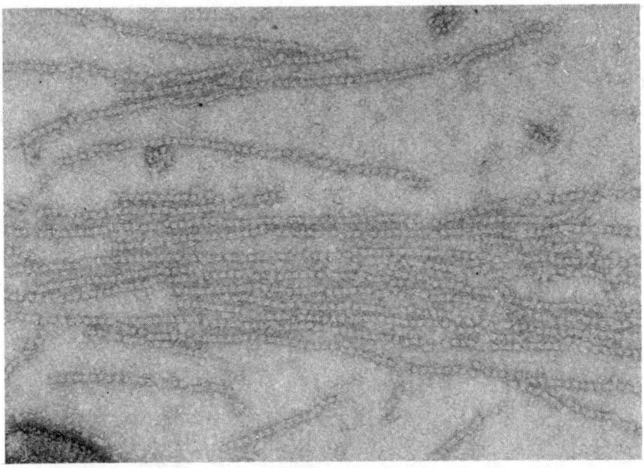

Figure 21-2 Rows of particles released from ileal membranes by $CaCl_2$ and negatively stained. Each row consists of individual 7.5-nm β-N-acetylhexosaminidase molecules joined together with a center-to-center separation of 14.5 nm. × 87,750. From Jakoi et al.[10] Reprinted with permission.

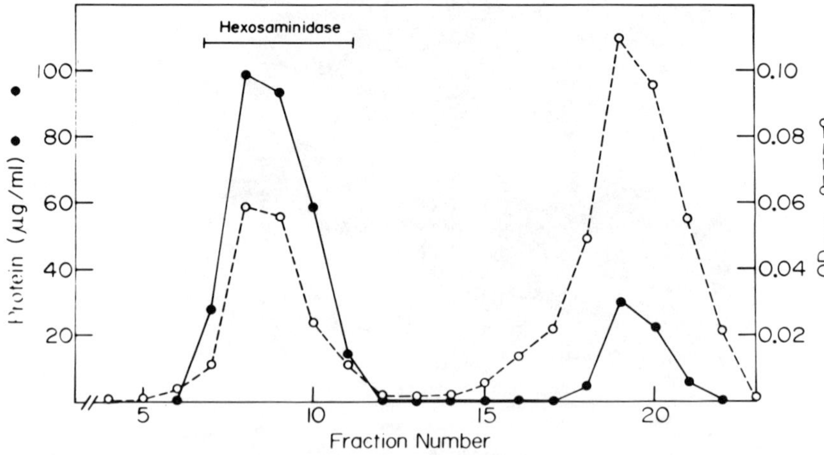

Figure 21-3 Elution profile from a Biorad P-60 separation of the protein components of the rows of particles released from endocytic membranes. Position of hexosaminidase activity is indicated. ●——● protein concentration. O——O absorbance at 280 nm.

Ligatin from Embryonic Chick Neural Retina

Some of the properties of ileal ligatin, eg, its plasma membrane location, its molecular weight of approximately 10,000 daltons, and its apparent inability to be stained in polyacrylamide gels, are identical to characteristics of the inhibitor of retinal adhesion chracterized by Merrell et al.[8] These physical similarities led us to an investigation in retina for a ligatin-like molecule.[11]

Retinal plasma membranes were examined by electron microscopy using negative staining techniques. Most of the membranes were covered in part with randomly organized filaments 4.5 nm in diameter (Figure 21-4). When the plasma membranes were treated with 40 mM Ca^{++}, individual filaments 4.5 nm in diameter were seen free of the membranes that were scattered over the grids (Figure 21-5). After dialysis against EGTA, at pH 8.0, filamentous materials were no longer seen. The release of filamentous material from plasma membrane surfaces after treatment with Ca^{++}, and the subsequent disappearance of the filaments after dialysis against EGTA, are identical to results obtained with ligatin-hexosaminidase filaments isolated from ileum. However, both the membrane-bound filaments and the free filaments differ in morphology from those seen in ileal preparations.

Subsequent molecular sieve chromatography on a Biorad P-60 column yielded profiles of optical density and protein concentration similar to those

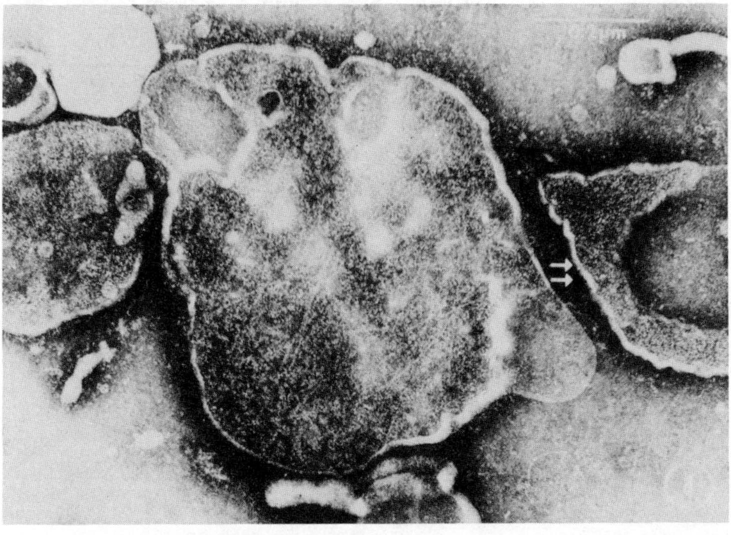

Figure 21-4 Negatively stained preparation of the isolated retinal plasma membranes. Most membrane surfaces are covered in part by filaments ~4.5 nm in diameter; others are completely covered by particles (double arrow) or are completely smooth (single arrow). × 114,000. From Jakio and Marchase.[11] Reprinted with permission.

shown for ileum (Figure 21-6). Again only two protein peaks were observed, one totally excluded and the other chromatographing as a globular protein of 10,000 daltons. Unlike ileum, however, the retinal proteins voided from the P-60 column were found to have no β-N-acetylhexosaminidase activity. They also had no morphologic resemblance to the ileal enzyme.

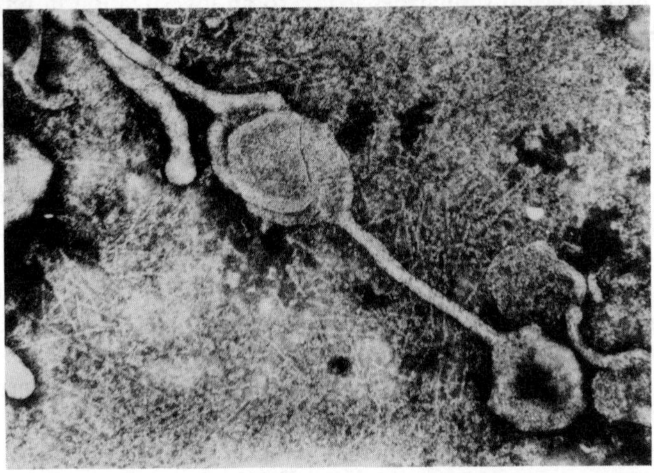

Figure 21-5 Negatively stained preparation of the Ca^{++}-treated membrane pellet. Both smooth and particulate-coated membranes are seen. In addition, filaments ∼4.5 nm in diameter are seen in the background, free from the membranes. × 102,000. From Jakoi and Marchase.[11] Reprinted with permission.

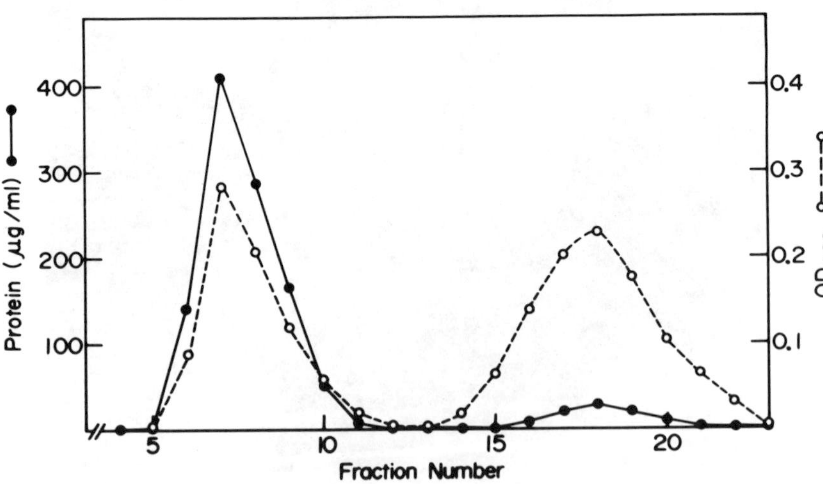

Figure 21-6 Elution profile from Biorad P-60 column of solubilized components from retinal plasma membranes after treatment with 40 mM Ca^{++} and dialysis vs 0.5 mM EGTA, pH 8. •——• protein concentration. O——O absorbance at 280 nm. From Jakoi and Marchase.[11] Reprinted with permission.

As with ileal ligatin, the included peak exhibited an optical density at 280 nm that greatly exceeds the molar extinction coefficient of a typical protein at that wavelength. The absorbance is maximal at ∿250 nm, and is attributable to conjugated phospholipids associated with the protein.

Rechromatography of the 10,000-dalton retinal protein in the presence of 0.5 mM EGTA did not alter its P-60 elution profile (Figure 21-7A). After dialysis against 5 mM Ca^{++}, however, the protein was completely excluded from the column (Figure 21-7B). Negative-stain electron microscopy of this material showed that the protein had polymerized into filaments ∿3 nm in diameter (Figure 21-8A). After freeze-thawing, this material formed tactoids (Figure 21-8B). The morphology of the repolymerized filaments is identical to that of repolymerized ileal ligatin.

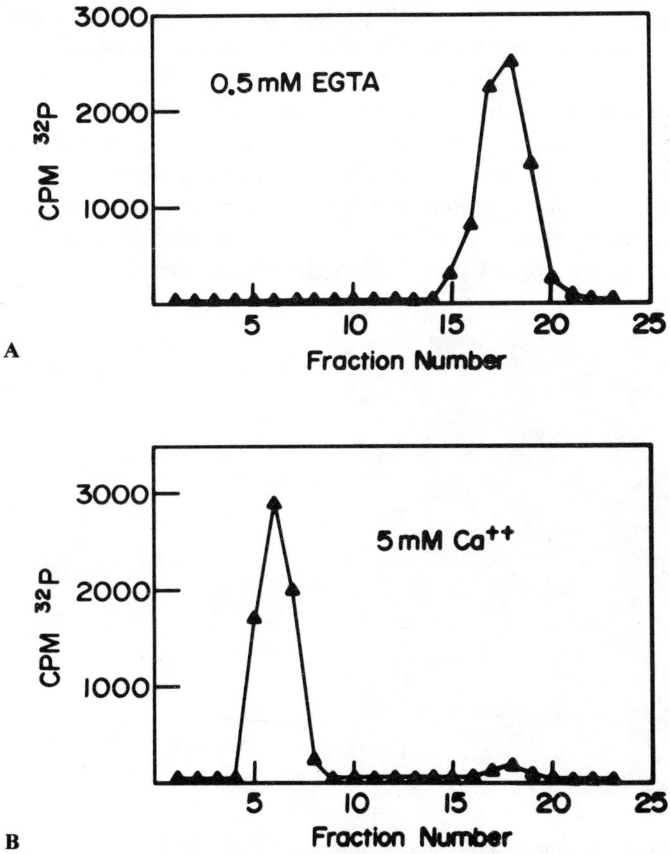

Figure 21-7 Effect of Ca^{++} on polymerization of ^{32}P-labeled protein from the included P-60 peak. **A** Chromatography in the presence of 0.5 mM EGTA. **B** Chromatography after dialysis against 5 mM Ca^{++}. In both cases, absorbance at 280 nm was superimposable with the distribution of radioactivity. From Jakoi and Marchase.[11] Reprinted with permission.

Polyacrylamide Gel Electrophoresis of Retinal Ligatin

Aliquots of retinal ligatin were subjected to electrophoresis under both nondenaturing and denaturing (SDS-urea) conditions and, as with ileal ligatin, no bands were visualized, although protein and glycoprotein standards in accompanying gels were clearly stained. Several explanations seemed possible: 1) the proteins did not react with either of the stains employed; 2) the proteins were heterogeneous and each polypeptide was present at too low a concentration to be observed; or 3) the proteins were soluble in the fixature or staining solutions and diffused from the gels. To examine these possibilities, duplicate preparations of retinal ligatin labeled with ^{32}P were electrophoresed under nondenaturing conditions. One gel was frozen and sliced without fixation. Its profile of radioactivity is shown in Figure 21-9. All of the counts per minute (cpm) applied to the gel were recovered in a single, discrete band migrating just ahead of the tracking dye. The duplicate gel was fixed and stained with sodium anazolene (Coomassie Blue) in methanol:acetic acid:trichloroacetic acid for 12 hours. No bands were visible after destaining. The gel was then sliced and the profile of radioactivity determined. No cpm were present in the fixed gel. Therefore, the native protein was apparently not retained in the gel using conventional fixation procedures. This interpretation was confirmed using ^{14}C-carboxymethylated protein and denaturing conditions. In these ex-

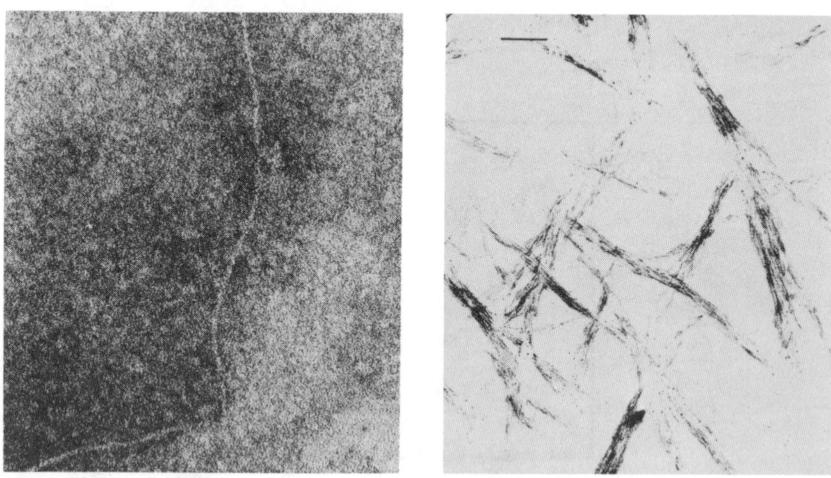

A **B**

Figure 21-8 **A** Negatively stained preparation of the included fraction from a Biorad P-60 column after the readdition of $CaCl_2$. Individual filaments are ∼3 nm in diameter. × 213,000. **B** Negatively stained preparation after being frozen (−20 C) and thawed. The filaments have aggregated into tactoids of varying lengths. No individual filaments are seen. × 75,000. From Jakoi and Marchase.[11] Reprinted with permission.

periments a single band of radioactivity with a mobility of 0.70 relative to the tracking dye was found before staining (Figure 21-10). Again, the cpm disappeared during the fixing and staining procedures.

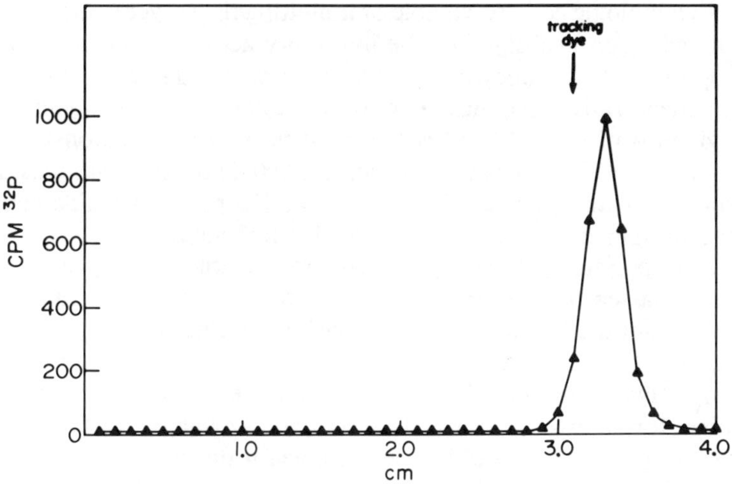

Figure 21-9 Profile of radioactivity of ^{32}P-labeled ligatin after electrophoresis under nondenaturing conditions and before fixation with acidic or acidified organic solvents. After fixation, no cpm remained in the gel.

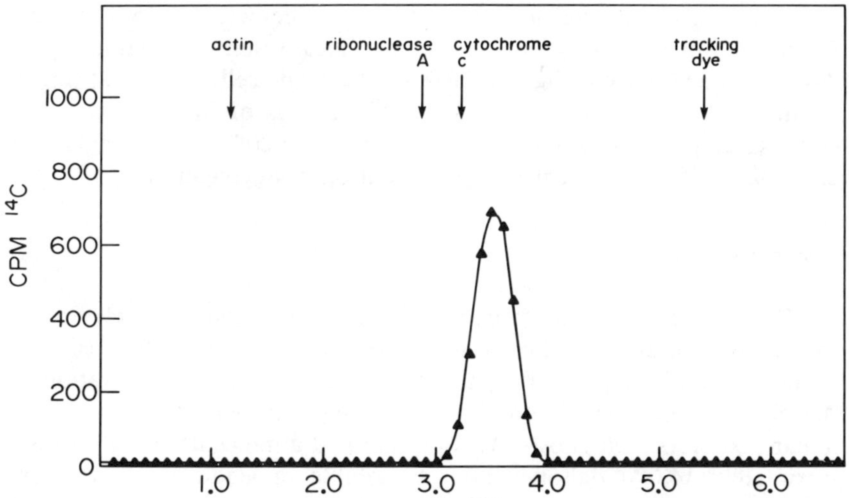

Figure 21-10 Profile of radioactivity of ^{14}C-carboxymethylated ligatin after electrophoresis in SDS-urea on 15% polyacrylamide gels and before fixation with acidic or acidified organic solvents. Arrows denote positions of standards: actin (43,000 daltons); ribonuclease A (13,600 daltons); cytochrome c (12,500 daltons). After fixation, no cpm remained in the gel. From Jakoi and Marchase.[11] Reprinted with permission.

Biological Activity of Retinal Ligatin: Inhibition of Single-Cell Disappearance

The primary stimuli for seeking ligatin in neural retina were similarities between a biologic activity capable of inhibiting the reaggregation of single retinal cells, and ileal ligatin. The inhibitory activity isolated by Merrell et al[8] possessed the following properties: it was extractable from retinal plasma membranes using lithium diiodosalicylate; it passed quantitatively through an Amicon PM 10 filter (nominal cutoff 10,000 daltons), but was retained by a UM 2 filter (nominal cutoff 2000 daltons); it was sensitive to trypsin; and it was not visualized after staining polyacrylamide gels with either a protein stain or the periodic acid-Schiff method.

Correspondingly, ileal ligatin was extractable from plasma membranes[10]; passed through PM 10 but was retained by UM 2 filters (E.R. Jakoi, unpublished data, 1978); was sensitive to cleavage by trypsin (E.R. Jakoi, unpublished data, 1978); and was not visualized in conventionally stained polyacrylamide gels.[10] In addition, ileal ligatin functions as a baseplate for an external cell-surface protein. An analogous function by a similar protein in retina could be responsible for the biologic activity found by Merrell et al[8]: if proteins that mediate intercellular adhesion are bound to retinal plasma membranes via ligatin, then exogenous, soluble ligatin might compete for this material and thus inhibit adhesion rates. In order to ascertain if retinal ligatin possessed similar biologic activity to the preparation of Merrell et al,[8] single-cell reassociation studies were carried out in the absence and presence of retinal ligatin. A typical result is shown in Figure 21-11. In the absence of ligatin, 70% of the single cells present at time 0 became associated with other cells by 30 minutes of incubation. When ligatin at 20 μg/ml was added to the assay, only 20% of the single cells reassociated. This represents a 70% inhibition of aggregation.

Ligatin from Other Tissues

Four proteins that fail to be visualized in polyacrylamide gels stained with sodium anazolene (Coomassie Blue) have been reported recently. Besides the initial report of ligatin from ileum[10] and the preparation of Merrell et al,[8] Gaetjens[12] found a nonstaining phosphorylated protein in human red blood cell ghosts. This protein had a molecular weight of approximately 14,000 daltons and was soluble in acidified chloroform: methanol. Also, Bárány et al[13] extracted a phosphorylated protein of 10,000 daltons from adult frog skeletal muscle with acidified chlorform: methanol.

The inability to visualize these proteins has been suggested to be due to a failure to bind sodium anazolene (Coomassie Blue),[10,13] or because the proteins were present at too low a concentration to be apparent.[8,12]

However, we propose that the proteins isolated from skeletal muscle and red blood cells were also ligatin or ligatin-like proteins and that their failure to be visualized was due to their solubility in acidic fixatives. This hypothesis is supported by the findings that the muscle[13] and red blood cell[12] proteins were in fact soluble in acidified chloroform:methanol, as are ileal and retinal ligatin. In addition, the molecular weights and amino acid compositions reported for the frog muscle[13] and human red blood cell[12] proteins agree closely with those obtained for ileal (E.R. Jakoi and J.D. Robertson, unpublished data, 1980) and retinal ligatin.[11] Lastly, using the Ca^{++} extraction described above, we have isolated proteins that satisfy the operational criteria for ligatin from skeletal muscle membranes and red blood cell ghosts (R.B. Marchase and E.R. Jakoi, unpublished data, 1979). These similarities thus suggest that ligatin is a plasma membrane protein or a family of proteins found in a wide variety of tissues and across a spectrum of vertebrate species.

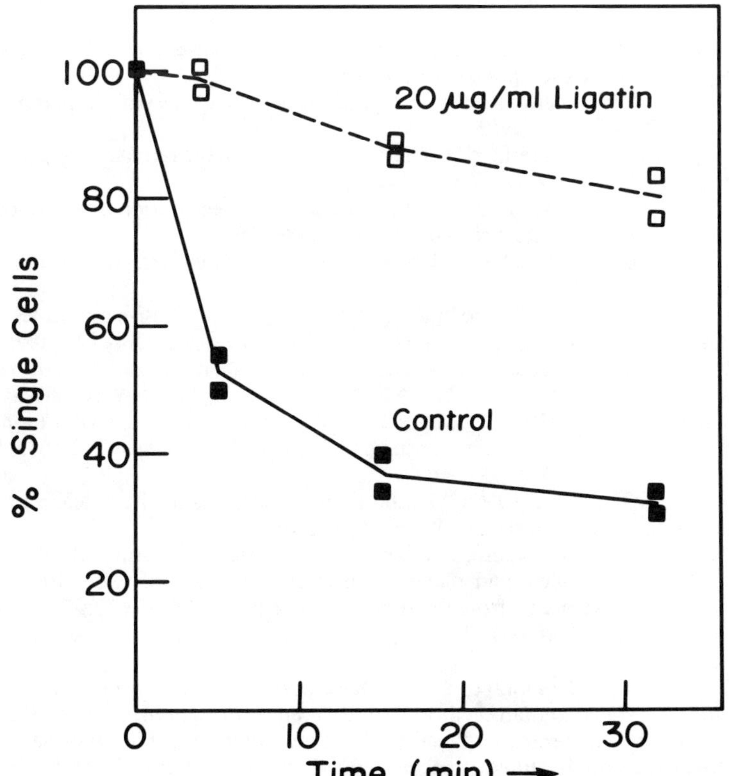

Figure 21-11 Number of dissociated retinal cells remaining as single cells in the absence (■——■) or presence (□----□) of 20 µg/ml retinal ligatin. Cells were shaken and at intervals aliquots of the suspension were removed; the numbers of single cells remaining were determined in hemocytomers.

While no information has been obtained regarding the association of ligatin with other proteins in either red blood cells or muscle, its putative roles in ileum and retina are suggestive of a generalized function. In ileum, ligatin secures a β-N-acetylhexosaminidase to the external surface of epithelial cells. The hexosaminidase participates in the partial digestion of polysaccharides prior to absorption by the ileum. In retina, our data suggest the possibility that ligatin serves as a baseplate for proteins active in intercellular adhesion. In both of these cases ligatin seems to be associated with proteins that help determine critical interactions with either neighboring cells or the biochemical milieu. It is intriguing to sepculate that ligatin might serve as a baseplate for other cell-surface proteins in other tissues, and that these associated proteins function in the social interactions of the respective cells.

REFERENCES

1. Sperry RW: Visuomotor coordination in the newt *(Triturus viridescens)* after regeneration of the optic nerves. *J Comp Neurol* 79:33–35, 1943.
2. Sperry RW: Chemoaffinity in the orderly growth of nerve fiber patterns and connections. *Proc Natl Acad Sci USA* 50:703–710, 1963.
3. Hamdi FA, Whitteridge D: The representation of the retina on the optic lobe of the pigeon. *Q J Exp Physiol* 39:111–119, 1954.
4. Barbera AJ, Marchase RB, Roth S: Adhesive recognition and retinotectal specificity. *Proc Natl Acad Sci USA* 70:2482–2486, 1974.
5. Marchase RB: Biochemical investigations of retinotectal adhesive specificity. *J Cell Biol* 75:237–257, 1977.
6. Balsamo J, Lilien J: The binding of tissue-specific adhesive molecules to the cell surface. A molecular basis for specificity. *Biochemistry* 14:167–171, 1975.
7. Hausman RF, Moscona AA: Purification and characterization of the retina-specific cell-aggregating factor. *Proc Natl Acad Sci USA* 72:916–920, 1975.
8. Merrell R, Gottlieb DI, Glaser L: Embryonal cell surface recognition. Extraction of an active plasma membrane component. *J Biol Chem* 250:5655–5659, 1975.
9. Wissig SL, Graney DO: Membrane modifications in the apical endocytic complex of ileal cells. *J Cell Biol* 39:564–579, 1968,
10. Jakoi ER, Zampighi G, Robertson JD: Regular structures in unit membranes. II. Morphological and biochemical characterization of two water-soluble membrane proteins isolated from the suckling rat ileum. *J Cell Biol* 70:97–111, 1976.
11. Jakoi ER, Marchase RB: Ligatin from embryonic chick neural retina. *J Cell Biol* 80:642–650, 1979.
12. Gaetjens E: Isolation of a ^{32}P-labeled polypeptide of low molecular weight from phosphorylated human erythrocyte membranes. *Biochemistry* 15:40–45, 1976.
13. Bárány M, Bárány K, Gaetjens E, et al: Isolation of phosphorylated acid chloroform/methanol-soluble proteins from live frog muscle. *Biochim Biophys Acta* 491:387–397, 1977.

22 A Plastic Deformation Artifact in the Freeze-Fractured Membranes of Mammalian Urothelium

J. David Robertson, MD, PhD
Juan A. Vergara, MD

Plastic deformation is frequently encountered as an artifact in the freeze-fracture method, but it is not always recognized. This artifact, resulting in various degrees of shape deformation of components of the frozen specimen during fracturing, is encountered at all ranges of temperature so far explored and is seen after a variety of fracturing procedures (for review see Southworth et al[1] and Sleytr and Robards[2]). It is an important factor to be assessed in determining the correct appearance of objects located in the fracture plane, because it is reasonable to expect that different macromolecular components will display different mechanical properties and hence will be deformed in different ways. This variation should be particularly evident in membranes in which hydrophobic and hydrophilic domains in ice must surely have different mechanical properties.

Plastic deformation can be a useful artifact because it often enhances the visibility of the fine structure of biologic domains. However, the

This work was supported by National Institutes of Health Grant No. 5 PO1 GM723911.

magnitude of plastic deformation has to be considered in a final interpretation of the size, shape, and localization of the structures observed in the freeze-fracture image. Any attempt to eliminate plastic deformation entirely is probably futile, but an effort to recognize it and better understand it is highly desirable. Membranes with regular arrays of particles provide a favorable system in which to understand plastic deformation, because the structure in the membrane, visualized by freeze-fracture, can also be seen by other methods of ultrastructural analysis such as sectioning, negative staining, shadowing of frozen and dried specimens, and possibly by x-ray diffraction and spectroscopic procedures. Urinary bladder hexagonal membranes, when fractured, give rise after certain technical procedures to heterogeneous structures in the fracture plane. The regular arrangement of particles as previously described by Robertson[3] is seen in some areas as relatively smooth surfaces, and in others as prominent globular discrete particle arrays like the ones previously described by Staehelin et al[4] and by Knutton and Robertson[5] using conventional methods. A hexagonal lattice can be discerned by the eye in the smooth areas, although the individual particles are difficult to make out. In the particulate areas they are, however, very distinct. The individual particles can be shown to be of different kinds, depending on the technique employed. The center-to-center spacing of particles in all areas, measured either directly or by optical diffraction, is the same: 160 Å. The vertical heights of the particles, however, are different in the fracture surfaces.

Materials and Methods

Isolation of hexagonal membranes Hexagonal membranes from cow urinary bladder were isolated by the procedure described by Vergara et al,[6] with slight modifications to enhance purity and improve the final yield. The procedure involved separation by sequential banding in two-step sucrose gradients. The material "zoning" on 32% wt/wt sucrose (d = 1.13 gm/ml) consisted of highly purified hexagonal membranes that were homogeneous in thin sections, negative-stain preparations, and freeze-fracture replicas. The procedure employed used only a sucrose solution in Tris-HCl buffer at pH 7.4 and avoided any extractive agent. Details of this method are presented in a more extensive paper.[7]

Freeze-fracture 0.25 ml of membrane suspension (0.5 mg/ml of membrane proteins) were deposited on the cationized surface of a glass coverslip by centrifugation as follows: The membrane suspension was layered on 0.5 M sucrose in 10 mM Tris-HCl buffer at pH 7.4 in a nitrocellulose tube. The bottom of the tube was plugged with a 15-mm high lucite cylinder, which at one end adapted to the concave bottom of the centrifuge tube, while at the other end a flat polished surface gave support to a round 13-mm diameter, 0.6-mm thick glass coverslip. After thorough cleaning and washing the surface of the coverslip was cationized with a solution of 0.1% alcian blue, as suggested by Sommer,[8] washed with deionized water, and dried with a jet of nitrogen to allow the formation of a thin film of dye on the face where the membranes were to be deposited. All this assembly (plug, cationized glass, sucrose solution, and the overlay membrane suspension contained in the nitrocellulose tube) was placed in a SW 41 Rotor (Beckman Corp, Palo Alto, Calif) and centrifuged at 10,500 g for 15 minutes. A thin film of membranes that strongly adhered to the glass was found to be formed with this procedure. The surface bearing the membranes was gently washed for 30 seconds with a jet of deionized water to get rid of buffer salts, sucrose, and excess membranes. It was then covered with a flat, thin, polished copper disc etched with nitric acid and washed just before use. Care was taken to maintain a film of water between the copper and glass surface.

The assembly thus prepared was mounted on the tip of fine forceps, the excess water blotted out with bibulous paper, and immediately frozen by mechanical insertion into propane at 80 to 82 K using a rapid freezing device constructed by Costello and Corless.[9] Fracture was performed by separating the coverslip from the copper disc at liquid-nitrogen temperature, either at atmospheric pressure or (to avoid surface contamination of the specimen) at 10^{-8} torr in the cooled stage of a Denton Vacuum Freeze-Fracture Apparatus. The prevacuum system comprised a mechanical pump and two Sorption pumps used in series. Very high vacuum was obtained with an ionization pump. The specimens were either etched at 170 K until no contamination could be seen on the glass surface (an average of 45 minutes), or if fractured under high vacuum they were sometimes replicated without etching. In all cases the replication was performed at 80 K and at a beginning vacuum of better than 10^{-8} torr. Platinum was deposited at an angle of 20° to the surface of the coverslip for 10 seconds, followed by carbon deposited vertically for 15 seconds. The replicated film was floated on 50% hydrofluoric acid, cleaned in two changes of chlorine bleach, and picked up from distilled water on a carbon film on an electron microscope grid.

Electron microscopy Micrographs were taken at magnification of about 40,000 using an AEI 801 or Siemens Elmiskop 102 electron microscope. Calibration of the microscopes was performed by using tropomyosin crystals at test specimens, and periodicities evaluated by means of a computerized density scanner built in this laboratory by Dr W. Longley.

Measurements of height To measure the height of the membrane structures, stereoscopic micrographs of selected areas were taken at tilt angles of 0°, +20°, and −20°. In making measurements it was necessary to carefully select areas in which the membrane edges were not obviously artifactually turned up from the substrate. This was done by examining the negatives or three-times-enlarged prints of the stereo pairs with a stereoscope. The measurements of the selected areas were performed on the untilted micrographs of each stereo set.

Results

In considering the results, it is germane to recall a few features of these membranes as seen in intact cells by thin-sectioning methods and in isolation by negative-stain techniques. The membrane, in transverse sections, after heavy metal staining, shows regions with the typical triple-layered unit membrane pattern ∼70 to 80 Å thick, which separate extensive regions ∼0.3 to 0.5 nm wide having arrays of regularly spaced particles. Each particle is ∼110 Å wide by ∼60 to 70 Å thick with the long axis in the plane of the membrane. The particles are spaced at a period of 160 Å. In the particulate regions the total membrane thickness is increased by the thickness of the particle to ∼150 Å. When observed in frontal sections the particle areas appear as plaques ∼0.3 to 0.5 μ in diameter with the ∼110-Å-diameter particles in a hexagonal array. The particle-free areas of membrane between plaques are referred to as hinge membrane. Negative-stain and conventional freeze-fracture preparations show the same arrangement of particle-covered plaques separated by smooth hinge membrane areas.

In our special freeze-fracture-etching (FFE) preparations of these membranes we again see particulate plaques separated by smooth hinge membrane areas. However, we find that the plaques display two distinctly different appearances. The differences are quite apparent in Figures 22-1 and 22-2. The area labeled GP (globular particles) is characterized by a globular appearance of amorphous particles rising as independent units from a smooth membrane matrix. This type of pattern is like the ones previously described by Knutton and Robertson.[5] In contrast, the area labeled SP (smooth pattern) is made up of smaller interlacing subunits, with the appearance of a smooth crystalline array. This appearance is a new one attributable, we believe, to the new techniques we have employed. No sharp boundaries seem to delimit either type of pattern and no change in the direction of the main rows of particles occurs when the globular and smooth patterns coexist over the same hexagonal plaque. In both structures the particles are organized in hexagonal close packing with a center-to-center spacing of 160 Å. Optical diffraction confirms the observation that both structures share the same dimensions for the unit cell. Each structure shows, however, a distinct and characteristic distribution of the intensities of the optical diffraction spots. While the coarse globular particles give a strong first and weak second order, the smooth pattern reverses this, giving stronger spots in the second order (Figure 22-2), a situation consistently found in optical diffraction of micrographs obtained from specimens negatively stained with PTA.

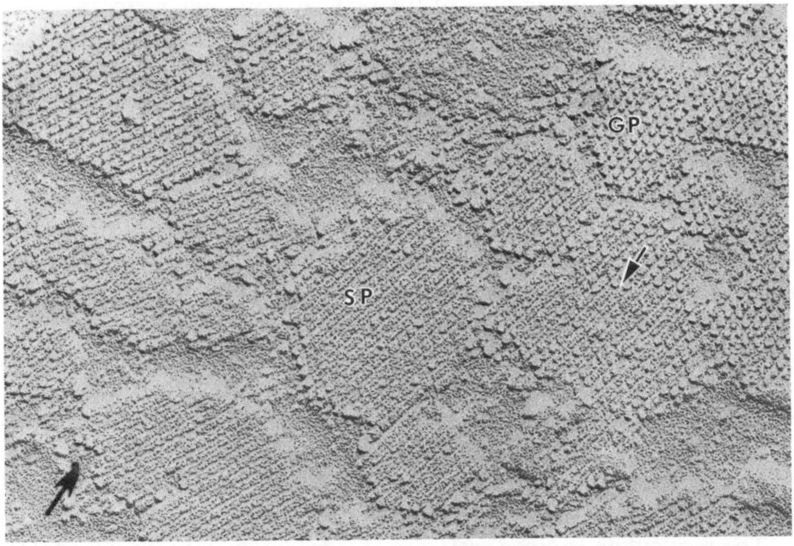

Figure 22-1 Isolated urinary bladder luminal membranes after freeze-fracture on cationized glass display two distinct appearances of the hexagonal array of particles. Smooth (SP) and globular (GP) particles occasionally coexist in the same hexagonal plaque (arrow). × 120,000. Large arrow in the lower left indicates the direction of shadowing.

Figure 22-2 Globular (GP) and smooth (SP) hexagonal patterns in the fracture plane of isolated urinary bladder luminal membranes. The insets show the distinct distribution of intensities of the optical diffraction patterns obtained from the globular (upper left corner) and from the smooth (upper right corner) appearance of the hexagonal plaques after freeze-fracture on glass. × 224,000. Large arrow in lower left indicates the direction of shadowing.

Figure 22-3 shows an area of the replica-containing membranes with some surfaces that we interpret as unfractured membranes. In analyzing Figure 22-3 we must take into consideration the fact that the isolated membranes most often occur as vesicles with the particle arrays either inside or outside the lumen. In the case of Figure 22-3 we see a vesicle in which the particle array was external and hence attached to the glass. The particulate arrays showing both the globular and smooth appearances represent the external fracture (EF) faces. In some regions designated PS (true protoplasmic surface) the unfractured protoplasmic surface of the membrane can be identified. To the upper left (arrowhead 1), this surface is roughened by irregular artifactual deposits. Because the substrate surface is flat we can measure heights with some confidence. Thus the height of the membrane here is ~150 Å, about right for the plaque membrane total thickness from thin-section studies. The thickness to the lower left (arrowhead 2) is ~ 85Å, about right for the hinge membrane as seen in thin sections. There is a transition to the lower right (arrowhead 3) in a plaque region where we see the PS surface of a plaque casting a shadow onto an EF face of a smooth lattice. The particulate EF face casts a serrated shadow onto the glass (g) surface at arrowhead 4. This serrated appearance is more clearly displayed in Figure 22-4 at arrowheads 1 and 2. Note that at arrowhead 3 in Figure 22-4 the serrations are absent. The shadows at arrowheads 1 and 2 are cast by the smooth lattice. The height of the smooth lattice at arrowhead 3 is ~ 0 Å,

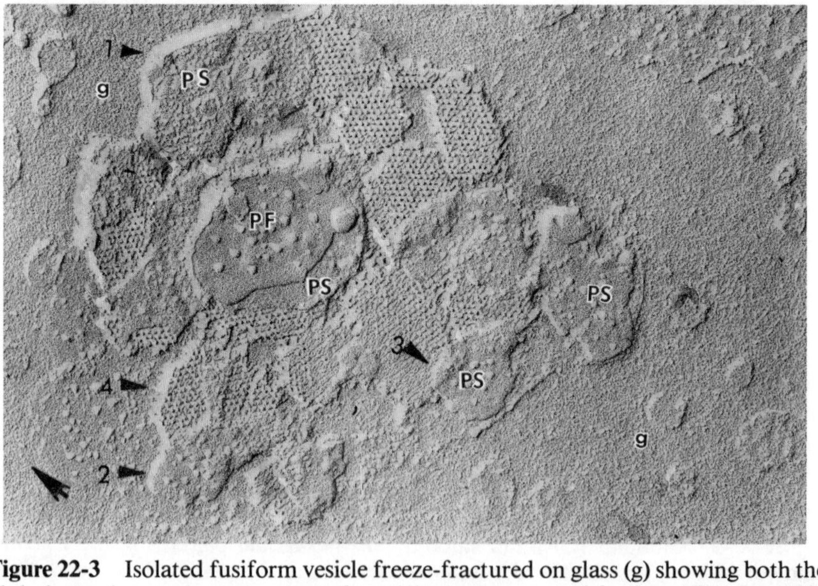

Figure 22-3 Isolated fusiform vesicle freeze-fractured on glass (g) showing both the globular and smooth appearance of particles in the external fracture (EF) face. PS: true protoplasmic surface; PF: protoplasmic fracture face. × 70,000. Large arrow in the lower left indicates the direction of shadowing.

while the serrations at arrowheads 1 and 2 in Figure 22-4 are caused by projections from a base of this height to as much as 230 Å. The projections are the shadows of the particles.

In the center of Figure 22-3 the surface labeled PS is partially covered by a membrane fragment that we interpret as the protoplasmic fracture (PF) face of a portion of the membrane on the side of the flattened vesicle; most of the other side of the vesicle, which was flattened by the centrifugal field, was fractured away. There are two small fragments that cast shadows onto this PF surface that are difficult to interpret, but may represent fragments of hinge membrane that did not fracture away. The PF surface is marred by artifactual deposits from the vacuum chamber. To the lower right (arrowhead 3) the EF surface of a hinge membrane can be made out. This is partially unfractured and extends onto a smooth lattice area where its shadow indicates a height of ∼45 Å.

We are currently collecting data for further analysis and have accumulated enough to allow us to say certain things about the two kinds of array that we see in the EF surfaces of the fractured plaques. The data support a finding from studies of stereo electron micrographs, which we shall present in a later paper. Studies of stereo micrographs show very clearly that the amorphous globular particles project above the plane of the smooth lattice areas to varying degrees, as shown quite clearly in Figure 22-4 at arrowheads 1 and 2. Indeed, in some cases they rise to heights above

Figure 22-4 Fracture plane of isolated urinary bladder luminal membranes freeze-fractured on glass. The globular particles project above the plane of the smooth lattice (arrowheads 1 and 2). × 104,000. Large arrow in the lower left indicates the direction of shadowing.

the smooth lattice area that greatly exceed the thickness of the unfractured membrane. The smooth lattice areas, on the contrary, are of a more uniform height, consistent with their representing a fracture plane proceeding along the center of the hinge membranes, and into and along the corresponding planes of the plaques. They thus represent, we believe, the true bilayer fracture surface. They display an order similar to that of the surface particles, but do not represent them directly. The globular particles are evidently plastic deformation artifacts of an unknown nature.

Our data supporting these conclusions is at present only preliminary, because there is much variability in the structures that makes precise measurements very difficult. There is, of course, an unavoidable subjective element involved in discarding certain measurements because they appear to be artifactual. We believe, however, that this can be overcome by making a very large number of measurements, far more than we now have. Nevertheless, the data that we have in hand follows a consistent pattern that supports the stereo studies. With these qualifications we present the highly preliminary figures, fully aware that the standard deviations are often so large as to make interpretation uncertain.

We have been able to measure several heights, defined as follows:

h_1 = height of fractured hinge membrane to glass (EF_{hinge} to g)
h_2 = height of fractured lattice bilayer to glass ($EF_{lattice}$ to g)
h_3 = height of smooth lattice to hinge fracture face ($EF_{lattice}$ to EF_{hinge})
h_4 = height of globular deformed lattice to glass
h_5 = height of individual deformed globules to underlying lattice
h_6 = height of deformed globular lattice to hinge EF face
h_7 = height of whole unfractured lattice membrane to glass

These values are presented in Table 22-1 with standard deviations, number of measurements, and the range. To the right are shown values to be expected if the fracture plane proceeds primarily along the center of the bilayer as defined by the clear zone in thin sections.

Discussion

Two distinct hexagonal arrays of intramembrane particles are present in freeze-fracture replicas of isolated urothelium luminal membranes. The terms "smooth" and "globular" have been used to roughly describe their appearance in the freeze-fracture replica.

We postulate that the globular pattern derives from the smooth pattern by a plastic deformation artifact. This assumption is based on the following considerations: 1) both lattices share a common center-to-center spacing of 160 Å; 2) in cases where both types of patterns coexist over the same hexagonal plaque (Figure 22-1, arrow), no sharp boundaries delimit the structures, as we would expect to find between two contiguous internal fracture planes; 3) in these cases the collinearity of the main rows of particles is maintained, as observed in individual hexagonal plaques studied by negative stain—this indicates that both structures belong to the same lattice

Table 22-1
Heights of Seven Identified Structures Found in Freeze-Fracture Replicas of Urinary Bladder Luminal Membranes

	No. Measurements	Value	SD	Range	Theoretical (Expected) Values
h_1	6	43	± 5	37–48	40
h_2	12	106	±14	92–120	110
h_3	22	85	±11	74–96	70
h_4					
Peaks	227	162	±22	96–23	
Valleys	90	120	±14	78–156	110
h_5	34	55	±14	26–104	56
h_6					
Peaks	9	123	±17	106–140	117
Valleys	8	84	±16	68–100	74
h_7	7	155	±13	142–168	150

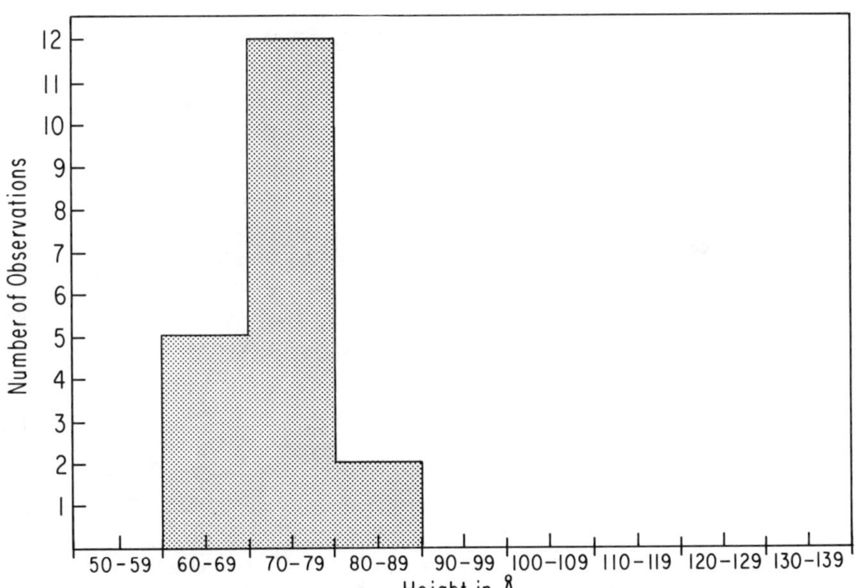

Figure 22-5 Distribution of values of heights for the smooth pattern.

system; and 4) in spite of these observations indicating that both structures are located in the same fracture plane, the dimensions of the vertical axes of the two structures are clearly different: the low relief of the smooth pattern contrasts with the prominent elevated appearance of the globular pattern. Inspection of stereoscopic pairs of micrographs make this observation more evident. Measurements of heights estimated by the lengths of the shadows cast onto the flat glass surface at the edges of both structures also confirm that the globular patterns rise higher than the smooth ones. Indeed the individual globules in the deformed lattice often rise to heights from the glass surface exceeding the total thickness of the whole membrane. The globules frequently rise about 60 Å above the level of the underlying smooth lattice. We have frequently been able to identify the complementary protoplasmic fracture faces, and they have never contained pits to correspond to these particles. This situation could be explained by plastic deformation of subunits, pulled up from the lattice during fracturing. Further evidence supporting this hypothesis is obtained from the characteristics of the optical diffraction patterns of both structures. The smooth pattern shows a distribution of intensities of the first and second orders similar to that obtained from micrographs of negatively stained specimens. This latter characteristic of the diffraction pattern is due to the six hollow fold rings that form the individual subunits of the lattice (Vergara et al[10]).

This modulation of the fine structure is lost in the pattern made up of globular particles. One-dimensional computer reconstruction shows that the profile of the component subunits of the globular pattern corresponds

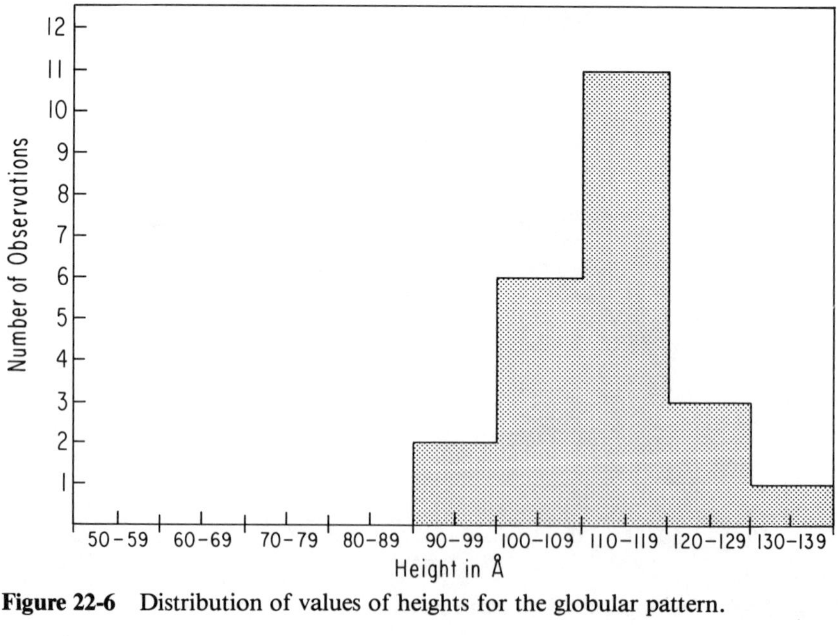

Figure 22-6 Distribution of values of heights for the globular pattern.

to a solid lump, which contrasts with the profile of the smooth pattern, in which the hollow ring structure of the individual subunits is maintained. A more detailed analysis of the fine structure is necessary to establish the mechanism of this filling-in process that occurs in the six fold rings during plastic deformation, but it is clear that a significant alteration is produced.

In conclusion, the globular pattern is an artifact resulting from mechanical stress during fracturing by pulling up and filling in the core of the individual rings of subunits. The plastic deformation that follows drastically modifies the fine structure of the hexagonal pattern, as shown by optical diffraction as well as by direct microscopy.

This observation emphasizes the fact that the size and shape of the intramembrane particles could be a misleading criterion to describe discrete areas of the membrane as morphologically distinct entities. Indeed, it suggests that any intramembrane particle should be examined and such artifacts as described here ruled out, before any interpretations are made about its nature.

The authors would like to thank Mrs Florence Stubbs and Susan Hester for excellent technical assistance. The photographic assistance of Mrs Wadeeah Beyah is gratefully appreciated.

REFERENCES

1. Southworth D, Fisher K, Branton D: Principles of freeze-fracturing and etching, in Glick D, Rosenbaum RM (eds): *Techniques of Biochemical and Biophysical Morphology,* vol 2. New York, John Wiley & Sons Inc, 1975, pp 247–282.

2. Sleytr UB, Robards AW: Plastic deformation during freeze-cleaving: a review. *J Microsc* 110:1–25, 1977.

3. Robertson JD: On the nature of intramembrane particles, in Bailey GW (ed): *Proceedings of the Electoral Microbiological Society of America. 35th Annual Meeting.* Baton Rouge, La, Claitor's Publishing Division, 1977, pp 680–683.

4. Staehelin LA, Chlapowski FJ, Booneville MA: Luminal plasma membrane of the urinary bladder. I. Three-dimensional reconstruction from freeze-etch images. *J Cell Biol* 53:73–91, 1972.

5. Knutton S, Robertson JD: Regular structures in membranes: the luminal plasma membrane of the cow urinary bladder. *J Cell Sci* 22:355–370, 1976.

6. Vergara JA, Zambrano F, Robertson JD: Isolation and characterization of luminal membranes from urinary bladder. *J Cell Biol* 61:83–94, 1974.

7. Robertson JD, Vergara JA: An analysis of the structure of intramembrane particles of the mammalian urinary bladder. *J Cell Biol* (In press)

8. Sommer JR: To cationize glass. *J Cell Biol* 75:245a, 1977.

9. Costello MJ, Corless JM: The direct measurement of temperature changes within freeze-fracture specimens during rapid quenching in liquid coolants. *J Microsc* 112:17–37, 1978.

10. Vergara JA, Longley W, Robertson JD: A hexagonal arrangement of subunits in membranes of mouse urinary bladder. *J Mol Biol* 46:593–598, 1969.

SECTION V
Membrane Transport

23 Coupling Between Epithelial Cells in Frog Skin

Hans H. Ussing, PhD
Robert Nielsen, PhD

The two-membrane theory for the origin of the frog skin potential[1,2] has been widely used as a model for epithelial ion transport. Its main elements, ie, an outward-facing sodium-selective and an inward-facing potassium-selective cell membrane, combined with a sodium pump that may be an electrogenic Na^+/K^+ exchange pump at the inner membrane, have turned out to be common features, especially in "high-resistance epithelia."[3,4]

As far as the frog skin itself is concerned, the morphologic equivalent of the model has given rise to lively discussion. Initially the active cells were thought to be those of the stratum germinativum. A careful microelectrode study[5] localized the sodium-selective membrane at the outside of the outermost living cell layer, and it was concluded that all the cell layers in the epithelium were electrically connected via resistive cell contacts. Thus the whole epithelium might function as a kind of syncytium, where all cells could participate in the pumping of cellular sodium into the pericellular spaces (which were closed toward the outside and open toward the inside). Simultaneously and independently a similar model was proposed by Farquhar and Palade.[6] The model gained credulity from the discovery by Kanno and Loewenstein[7] of coupling between gland cells and several other cell types. The coupled-epithelium model explains the electrochemical properties of the skin reasonably well. It explains why lithium, being pumped much less efficiently than sodium, piles up in all cell layers in the epithelium.[8-10] It is in agreement with the fact that most of the sodium ATPase of the epithelium is localized in the stratum spinosum (the middle layer of the epithelium,[11] and it is strongly supported by the electron microprobe analysis,[12] which shows that the sodium and potassium concentrations of the different epithelial layers vary in parallel, even under very extreme changes of the bathing solutions. (We shall return to this important paper in the discussion.) There are, however, observations that are hard to reconcile with a simple syncytium model.

Thus Zerahn,[13] studying the kinetics of the sodium uptake with ^{24}Na, concluded that the transport pool of sodium must be quite small, only a very small fraction of the total epithelial sodium. Similarly, Aceves and Erlij[14] found that the pool from which radio-sodium could be washed back into the outside bath was much smaller than the total epithelial sodium pool. Similar results were obtained in several other studies.

The finding by Voute and Ussing[15] that changes in the transport rate for sodium influenced mainly or only the outermost living cell layer has also been cited as evidence against coupling between cell layers, and has been used to support the view that only this outer cell layer is performing the transport. It should be stressed, however, that the authors of that paper themselves explicitly state that the observations are compatible with the concept of coupling between cell layers.

Still, it is surprising that excess sodium forced into the cells by short-circuiting or by ingoing currents, even higher than that necessary for short-circuiting, does not give rise to swelling of the whole epithelium,[16] but only to swelling of the outermost living cell layer. In fact, since increased active sodium transport leads to a pronounced expansion of the intercellular spaces,[17] short-circuiting seems rather to decrease the total cell volume.

Similarly, when the sodium transport is inhibited by ouabain the theory demands that sodium chloride should enter the cells, and once sodium has replaced all cellular potassium, one would expect colloid osmotic swelling. In fact, in such experiments one only observes a slight shrinkage and no swelling, even in prolonged experiments. This should be contrasted to the excessive swelling of the epithelium that takes place if sodium on the inside of the skin is replaced by potassium.

Working Hypothesis

In order for all or most of the apparently contradictory findings cited above to be correct, the simple syncytium theory has to be modified. We have been concerned with developing a possible solution to this problem for some time, and have formed the following working hypothesis: 1) The degree of coupling between the cell layers is variable, and determined mainly by the possibility for close contact between neighboring cells. Intense sodium transport will tend to expand the interspaces and thus reduce cell contacts. Conversely, inhibition of sodium transport with ouabain will lead to collapse of the interspaces and increase cell contacts. 2) The entry of sodium through the sodium-selective outward-facing membrane is regulated not only by the sodium-sensitive regulatory site at the outer end of the sodium channel,[18] but also by some regulatory site at the cytoplasmic side of the membrane, opening the channel when the cell volume increases[19] and/or when the cellular sodium concentration decreases.[20]

Experimental Evidence

In order to test the hypothesis outlined here, the epithelial volume changes have been studied during well-defined changes in the electrolyte composition of the media, with and without ouabain inhibition and with and without current flow through the system. The histologic changes have been followed on samples osmium-fixed at appropriate times.

As already mentioned, frog skin epithelia do not show the swelling predicted from the two-membrane theory when sodium pumping is inhibited by ouabain. Conceivably this lack of volume response might be due to, for instance, a secondary effect of ouabain poisoning on the chloride permeabilities of the cell membranes. If, however, sodium is replaced by potassium in the ringer solution bathing the inside of a ouabain-poisoned skin, the epithelium swells rapidly, indicating normal permeabilities to both potassium and chloride. Alternatively, the absence of swelling of the ouabain-poisoned epithelium with sodium Ringer on both sides might be due to a reduced sodium permeability of the outward-facing membrane, possibly due to an inhibitory effect of increased cellular sodium.[20]

It has been shown[19] that the sodium entry is positively correlated to epithelial volume. Therefore, a series of experiments was performed where ouabain-poisoned skins were exposed to half-strength Ringer on the inside. This should lead to some cell swelling, which in turn ought to increase the sodium permeability of the outer membrane. The prediction turned out to be correct (Figure 23-1).[21] The combined effects of ouabain poisoning and dilution to half in the inside medium leads to an excessive swelling of the epithelium when the outside medium is normal sodium-chloride Ringer.

If, on the other hand, the outside medium is sodium-free potassium-chloride Ringer, the swelling is only that expected from the change in osmolarity of the inside medium. Thus the excessive swelling in the first experiments must be due to the entry of sodium from the outside bathing solution. The epithelium indeed swelled to such an extent (often by more than 50μ) that several cell layers must have contributed to the swelling. This conclusion was verified histologically. Thus, the experiments indicated that coupling between cell layers is possible.

The light-microscopic appearance of skins that had undergone the excessive swelling in the first type of experiment indicated that the interspaces were completely compressed and that all cell layers showed pronounced swelling. Even electron-microscopically the interspaces were toally collapsed (Figure 23-2), so that the cell-cell contact was maximal. In agreement with our working hypothesis this should give maximal coupling.

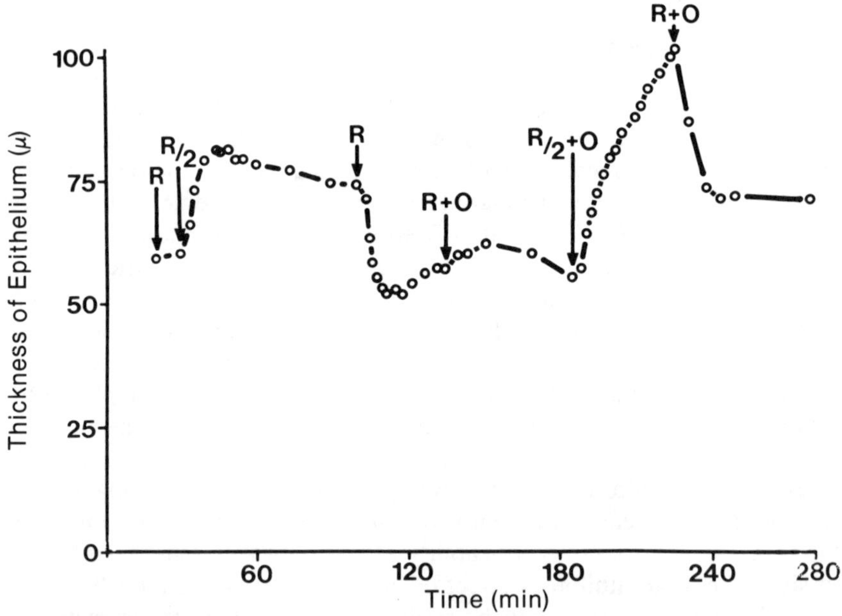

Figure 23-1 Volume changes of epithelium of skin of *Rana temporaria* in response to changes in inside osmolarity and to ouabain poisoning. Outside medium: NaCl Ringer throughout. Inside medium: initially also NaCl Ringer (first arrow, marked R). Half-strength Ringer was applied (second arrow, R/2). At third arrow (R), NaCl Ringer was again applied. At arrow marked R + O ouabain was added to the inside solution to give a 10^{-5} M solution. At arrow marked R/2 + O, half-Ringer, made 10^{-5} M with respect to ouabain, was introduced, and finally at the last arrow Ringer with ouabain was applied. Procedure as described by MacRobbie and Ussing.[16] From Ussing.[21] Reprinted with permission.

According to Zeiske and Lindemann,[22] the inhibitory effect on sodium entry exhibited by outside sodium can be counteracted by certain drugs, eg, benzimidazole guanidine (BIG). Therefore, we tested if the combination of ouabain on the inside (to stop the sodium pump) and BIG on the outside (to increase sodium entry) would lead to swelling of the epithelium. In itself this combination of drugs did not give rise to swelling. If, however, sodium was now pushed into the epithelium from outside by a clamping current two times the normal short-circuit current (about 30 μamp/cm_2), excessive swelling ensued, involving all cell layers of the epithelium (Figure 23-3). It should be stressed that the swelling depends on the combination of ouabain and BIG treatment with an ingoing electric current, carried by sodium ions. Also it should be noted that this experiment can be done with skins from *Rana esculenta* but not with skins from *R. temporaria,* which are known to be insensitive to BIG.

Figure 23-4 shows that the swelling induced by the combined effects of ouabain, BIG, and an ingoing sodium current did indeed lead to swelling of all cell layers associated with a collapse of the interspaces.

The collapse of the interspaces in these experiments is mainly due to

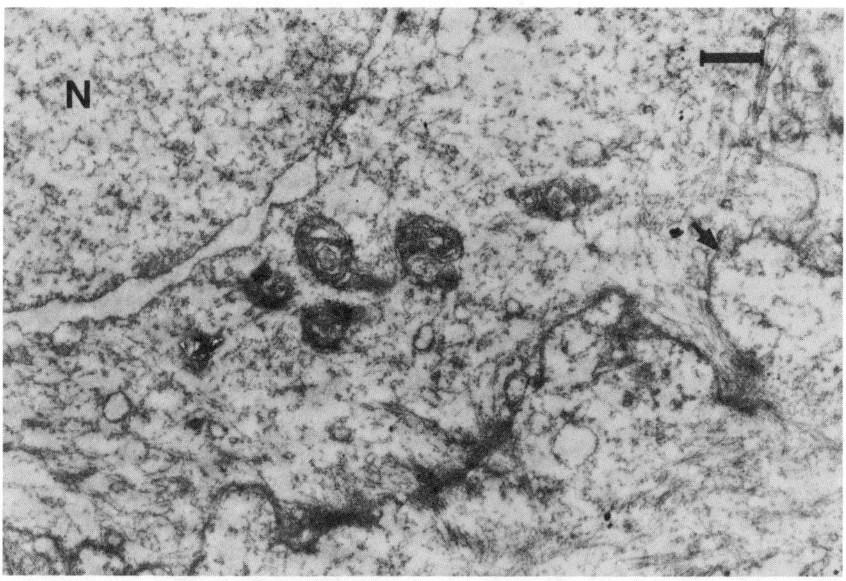

Figure 23-2 Electron micrograph showing excessive swelling of epithelium cell (stratum spinosum) from frog skin, where sodium pumping was inhibited by ouabain while sodium entry from outside was stimulated by dilution of inside solution *(R temporaria)*. Outside solution: NaCl Ringer. Inside solution: half NaCl Ringer made 5×10^{-4} M with respect to ouabain. N: nucleus. Arrow shows collapsed interspace. Length of bar = 4000 Å. Fixed 60 minutes after application of half-Ringer to inside (see Figure 23-1). Fixation and general histologic procedure in this and the following experiments as described by Voûte and Ussing.[15]

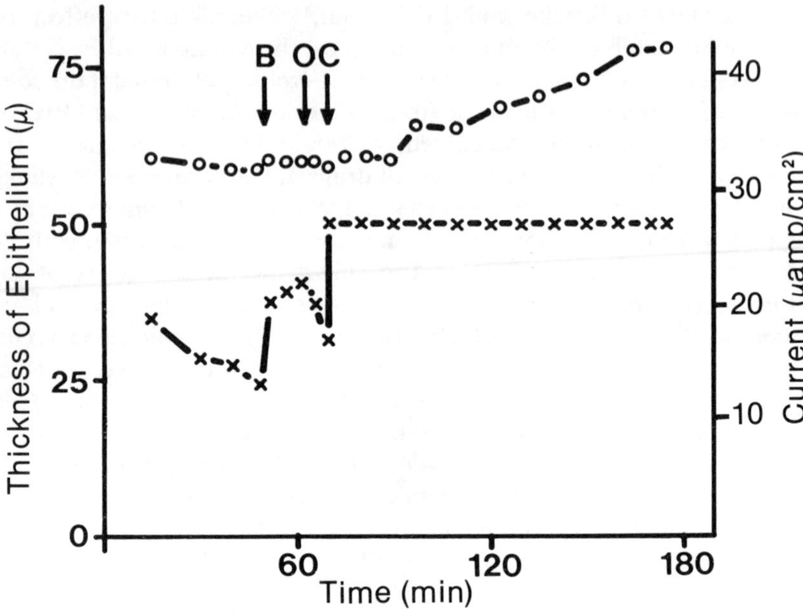

Figure 23-3 Volume changes (upper curve) of frog skin epithelium *(R esculenta)* (method of MacRobbie and Ussing[16]) exposed to the combined effects of BIG (1 mM in outside solution), ouabain (5×10^{-4} M in inside solution), and an applied ingoing electric current (lower curve). BIG was added at B, ouabain at O, and ingoing current applied at C. Both bathing solutions: NaCl Ringer.

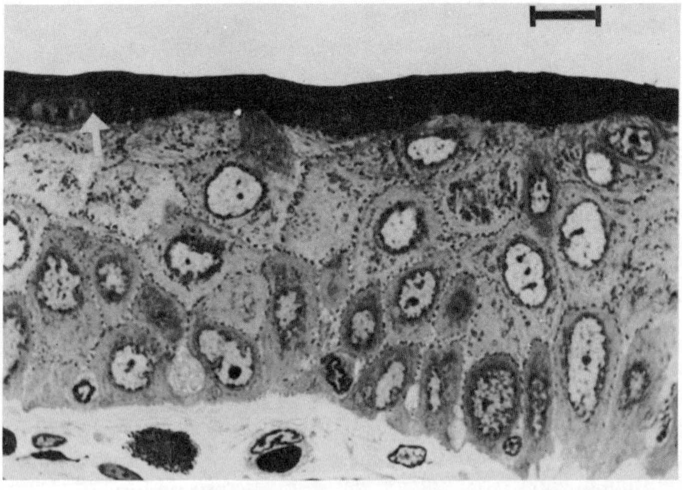

Figure 23-4 Excessive swelling of epithelium *(R esculenta)* when sodium pumping was stopped by ouabain, while sodium entry from outside was stimulated by BIG in combination with an ingoing electric current. Same epithelium as in Figure 23-3, fixed at the end of the experiment. Arrow shows outermost living cell layer. Length of bar = 10 μ.

the absence of active sodium transport from cells to interspaces. However, increased intercellular pressure and expansion of the interspaces can also be achieved by increasing the hydrostatic pressure head on the inside of the skin. This, according to Voute and Ussing,[17] gives rise to an intraepithelial edema, even with pressures as low as 10 cm H_2O.

If our working hypothesis is correct the separation of the cells caused by such a pressure head should decrease the coupling between cell layers by breaking the gap junctions. Figure 23-5 shows the result of such an experiment. The pressure of the inside bathing solution was raised 15 cm H_2O over that of the outside solution. The sodium transport was stopped with ouabain added to the inside solution, and sodium entry increased by BIG combined with an ingoing current. It will be seen that the swelling is limited to the outermost living cell layer, whereas the other layers look normal.

The fact that the epithelium during the combined effects of ouabain, BIG, and ingoing current swells less when the interspaces are expanded than when they are not seems clear just from inspection of the figures. One can get a quantitative estimate of the difference by deducting the area of the interspaces from the total area per unit length of the sections. This yields a ratio of pressure-treated to non-pressure-treated skin of 0.32. In order to estimate the stretch of the epithelium brought about by the pressure head, the number of stratum germanativum cells per unit length of the section has been determined. The stretch turned out to be on the order of 5% on a linear basis, which corresponds to some 10% calculated as increases in epithelial area. Thus, the true ratio between the volumes with and without pressure is about 0.22.

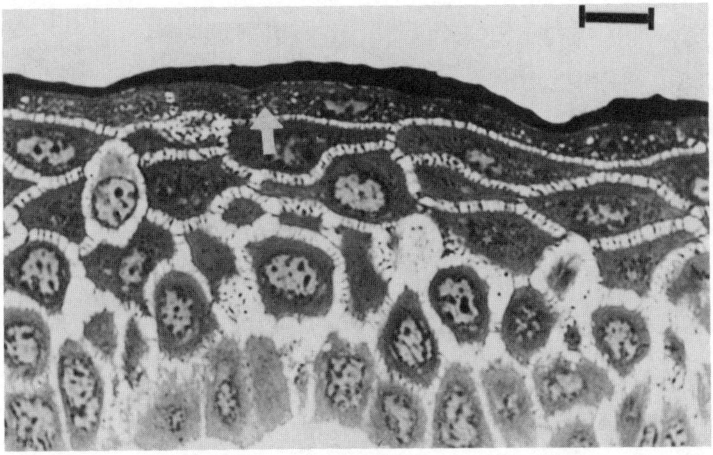

Figure 23-5 Frog skin epithelium *(R esculenta)* did not undergo excessive cellular swelling under the conditions of the experiment shown in Figure 23-4, if the interspaces were expanded by application of a hydrostatic pressure head of 15 cm H_2O in the inside bathing solution. Only the cells of the outermost living cell layer, indicated by arrow, were swollen. Length of bar = 10 μ.

In the types of experiments so far discussed the close contact between cells was obtained by inhibition of the sodium pump, and thus by a drop of the interepithelial pressure. One might think that good cell-cell contact could also be obtained by swelling the epithelial cells. One way of doing this is to replace sodium chloride in the inside bathing solution with potassium chloride. This did indeed lead to swelling of all cell layers (except the dead cells of the cornified layer).[16,23] As seen in Figure 23-6 the swelling was particularly pronounced in the first living cell layer. These cells evidently were taking up sodium ions from the outside medium and potassium from the inside. Most interspaces seem compressed. If a similar experiment was performed with potassium chloride Ringer on both sides, the picture was slightly different. Again there was excessive swelling of the epithelium, but here it will be noticed that the outermost living cell layer did not take part in the swelling (Figure 23-7). In other words, these cells could not be tightly coupled with the deeper cell layers that were swelling uniformly. One might argue that the KCl treatment might have induced a change in these cells, so that they lost their characteristic properties. This is not the case, however. If the potassium chloride of the outside medium was replaced by ordinary sodium chloride Ringer, the cells of the first living cell layer swelled (Figure 23-8), indicating that their outward-facing membrane is still the sodium-selective barrier of the epithelium. Thus the high potassium/low sodium medium on the inside of the skin did not lead to increased coupling between the cells. Indeed the outermost living cell layer seemed to be uncoupled from the deeper layers, even when the interspaces were collapsed.

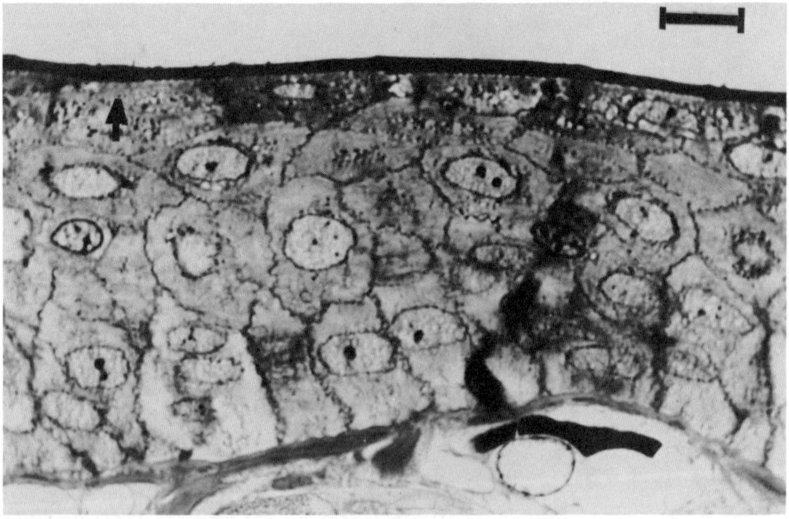

Figure 23-6 Excessive swelling of frog skin epithelium *(R temporaria)* when skin was exposed to NaCl Ringer on the outside and KCl Ringer on the inside for 100 minutes. Arrow shows outermost living cell layer, which is highly swollen. Length of bar = 10 μ.

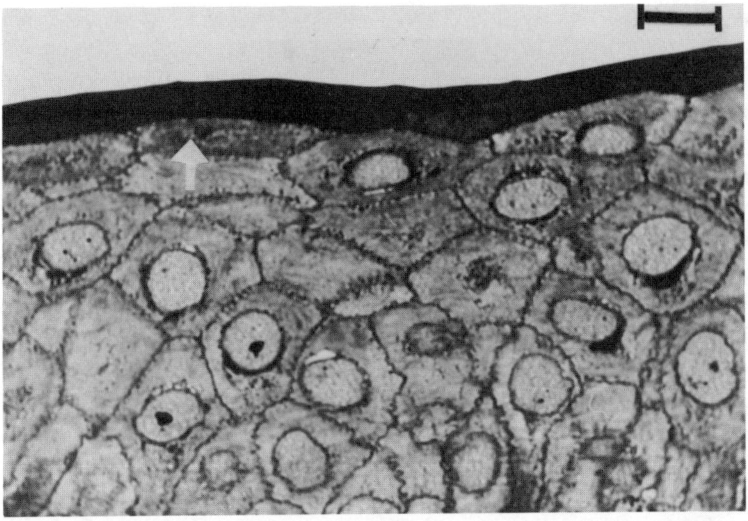

Figure 23-7 Excessive swelling of frog skin epithelium *(R temporaria)* when the skin was exposed to KCl Ringer on both sides for 100 minutes. Arrow shows outermost living cell layer, which is totally collapsed. Length of bar = 10 μ.

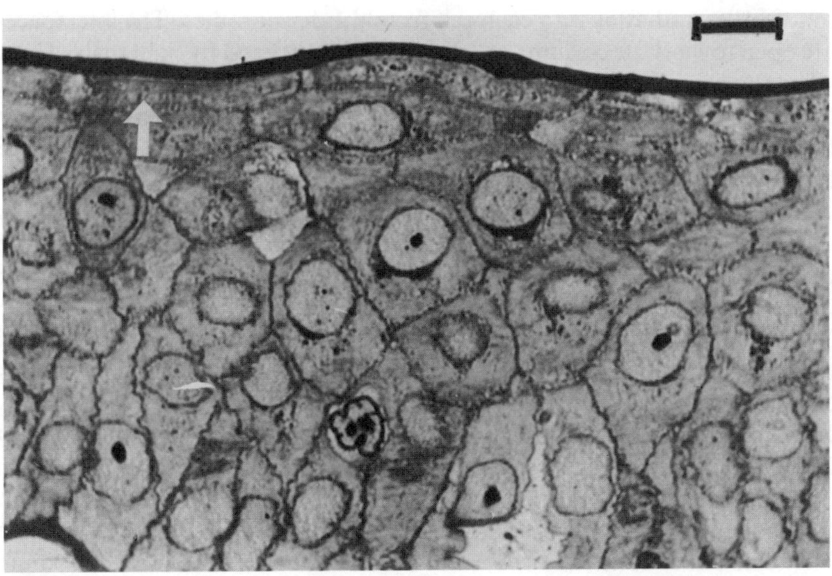

Figure 23-8 Excessive swelling of frog skin epithelium *(R temporaria),* which was first exposed to KCl Ringer on both sides for 80 minutes and then NaCl Ringer on the outside (but still KCl Ringer on the inside) for 30 minutes. Arrow shows outermost living cell layer that is now swollen, indicating that the outward-facing cell membrane is still sodium-selective. Length of bar = 10 μ.

Discussion

The experiments presented here indicate that the cell-cell coupling between adjacent cell layers in the frog skin epithelium may vary, depending on the functional state of the tissue. Thus one may say that both the proponents of the view that most of the sodium transport is performed by the outermost laying cell layer, and those in favor of the view that the epithelium acts like a syncytium, may be right.

Our hypothesis at first sight seems to contradict the conclusion by Rick et al[12] that the epithelium under all situations studied behaves like a functional syncytium, as far as sodium and potassium concentrations are concerned. Rick et al, however, studied only situations where the rate of active sodium transport was small (ouabain poisoning, amiloride inhibitions, low outside sodium concentrations), whereas the present study also includes situations where the interspaces are expanded either by massive active sodium transport or by hydrostatic pressure. Here it may be appropriate to discuss the question: Why does a hydrostatic pressure head on the inside of the skin lead to expansion of the interspaces?

To understand this one must consider the physical properties of the different tissues in the skin. The corium is mechanically very strong, but provides hardly any resistance to filtration of water. The epithelium has a resilient sponge-like structure, being composed of cells held together by tonofibrils irradiating into each cell from the desmosomes. The interspaces are open toward the corium, but closed at the outer end by tight seals. Thus, when the pressure of the inside bathing solution is increased, fluid will filter through the corium, through small holes in the basement membrane, into the epithelial interspaces, which will expand until the elastic tension of the tonofibrils balancs the hydrostatic pressure. In short-term experiments the expansion is fully reversible when the pressure is released. Histologically one cannot tell the difference between a skin whose interspaces have been expanded as in Figure 23-4, and a control specimen that has never been exposed to hydrostatic expansion.

The working hypothesis advanced in the beginning of this chapter, ie, that the hydrostatic pressure in the interspaces (which is normally a function of the rate of active sodium transport) regulates the degree of coupling, can explain most of our observations. The uncoupling of the outermost living cell layer in experiments with potassium Ringer on both sides of the skin does not quite fit into the pattern. Possibly the complete absence of sodium combined with a depolarization of the cells make them unable to regulate their calcium concentration. Several observations indicate that the sodium gradient is in some way involved in the expulsion of calcium from the cells (it is known that cells uncouple when the cytoplasmic concentration increases above a critical value).

If these considerations are correct we must assume that the ouabain-inhibited skin still retains the capacity to regulate cellular calcium, since it seems to couple excellently. This coupling obviously is favored by the close contact between the cells, which again is due to the complete collapse of the interspaces in the absence of sodium pumping (see Figure 23-2). An interesting implication of this view is that the interspaces in the epithelium do not contain significant amounts of intercellular material (mucopolysaccharides, protein, etc), which would impede cell-cell contact.

Finally, it should be mentioned that gap junctions have been observed between the cells of the frog skin epithelium.[6] The present study indicates that the number of active gap junctions may vary tremendously depending on the conditions of the experiment, so that future search for gap junctions and other types of conducting junctions should be correlated to the functional state at the moment of fixing or freezing the specimen.

REFERENCES

1. Ussing HH, Koefoed-Johnsen V: Nature of the frog skin potential, abstracted, in *Commerce of the 20th International Physiological Congress.* Brussels, 1956.
2. Koefoed-Johnsen V, Ussing HH: The nature of the frog skin potential. *Acta Physiol Scand* 42:298–308, 1958.
3. Ussing HH, Erlij D, Lassen U: Transport pathways in biological membranes. *Annu Rev Physiol* 36:17–49, 1974.
4. Ussing HH, Leaf A: Transport across multimembrane systems, in Giebisch G, Tosteson DC, Ussing HH (eds): *Membrane Transport in Biology,* vol 3. Berlin, Springer Verlag, 1978, pp 1–26.
5. Ussing HH, Windhager EE: Nature of shunt path and active sodium transport path through frog skin epithelium. *Acta Physiol Scand* 61:484–504, 1964.
6. Farquhar MG, Palade GE: Functional organization of amphibian skin. *Proc Natl Acad Sci USA* 51:569–577, 1964.
7. Kanno Y, Loewenstein WR: Intercellular diffusion. *Science* 143:959, 1964.
8. Hansen H, Zerahn K: Concentration of lithium, sodium and potassium in epithelial cells of the isolated frog skin during active transport of lithium. *Acta Physiol Scand* 60:189–196, 1964.
9. Morel F, Leblanc G: Kinetics of sodium and lithium accumulation in isolated frog skin epithelium, in Ussing HH, Thorn NA (eds): *Transport Mechanisms in Epithelia. Alfred Benzon Symposium V.* Copenhagen, Munksgaard, International Booksellers & Publishers Ltd, 1973, pp 73–85.
10. Zerahn K: Studies on the active transport of lithium in the isolated frog skin. *Acta Physiol Scand* 33:347–358, 1955.
11. Mills JW, Ernst SA, DiBona DR: Localization of Na^+-pump sites in frog skin. *J Cell Biol* 73:88–110, 1977.
12. Rick R, Dörge A, von Arnim E, et al: Electron microprobe analysis of frog skin epithelium: evidence for a syncytical sodium transport compartment. *J Membr Biol* 39:313–331, 1978.
13. Zerahn K: Nature and localization of the sodium pool during active transport in the isolated frog skin. *Acta Physiol Scand* 77:272–281, 1969.

14. Aceves J, Erlij D: Sodium transport across the isolated epithelium of the frog skin. *J Physiol* 212:195–210, 1971.

15. Voute CL, Ussing HH: Some morphological aspects of active sodium transport. The epithelium of the frog skin. *J Cell Biol* 36:625–638, 1968.

16. MacRobbie EAC, Ussing HH: Osmotic behaviour of the epithelial cells of frog skin. *Acta Physiol Scand* 53:348–365, 1961.

17. Voute CL, Ussing HH: The morphological aspects of shunt-path in the epithelium of the frog skin *(Rana Temporaria). Exp Cell Res* 61:133–140, 1970.

18. Fuchs W, Hviid Larsen E, Lindemann B: Current-voltage curve of sodium channels and concentration dependence of sodium permeability in frog skin. *J Physiol* 267:137–166, 1977.

19. Ussing HH: Relationship between osmotic reactions and active sodium transport in the frog skin epithelium. *Acta Physiol Scand* 63:141–155, 1965.

20. Erlij D, Smith MW: Sodium uptake by frog skin and its modification by inhibitors of transepithelial sodium transport. *J Physiol* 228:221–239, 1973.

21. Ussing HH: Physiology of transport regulation. *J Membr Biol* 40S:5–14, 1978.

22. Zeiske W, Lindemann B: Chemical stimulation of Na^+ current through the outer surface of frog skin epithelium. *Biochim Biophys Acta* 352:323, 1974.

23. Ussing HH, Biber T, Bricker NS: Exposure of the isolated frog skin to high potassium concentrations at the internal surface. II. Changes in epithelial cell volume, resistance, and response to antidiuretic hormone. *J Gen Physiol* 48:425–433, 1965.

24 On Water Transport and Amphibian Epithelia

Rui C. de Sousa, MD
Alda Grosso, DSc

The interplay between biologic membranes and the environment is particularly well illustrated by amphibian epithelia. Among these, the skin and the urinary bladder of anurans have been extensively investigated and are generally considered as archetypal examples of osmoregulatory epithelia. Such membranes also play important roles in other mechanisms that preserve the constancy of the *milieu intérieur,* namely those implicated in acid-base equilibrium and in volume regulation. For conciseness, however, we shall restrict ourselves to the discussion of the osmoregulatory aspects of their function.

Epithelia and Their Environment

The major challenges faced by animal organisms in the course of their evolution were admirably summarized by M.B. Strauss in two verses written in 1951 and quoted by Smith.[1] The challenges went from "profligacy of

This work was supported by Grant No. 3.043-0.76 from the Swiss National Foundation.

salt and water" in the marine habitat, to overabundance of water and scarcity of salt in freshwater life, and culminated in the need to save both salt and water in the terrestrial habitat. Different phyletic groups of amphibians adapted to one or more of these three forms of life. In each instance, the permeability of the skin of these animals to salt and water varied according to the osmoregulatory stress(es) imposed on them.[2,3] Let us review some typical examples.

The epidermis of the freshwater species *Xenopus laevis* has a very low permeability to water, a *sine qua non* condition to avoid lethal hemodilution. On the other hand, the same skin transports NaCl actively, which will increase the input of salt in very adverse conditions to equilibrate the body budget of sodium. In terrestrial toads (eg, the tropical toad *Bufo marinus*), the ventral epidermis also transports NaCl but has in addition a much higher permeability to water.

A particularly interesting example of adaptation of epithelia is seen in euryhaline amphibians. Such animals tolerate a wide variety of external media, ranging from freshwater to media of high salinity. They include two anurans, *B viridis* and *Rana cancrivora*[4,5] and the urodele *Batrachoseps*.[6] Without going into details concerning humoral and metabolic aspects of the adaptation to hyperosmotic media,[2,3,7] mention should be made here to the reduction in sodium permeability of the skin when the animals are exposed to hypersaline environments. This phenomenon is particularly evident in *B viridis* and represents an effective protection against a massive input of sodium.[4] It has also been reported that increases in intracellular sodium concentration decrease the number of sodium entry sites at the external barrier of epithelia, as evidenced by the lesser number of receptors that bind amiloride, a well-known blocker of sodium channels in tight epithelia.[8]

It is thus apparent that the apical membrane of epithelia reacts to environmental conditions in order to become an effective barrier against chemical species threatening the economy of the organism. This is done, however, without losing the capacity to transport other species. The examples discussed above were taken from amphibians, but similar adjustments have also been found in other animal groups submitted to comparable osmotic stresses.[2] Obviously, the permeability changes are not restricted to the integument but involve different epithelia in gills, intestine, and other organs.[2,3] Collectively, they represent a challenging problem of cell and tissue physiology, and must implicate subtle changes in the structure and function of the plasma membrane. We still ignore most of the cellular mechanisms underlying such permeability changes. In this regard, a particular effort should be made to distinguish between changes resulting from local regulatory processes of the epithelium itself, and changes resulting from the overall homeostatic reaction of the whole organism.

Osmoregulation and Long-Range Cell Communication

In the previous section we analyzed very briefly some changes in membrane function that can be observed in vitro in the resting or nonstimulated state of a transporting epithelium. We shall consider now a second type of osmoregulatory phenomena in epithelia: the modulation of membrane permeability by humoral agents. This is part of a major theme of physiology concerning biologic regulation and long-range cell communication.[9] Changes in water permeability induced by neurohypophyseal hormones are a good example of this kind of phenomena. Such a topic leads us to the consideration of another interesting epithelium—the anuran urinary bladder.

It is well known that the amphibian kidney can not excrete urine with an osmolality higher than that of plasma, since its nephrons lack a critical segment, the loop of Henle, an observation already made by Crane in 1927.[10] This serious drawback, from the viewpoint of osmoregulation, is circumvented to a large extent by the presence of a urinary bladder, functioning as a large and dependable reservoir of water. In amphibians, the major input of water goes through the skin since, in normal conditions, these animals apparently do not drink. Part of the water reabsorbed through the integument is stored inside a urinary bladder that, in some species, can hold an amount of fluid representing up to 50% of the body weight. Two contradictory demands are imposed on the epithelium lining the bladder: 1) that of being almost water-impermeable, in order to function as a reservoir of water; and 2) that of being highly water-permeable, whenever fluid must be quickly released in the extracellular space to maintain homeostasis.[11] The transition from the first to the second state is triggered by neurohypophyseal hormones.

The modulation of membrane permeability in the anuran urinary bladder is paradigmatic of similar changes taking place at the apical membrane of the mammalian collecting duct. This segment of the urinary tubule functions as an osmotic exchanger and plays a critical role in the excretion of both a diluted and a concentrated urine in mammals. The portion of the collecting duct traversing the kidney medulla has a unique extracellular environment in the whole body: the stratified, cortico-papillary, osmotic gradient, generated by the countercurrent system of the loop of Henle. To excrete a diluted urine, the tubule must have, like the anuran bladder in the resting state, a very low basal water permeability. To excrete a concentrated urine, advantage must be taken of the peculiar surroundings of the tubule, by allowing osmotic equilibration between the tubular fluid and the hypertonic interstitium. This is achieved by reabsorption of water following permeabilization of the apical membrane by vasopressin.

The amphibian skin would be a rather ineffective osmoregulatory organ if, for a given species, its water permeability were to just fluctuate around a relatively fixed value. In contradistinction this epithelium, despite having a higher basal permeability than that of the amphibian bladder or that of the mammalian collecting duct, also markedly increases its conductance to water under the effect of neurohypophyseal hormones. Such is the basis of the so-called Brunn reaction of amphibians.[12] As discussed later in this chapter, water flows across the pelvic skin of *B marinus* may reach values as high as those recorded in the bladder.

Osmosis: A Historical Sketch

In a recent note to a famous medical journal, a leading mathematician emphasized how little mathematics, and even physics and chemistry, influenced the development of medicine and biology during most of their history.[13] Fortunately, this situation changed radically in the last century and a half. If we were to look for an early example of interdisciplinarity among these sciences, osmosis would be, we believe, one of the best. From the very beginning of this concept one finds a physicist, J.A. Nollet, studying a bladder separating two chambers filled with water and wine, respectively, while a botanist, Pfeffer, develops a shrewd procedure to manufacture a synthetic semipermeable membrane for measuring osmotic pressure.[13,14]

Osmosis has been an important topic in the study of amphibian epithelia and has had the virtue of polarizing the attention of both the basic scientist and the clinician on some fundamental aspects of cell physiology. References to early work with frog skin usually go back to the observation of a skin electrical potential by Du Bois-Reymond in 1848.[16] Three years earlier however, Matteucci and Cima[17] were already using frog skin to study osmotic flow and found an easier transfer of fluid from the inside outwards. Almost 50 years later, Reid[18] repeated some of the experiments performed by Matteucci and Cima and found, in contrast, a facilitation of water transfer in the opposite direction. Leaving aside the reasons for this discrepancy of results, the important point to make here is that Reid considered this asymmetry of fluxes to be caused by "an absorptive force dependent on protoplasmic activity."[18]

In 1913, Maxwell,[19] from the University of California, challenged the conclusions of Reid concerning the presence of water flows against an osmotic gradient. After a series of careful experiments, Maxwell stated very clearly that "the assumption of a vital activity on the part of the frog skin in transferring water through itself is shown to be unnecessary."[19]

By 1921, a connection was made between the phenomenon of osmosis and endocrinology, or, more precisely, an incipient neuroendocrinology. Fritz Brunn, from the University of Prague, made the critical observation that a "water-balance principle" from the neurohypophysis induced a marked weight gain in frogs immersed in water.[12] That the skin was the site of action of such "principle" was deduced from experimental evidence derived from skinless frogs.[20] Twenty years later, Sawyer[21] provided indisputable evidence for a direct action of pituitrin on frog skin by using an in vitro preparation. Similar results were reported by Fuhrman and Ussing.[22]

Meanwhile, the fruitful collaboration between the Institute of Theoretical Physics and the Laboratory of Zoophysiology in Copenhagen was working to unravel a major problem of water transport in biologic membranes: the discrepancy between permeability measurements made with isotope tracers and with a hydrosmotic gradient.[23] Von Hevesy and coworkers expressed their results by using rather curious units abbreviated as DN, "meaning the number of days necessary for 1 mole to pass through 1 cm^2 by constant concentration difference of 1 mole."[23] Differences of almost one order of magnitude were found between D_2O and ordinary water. This problem has been re-examined from the viewpoint of non-equilibrium thermodynamics.[24,25] Physical distinguishability of isotope species and kinetic considerations based on Onsager relations led to the notion of interaction of isotope flows in some membranes.[25] The discrepancy noticed by von Hevesy and associates[23] was confirmed by others, and is compatible with single-file diffusion in narrow pores. Such a model is consistent with current views on water transport and on the microstructure of the plasma membrane.[11,26-28] Interestingly, von Hevesy et al[23] put forward an explanation based on the effects of a complex unstirred layer situated between the surface of the skin and the stratum germinativum, which they assumed to be the quasi-semipermeable membrane of frog skin. Although this morpho-functional correlation is wrong according to our present knowledge, the problems introduced by Nernst layers are a real issue and constitute an artifact pervading most studies on water transport.[29] Use of isotopes led Koefoed-Johnsen and Ussing[30] to propose the famous pore theory to account for the effects of neurohypophyseal hormones on anuran skin. We shall come back to this model later.

Overall, the study of osmosis in amphibian skin led to the characterization of a purely passive transport process across a biologic membrane, without implying a necessary coupling to solute flow, as evidenced by the osmotic flows obtained when distilled water was the external medium. This does not exclude, however, a small salt-linked flow in frog skin.[31] Besides, nonlinearities have been reported in the osmotic force-flow relationship.[31,32] They are not yet fully understood, but in any case they do not challenge the passive nature of water flow in epithelia.

In this historical sketch we have focused our attention on water transport across amphibian skin. However, following the work of Hays[29] and Leaf,[33] the toad urinary bladder became the favorite experimental model for that purpose. In our laboratory we have been systematically comparing the results obtained with the skin and the bladder in a variety of experimental conditions, in an attempt to get some insight into the general features of water transport in tight epithelia. This became possible by virtue of a technical improvement, as discussed in the next section.

Measurement of Water Flow

Accurate measurements of volume flow across biologic membranes are not easy to obtain. With amphibian epithelia two major approaches have been used: 1) the bag or sac method, and 2) the chamber or diaphragm method.

The bag technique of Bentley[34] is particularly useful for working with the urinary bladder, but very difficult to adapt to the amphibian skin. It is a gravimetric method based on iterative weighings of the bladder mounted as a bag at regular time intervals (usually 30 to 60 minutes), following a procedure designed to minimize tissue damage, evaporative losses, and contamination with residual fluid on the bladder surface.[35] Despite its apparent crudity, the technique provides sensitive and reproducible results in experienced hands, and has the advantage of keeping the bladder in a physiologic conformation.[35,36] Its obvious drawbacks are the changing geometry of the membrane surface, with poor definition, and inadequate time resolution for kinetic studies.

The use of membranes as diaphragms is applicable to both the bladder and the skin. Many variants of this method have been used in the past with relevance for the setups described by Edelman et al[37] and by Lichtenstein and Leaf.[38] However, the major breakthrough in this particular field was the automatic technique described by ourguet and Jard.[39] Theirs is a volumetric method, in which the loss of fluid from the external compartment is compensated by an automatic injection of liquid with an electrically driven syringe.

Over the last few years we developed in our laboratory automatic techniques of the volumetric type for measuring water flow. The membrane separates two hemichambers, one of which is prolonged by a horizontal pipette. The basic principle of the method is the continuous tracking of the meniscus, by means of a carriage moving alongside the pipette. Either an optical or a capacitance signal has been used to trigger the motor driving the carriage.[40–42] The original optical version of the method is semiautomated.[40] Attached to the carriage there is a light source, a photosensitive element, and a pen. Interception of the light by the meniscus sharply reduces the illumination of the photoresistance, which triggers the motor and drives the carriage until the initial conditions of illumination are reestablished. The pen attached to the carriage simply and directly records its displacements on a moving chart, thus giving a continuous recording of the water flow, as shown in the upper part of Figure 24-1. This method, however, has the disadvantage of requiring the calculation of the actual water fluxes from the slopes of the record, which is tedious and often inaccurate. We achieved full automation of this method by adding an electronic device that processes the information and calculates average water fluxes (in μl/min/cm²) during a preset time interval. This interval is chosen according to the experimental conditions and usually is one minute. Recently we obtained excellent results by integrating the flux over periods as short as 12 seconds.

The basic features of the optical method are also found in the variant using a capacitance signal, generated by the position of the meniscus between two metal pieces attached to the carrier and surrounding the pipette.[41] Figure 24-2 shows a record obtained with this fully automated method.

Very accurate techniques for measuring water flow have been reported quite recently.[43,44] One of them, also based on a capacitance signal, appears to be exceedingly interesting for continuous recording of flow rates in renal collecting tubules.[44]

Before ending this section, we would like to mention the ingenious method described by Reid.[18] This author used a classic form of the osmometer of Dutrochet, and a second instrument that he named "oil

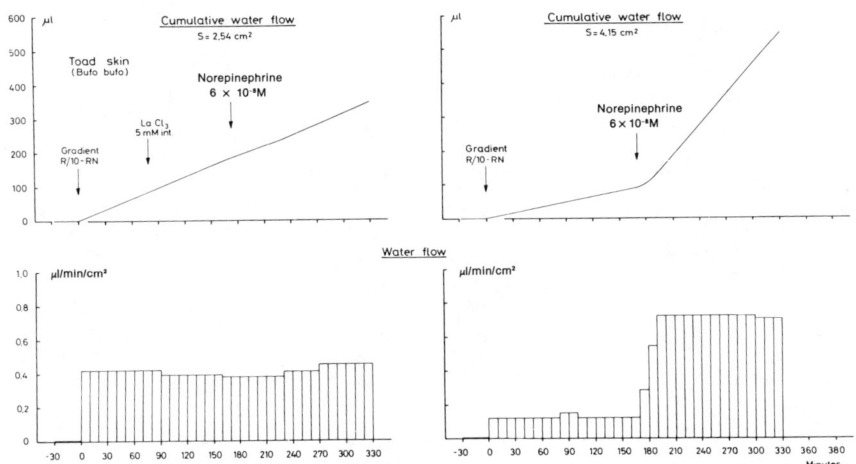

Figure 24-1 Typical recording obtained with the semi-automated optical method described in reference 38. The fluxes shown in the lower part of the figure were calculated from the slopes above. Two segments of the same toad skin *(B bufo)* were used. Note the clear hydrosmotic effect induced by norepinephrine and its total inhibition by lanthanum.[22] RN = normal Ringer; R/10 = normal Ringer diluted tenfold; s = membrane surface area.

discharging osmometer." With this equipment, Reid obtained a continuous record of water flow by means of an optical system. The following is his description[18]: "any rise in fluid in the recording tubes shows itself on development of a photographic paper by means of a white line, produced by the interception of the light of a paraffine lamp or Argand gas burner by the meniscus at the surface of the fluid." And he adds a little further, "I have to thank Prof. Schäfer for pointing out to me that the meniscus was sufficient to intercept the light." We find the venerable tracings on Plate IX of Reid's article an astounding match to the recordings we obtained 82 years later.

New Data on Water Transport

The technique described in the previous section allowed for the collection of a considerable amount of data on water flow, both in the skin and the bladder of amphibians. Since the skin in particular has received little attention over the years, we shall review here some recent data obtained with the epithelium in our laboratory.

The skin has at least two receptors that trigger a hydrosmotic response: 1) the receptors to neurohypophyseal hormones; and 2) the receptors to

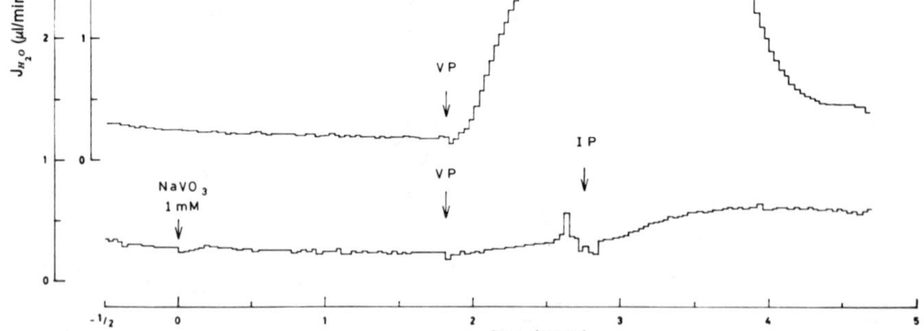

Figure 24-2 Hydrosmotic flow in toad skin *(B marinus)*. Block by metavanadate (Na VO_3) internal. Typical recording obtained with the fully automated method, in which the tracking of the meniscus inside the pipette is triggered by a capacitance signal. Time interval for computation of average water flow (J_{H_2O}) is two minutes. Note that vanadate strongly inhibits the combined stimulation of J_{H_2O} induced by vasopressin and isoproterenol, but does not affect basal J_{H_2O}.

catecholamines of the β-type. Both of them are in series with membrane adenylate cyclase, but we still do not know whether they converge on the same enzyme unit or if they are linked to spatially distinct enzyme molecules.[45] Recent work with methohexital and propranolol revealed that these drugs selectively inhibit the responses to vasopressin and to isoproterenol.[45] since the inhibition appears to occur prior to cyclic AMP generation, these drugs offer an interesting possibility for dissecting the hormonal responses at a very early phase of the stimulus-effect coupling.

A prominent feature of the hydrosmotic response of the *B marinus* skin was the magnitude of the fluxes measured across it. The fluxes were as large as those recorded in the bladder of the same animal, and sometimes even exceeded them.[11,45]

A second feature of great interest was the discovery of a series of new hydrosmotic agents. These substances may have some impact on pathophysiologic and clinical problems. For example, we found that several psychotropic drugs induce conspicuous increases of water flow in toad skin.[42] There have been numerous reports, in recent years, on episodes of water intoxication in patients taking the same drugs.[46] Admittedly, our findings and the clinical reports are not necessarily related. A formal link exists between them however, and the subject certainly deserves further investigation. Harmaline, a hallucinogen, was the psychotropic drug that we studied in greater detail.[47] We examined interaction between the drug and the cyclic nucleotide system. Mutual inhibition was seen with hormones (vasopressin, catecholamines), exogenous cAMP, and theophylline. An example is given in Figure 24-3. The vasopressinomimetic effects of harmaline suggest a very general remark on an intriguing dualism: psychotropic drugs, like harmaline, markedly stimulate water transport, while the antidiuretic hormone par excellence, vasopressin, interacts with complex functions of the brain, such as memory and probably behavior. This reflects a sort of ill-defined, intuitive feeling about the unit of biology that persists even when we are in the presence of such dissimilar phenomena as antidiuresis and brain function.

A third feature of interest concerns the chemistry of several inhibitors of water transport and the nature of their inhibition. Biamide is a thiol-oxidizing agent that probably affects the last steps of the stimulus-effect coupling. In fact the drug is effective only when it is in contact with the external surface of the epidermis.[41] Lithium is another powerful inhibitor of water flow. The results obtained with tissues like the skin or the bladder are the counterpart in vitro of the nephrogenic diabetes insipidus found in manic patients treated with this cation. The inhibition of water flow apparently results from a dual action of lithium: inhibition of adenylate cyclase and stabilization of microtubules.[42] Elements of the lanthanide series also block the hydrosmotic flow induced by vasopressin, norepinephrine (see Figure 24-1), or cAMP.[24] This action is particularly interesting, since these elements are specific membrane calcium probes. Finally, we reported quite recently that vanadate, an inhibitor of Na^+/K-ATPase that has been intensively investigated in the last year and a half, strongly inhibits cAMP-induced stimulation of water flow.[48] An example is shown in Figure 24-2. This new and unexpected finding might be related to a vanadate-induced inhibition of ATPases linked to mechanochemical proteins of the epithelial cell.

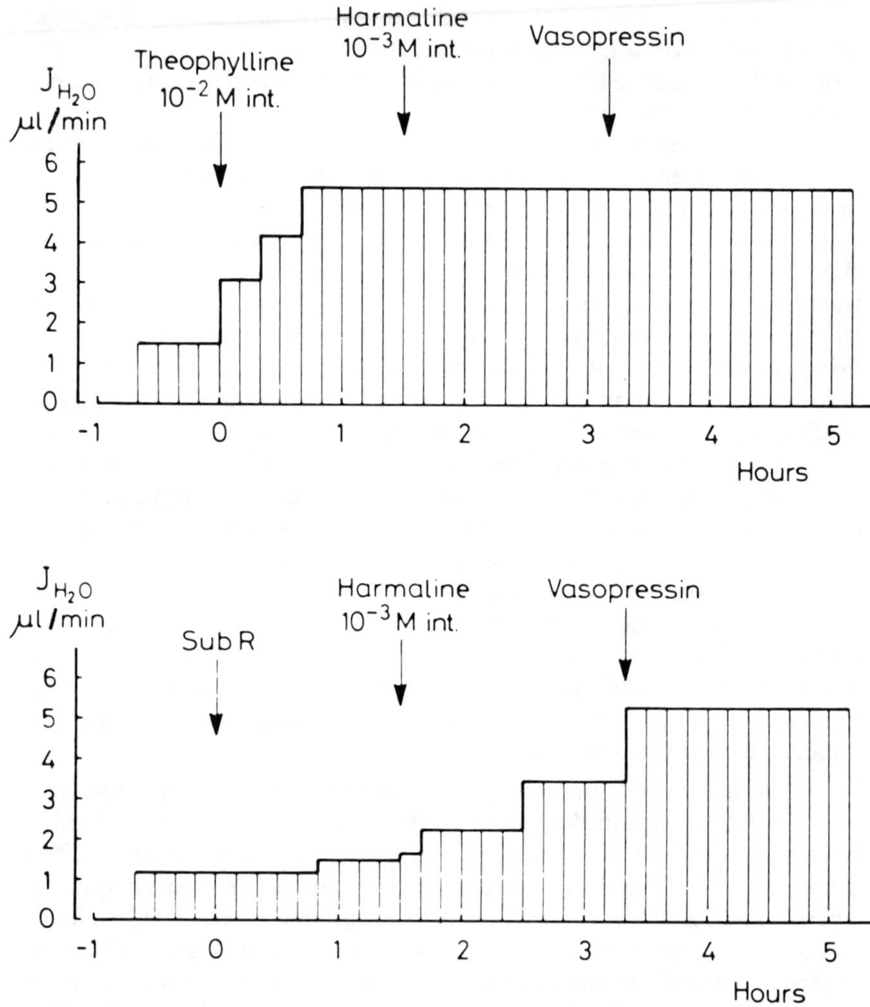

Figure 24-3 Vasopressinomimatic effect of harmaline: water flow (J_{H_2O}) in toad skin *(B bufo)* roughly doubled upon addition of the hallucinogen. Note the mutual inhibition between the hydrosmotic effects of harmaline, theophylline, and vasopressin. Sub R = substitution of Ringer solution. Membrane surface area is 4.15 cm_2. Two pieces of the same skin were used.

Water Transport and Membrane Structure

Water transport has been a strong incentive for the study of some very challenging correlations between structure and function of the plasma membrane. In tight membranes, like amphibian epithelia, theoretical considerations and experimental results indicate that the water flux is mainly transcellular.[24,49,50] Work by DiBona et al[51] showed that, at least in the toad bladder, there is cellular specificity for the transport of water, the granular cell being the epithelial element involved in this transfer. Work with the scanning electron microscope[50,52] confirmed the findings of DiBona et al[51] and showed, in addition, alterations of microvilli in the apical membrane of the granular cells. Such alterations were seen if, and only if, the bladders were challenged with vasopressin in the presence of an osmotic gradient. Finally, some organelles have also been implicated in water transport, namely microtubules and microfilaments[53,54] and lysosomes.[55]

A very important observation was made by Bourguet's group with the freeze-fracture technique: the appearance of aggregates of intramembrane particles in frog bladders exposed to oxytocin.[56,57] This finding has been confirmed by other groups in toad bladders stimulated with vasopressin.[58,59] The aggregates do not depend on the presence of an osmotic flow, and appear in the apical membrane of the granular cell but not in other cell types. More recently, similar aggregates in intracytoplasmic structures were reported simultaneously by two independent groups.[59,60] The presence of particle aggregates in other osmoregulatory, vasopressin-sensitive epithelia would strengthen considerably the biologic significance of this finding. Such is indeed the case in the rat collecting duct[61] and, most importantly, in a nonurinary epithelium, the amphibian skin.[62]

The schematic diagram of Figure 24-4 can be used to summarize our current views on water transport in tight epithelia. At the onset of the stimulus-effect coupling there is binding of the hormone to its receptors, activation of membrane adenylate cyclase, and diffusion inside the cell of the cAMP generated by the stimulus. Prior to the final steps taking place at the apical plasma membrane, there is strong evidence, at least in the toad bladder, for a role of microtubules and microfilaments in water transport. At the level of the apical membrane, two features are reasonably well established: an increase in fluidity[63] and the appearance of aggregates.[54–58] The latter would represent "permeable patches" or "osmotic shunts" across the apical membrane. Their formation would account for the sharp transition from a state of very low to a state of high permeability to water.[11]

According to one view the appearance of aggregates or clusters of intramembrane particles would be a direct consequence of increased fluidity and lateral diffusion of proteins, leading to an increased number of their collisions and formation of aggregates (hypothesis A). The water channels could result either from the alignment of peripheral proteins containing hemichannels, or from the juxtaposition of integral proteins (Figure 24-4). According to another view, the aggregates are preformed in the cytoplasm and inserted as such in the apical membrane (hypothesis B). Further work is needed to distinguish between these two possibilities. Finally, the water pathway across the apical membrane appears to go through membrane proteins containing or forming extremely narrow channels through which the water molecules proceed by single-file diffusion.[26-28]

As seen in the past, osmosis continues to favor interdisciplinarity. From what we have discussed here, transepithelial osmosis is at the confluence of three important lines of thought in modern biology: the cAMP system of Sutherland, the model of Singer and Nicolson, and Onsager's thermodynamics.

We thank Mrs A. Cergneux for secretarial assistance.

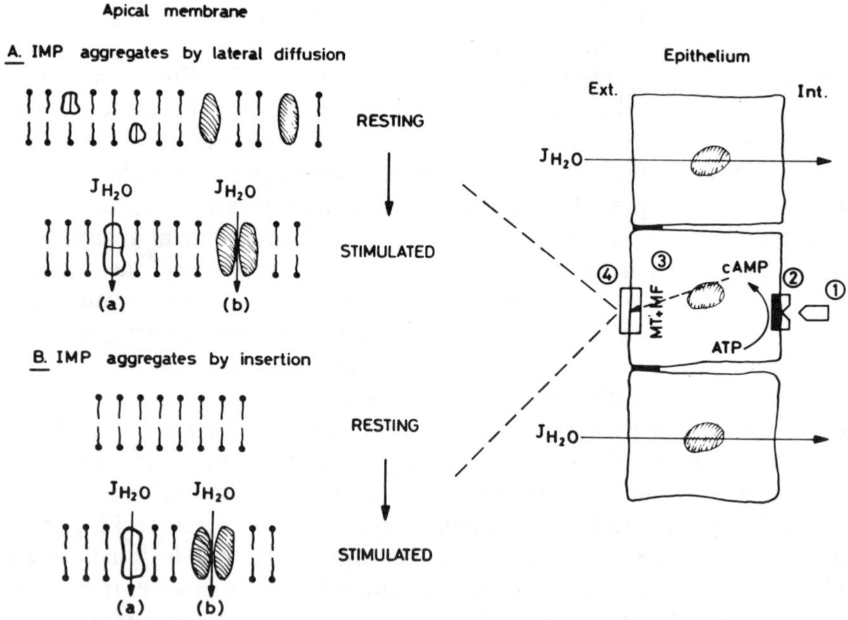

Figure 24-4 Schematic view of current concepts on hydrosmotic flow in vasopressin-sensitive epithelia. On the right, three cells are represented, with tight junctions between them. Water flow (J_{H_2O}) is thought to be transcellular. 1: Hormone (vasopressin or catecholamine); 2: membrane complex formed by the specific hormone receptor and adenylyl cyclase; 3: microtubules (MT) and microfilaments (MF); 4: apical membrane. On the left, two hypotheses to explain the change from a low (resting) to a high (stimulated) state of water permeability of the apical membrane. (a) passive channel for water flow through an intramembrane particle (IMP); (b) between particles. Hypothesis A implies lateral diffusion of membrane components; hypothesis B, insertion of preformed cytoplasmic aggregates.

REFERENCES

1. Smith HW: *From Fish to Philosopher.* Edited by The Natural History Library Anchor Books. New York, Doubleday & Company, Inc, 1961, p 33.

2. Bentley PJ: *Endocrines and Osmoregulation. A Comparative Account of the Regulation of Water and Salt in Vertebrates,* Farner DS (ed): Berlin, Springer Verlag, 1971.

3. Shoemaker VH, Nagy KA: Osmoregulation in amphibians and reptiles. *Annu Rev Physiol* 39:449–471, 1977.

4. Katz U: Salt-induced changes in sodium transport across the skin of the euryhaline toad, *Bufo viridis. J Physiol* 247:537–550, 1975.

5. Dicker SE, Elliott AB: Neurohypophyseal hormones and homeostasis in the crab-eating frog, *Rana cancrivora. Horm Res* 4:224–260, 1973.

6. Licht P, Feder ME, Bledsoe S: Salinity tolerance and osmoregulation in the salamander *Batrachoseps. J Comp Physiol* 102:123–134, 1975.

7. Romspert AP: Osmoregulation of the African clawed frog, *Xenopus laevis,* in hypersaline media. *Comp Biochem Physiol* 54A:207–210, 1976.

8. Cuthbert AW, Shum WK: Interdependence of the two borders in a sodium transporting epithelium. Possible regulation by the transport pool. *J Membr Biol* 40(special issue):221–245, 1978.

9. De Sousa RC: La membrane cellulaire: une frontière entre deux mondes. *Schweiz Med Wochenschr* 107:1605–1612, 1977.

10. Crane MM: Observations on the function of the frog's kidney. *Am J Physiol* 81:232–243, 1927.

11. De Sousa RC: On membrane transport and amphibian epithelia: some recent acquisitions, in Bourguet J, Chevalier J, Parisi M, Ripoche P (eds): *Hormal Control of Epithelial Transport.* Paris, Editions INSERM, 1979, pp 337–350.

12. Brunn F: Beitrag zur Kenntnis der Wirkung von Hypophysenextrakten auf den Wasserhaushalt des Frosches. *Z Gesamte Exp Med* 25:170–175, 1921.

13. Kac M: A mathematician looks at medicine. *Am J Med* 66:725–726, 1979.

14. Schafer JA, Andreoli TE: Water transport in biological and artificial membranes. *Arch Intern Med* 129:279–292, 1972.

15. West ES: *Textbook of Biophysical Chemistry,* ed 3. New York, Macmillan, Inc, 1963, pp 141–143.

16. Du Bois-Reymond E: *Untersuchungen über Thierische Elektricität,* vol I. Berlin, Verlag von G Reimer, 1848, p 743.

17. Matteucci C, Cima A: Mémoire sur l'endosmose, in *Annales de Chimie et de Physique,* vol 13. 1845, pp 63–86.

18. Reid EW: Osmosis experiments with living and dead membranes. *J Physiol* 11:312–351, 1890.

19. Maxwell SS: On the absorption of water by the skin of the frog. *Am J Physiol* 32:286–294, 1913.

20. Steggerda FR: The relation of pitressin to water interchange in frogs. *Am J Physiol* 98:255–261, 1931.

21. Sawyer WH: Effect of posterior pituitary extract on permeability of frog skin to water. *Am J Physiol* 164:44–48, 1951.

22. Fuhrman FA, Ussing HH: A characteristic response of the isolated frog skin potential to neurohypophyseal principles and its relation to the transport of sodium and water. *J Cell Comp Physiol* 38:109–130, 1951.

23. von Hevesy G, Hofer E, Krogh A: The permeability of the skin of frogs to water as determined by D_2O and H_2O. *Arch Physiol* 72:199–214, 1935.

24. De Sousa RC: Mécanismes de transport de l'eau et du sodium par les

cellules des epithelia d'amphibiens et du tubule rénal isolße. *J Physiol* 71:7A–71A, 1975.

25. Kedem O, Essig A: Isotope flows and flux ratios in biological membranes. *J Gen Physiol* 48:1047–1070, 1965.

26. Finkelstein A: Nature of the water permeability increase induced by antidiuretic hormone (ADH) in toad urinary bladder and related tissues. *J Gen Physiol* 68:137–143, 1976.

27. Finkelstein A, Rosenberg PA: Single-file transport: implications for ion and water movement through gramicidin A channels, in Stevens CF, Tsien RW (eds): *Membrane Transport Processes,* vol 3. New York, Raven Press, 1979, pp 75–88.

28. Rosenberg PA, Finkelstein A: Water permeability of gramicidin A-treated lipid bilayer membranes. *J Gen Physiol* 72:341–350, 1978.

29. Hays RM: The movement of water across vasopressin-sensitive epithelia, in Bronner F, Kleinzeller A (eds): *Current Topics in Membranes and Transport,* vol 3. New York, Academic Press, 1972, pp 339–366.

30. Koefoed-Johnsen V, Ussing HH: The contributions of diffusion and flow to the passage of D_2O through living membranes. Effect of neurohypophyseal hormone on isolated anuran skin. *Acta Physiol Scand* 28:60–76, 1953.

31. Lau YT, Parsons RH, Brady RJ, et al: Water movement across split frog skin. *Biochim Biophys Acta* 551:448–451, 1979.

32. Celentano F, Monticelli G, Orsenigo MN: Non linear volume flow dependence on osmotic pressure difference in frog skin. *J Physiol* 74:365–367, 1978.

33. Leaf A: Transepithelial transport and its hormonal control in toad bladder. *Ergeb Physiol* 56:216–263, 1965.

34. Bentley PJ: The effects of neurohypophyseal extracts on water transfer across the wall of the isolated urinary bladder of the toad *Bufo marinus. J Endocrinol* 17:201–209, 1958.

35. Sharp GWG: Water flow studies in toad bladder and the effect of antidiuretic hormones. *Meth Enzymol* 36(part B):251–256, 1975.

36. Eggena P, Schwartz IL, Walter R: A sensitive hydroosmotic toad bladder assay. Affinity and intrinsic activity of neurohypophyseal peptides. *J Gen Physiol* 52:465–481, 1968.

37. Edelman IS, Petersen MJ, Gulyassy PF: Kinetic analysis of the antidiuretic action of vasopressin and adenosine-3′,5′-monophosphate. *J Clin Invest* 43:2185–2194, 1964.

38. Lichtenstein NS, Leaf A: Effect of amphotericin B on the permeability of the toad bladder. *J Clin Invest* 44:1328–1342, 1965.

39. Bourguet J, Jard S: Un dipositif automatique de mesure et d'enregistrement du flux net d'eau à travers le peau et la vessie des amphibiens. *Biochim Biophys Acta* 88:442–444, 1964.

40. Rüphi M, De Sousa RC, Favrod-Coune E, et al: Optical method for measuring water flow with automatic recording. *Experientia* 28:1391–1393, 1972.

41. Grosso A, De Sousa RC: Sidedness of the inhibitory effects of diamide on Na and water transport in amphibian skin. *Experientia* 34:593–595, 1978.

42. Grosso A, De Sousa RC: Vasopressin-like effects of psychotropic drugs in amphibian epithelia. *J Membr Biol* 40(special issue):305–321, 1978.

43. Fischbarg J, Lim JJ, Bourguet J: Adenosine stimulation of fluid transport across rabbit corneal endothelium. *J Membr Biol* 35:95–112, 1977.

44. Jensen PK: Continuous measurement of flow rate and volume in the nanoliter range. *Acta Physiol Scand* 106:5–9, 1979.

45. Grosso A, De Sousa RC: Chemical dissection of transport processes in epithelia, in *Hormonal Control of Epithelial Transport*. INSERM Symposia series. (In press).

46. Grosso A, De Sousa RC: Vasopressinomimetic action of psychotropic drugs. *Eur J Clin Invest* 8:333, 1978.

47. De Sousa RC, Grosso A: Vasopressin-like effects of a hallucinogenic drug—harmaline—on sodium and water transport. *J Membr Biol* 40:77–94, 1978.

48. De Sousa RC, Grosso A: Vanadate blocks cyclic AMP-induced stimulation of sodium and water transport in amphibian epithelia. *Nature* 279:803–804, 1979.

49. Civan MM: Path of bulk water movement through the urinary bladder of the toad. *J Theor Biol* 27:387–391, 1970.

50. Spinelli F, Grosso A, De Sousa RC: The hydrosmotic effect of vasopressin: a scanning electron-microscopy study. *J Membr Biol* 23:139–156, 1975.

51. DiBona DR, Civan MM, Leaf A: The cellular specificity of the effect of vasopressin on toad urinary bladder. *J Membr Biol* 1:79–91, 1969.

52. Mills JW, Malick LE: Mucosal surface morphology of the toad urinary bladder. Scanning electron microscope study of the natriferic and hydro-osmotic response to vasopressin. *J Cell Biol* 77:598–610, 1978.

53. Taylor A, Mamelak M, Reaven E, et al: Vasopressin: possible role of microtubules and microfilaments in its action. *Science* 181:347–350, 1973.

54. Grosso A, Spinelli F, De Sousa RC: Cytochalasin B and water transport. A scanning electron microscope study of the toad urinary bladder. *Cell Tissue Res* 188:375–388, 1978.

55. Pietras RJ, Naujokaitis PJ, Szego CM: Differential effects of vasopressin on the water, calcium and lysosomal enzyme contents of mitochondria-rich and lysosome-rich (granular) epithelial cells isolated from bullfrog urinary bladder. *Mol Cell Endocrinol* 4:89–106, 1976.

56. Chevalier J, Bourguet J, Hugon JS: Membrane associated particles: distribution in frog urinary bladder epithelium at rest and after oxytocin treatment. *Cell Tissue Res* 152:129–140, 1974.

57. Bourguet J, Chevalier J, Hugon JS: Alterations in membrane-associated particle distribution during antidiuretic challenge in frog urinary bladder epithelium. *Biophys J* 16:627–639, 1976.

58. Kachadorian WA, Wade JB, DiScala VA: Vasopressin: induced structural change in toad bladder luminal membrane. *Science* 190:67–69, 1975.

59. Humbert F, Montesano R, Grosso A, et al: Particle aggregates in plasma and intracellular membranes of toad bladder (granular cell). *Experientia* 33:1364–1367, 1977.

60. Wade JB: Membrane structural specialization of the toad urinary bladder revealed by the freeze-fracture technique. *J Membr Biol* 40(special issue):281–296, 1978.

61. Harmanci MC, Kachadorian WA, Valtin H, et al: Antidiuretic hormone-induced intramembranous alterations in mammalian collecting ducts. *Am J Physiol* 235:F440–F443, 1978.

62. Brown D, Grosso A, De Sousa RC: Isoproterenol-induced intramembrane particle aggregation and water flux in toad epidermis. *Biochim Biophys Acta* 596:158–164, 1980.

63. Pietras RJ: Vasopressin-induced redistribution of binding sites for concanavalin A at the surface of epithelial cells from urinary bladder. *Nature* 264:774–776, 1976.

25 Physical Basis of Water Transport in *Valonia utricularis*

Francesco S. Gaeta, PhD
Damiano G. Mita, PhD
Angelo D'Acunto, PhD
Paulo Canciglia, PhD

Cell membranes are generally crossed by fluxes of thermal energy, owing to the different rates of metabolic activity in the two compartments separated by the membrane. Naturally enough one is led to ask whether the simultaneous existence of heat and matter transmembrane fluxes involves their mutual interaction, or, more specifically, to explore the possibility that the heat flux might provide the driving force for some energy-requiring form of biologic membrane transport. Even if the thermodynamic efficiency of the known processes of nonisothermal membrane transport is low, still one of these forms of transport might have been adopted in living organisms for want of anything better.

Nonequilibrium thermodynamics assumes that the temperature gradient is a generalized force,[1,2] and accordingly predicts coupling between thermal energy and matter fluxes in all suitable physical systems. On the other hand, being a phenomenologic theory, it cannot predict what will be the actual value of the coupling coefficient in any real circumstance. Experimental evidence is fortunately available to supply the quantitative information. Already known for a century, the effect of thermal diffusion[3] has shown that heat flux causes appreciable matter fluxes in all kinds of solutions. Also known for some decades is the effect of thermoosmosis,[4-11] consisting of the production of matter flux induced by a temperature gradient applied across a dense, semipermeable membrane. Such membranes do not possess anastomosing liquid-filled channels, and thermoosmosis accordingly consists of a heat-flux-driven solvent and/or solute transport through a solid or solid-like structure. In this case the values of coupling coefficients among heat and matter fluxes are generally very small. The small extent of observed transport, the exceedingly slow time course of the effect, and the awkward all-or-none permeability characteristics of dense membranes, have made thermoosmosis improbable as a model of biologic transport. Probably owing to these circumstances, debate on the biologic significance of transmembrane heat fluxes has never passed the subthreshold level.

The possibility of a new approach to this problem, however, stemmed from studies on the recently discovered effect of thermodialysis.[12-18] Akin to, but also remarkably different from thermoosmosis, this effect consists of nonisothermal transport of matter and charge across unselective porous partitions. Interestingly enough, such partitions, crossed by anastomosing liquid-filled pores and unable to discriminate per se between solvent and solute, or among different solutes, when permeated by a solution and crossed by a flux of thermal energy, acquire a very pronounced selectivity and exhibit remarkable vectorial properties.

Provoked by the many analogies between biologic membrane transport and the effect of thermodialysis, we tried to adapt the current biologic membrane model to the prerequisites of thermodyalysis, and sought direct experimental evidence confirming the relevance of heat flux to biologic transport.

Nonisothermal Membrane Transport Through Porous Partitions: The Effect of Thermodialysis

Since thermodialysis is less generally known than thermoosmosis, it would be useful at this point to give a concise description of the main features that might be relevant to biologic membrane transport.

Thermal energy, propagating in a nonisothermal liquid, gives rise to radiation forces (Gaeta, unpublished data, 1979).[19-21] If K is the thermal conductivity, v the velocity of propagation of phononic excitations in the medium, and dT/dy is the temperature gradient, then:

$$(1) \qquad F_{rad,th} = H \frac{K}{v^2} \frac{dv}{dT} \left(\frac{dT}{dy} \right)^2$$

is the thermal radiation force acting on the unit of volume of the liquid; H is a numerical constant not much different from unity. It is easy to calculate that, for instance, in water at room temperature $F_{rad,th}$ will amount to about 10^{-3} dines/cm^3 when $dT/dy = 1°C\text{-}cm^{-1}$.

If thermal energy propagates in a solution containing different molecular species, then:

$$(2) \qquad \left[F_{rad,th}\right]_j = -2\left[\left(H\,\frac{K}{v}\frac{dT}{dy}\right)_{1^*} - \left(H\,\frac{K}{v}\frac{dT}{dy}\right)_j\right]\sigma_j$$

will be the force acting on a molecule of the *j* species (Gaeta, unpublished data, 1979).[19-21] Symbols here have the same meaning as in equation (1); σj is molecular cross section; subscript (1*) indicates that the quantities enclosed in this term are those pertaining to the solution as a whole; the asterix then recalls the possibility that some of these physical parameters may be modified by special circumstances—as, for instance, by interactions with pore walls when the fluid is contained in a narrow pore (see point C below).

To give us a feeling for the order of magnitude of this force, let us consider a dilute aqueous solution (where the quantities in the term with subscript (1*) are practically the same as in pure water) contained in a vessel (or pore) large enough to make interactions with the walls unimportant. In the case of spherical hydrate solute particles 3 Å in radius, the drift velocity produced by the thermal radiation force of equation (2) would amount to $10^{-7} \div 10^{-8}$ cm/sec in a temperature gradient of $1°C\text{-}cm^{-1}$.

Equations (1) and (2) predict all the principal characteristics of thermodialysis, illustrated by the examples summarized below in points A through H.

A. Transport of ions and molecules by thermodialysis can proceed against gradients of electrochemical potential, since thermal radiation forces do the required work at the expense of the external thermal energy source. The simplest way to illustrate this point is by showing experimental results on water transport through an AP-20 Millipore fiberglass porous partition, separating two chambers containing pure water at different temperatures in an apparatus of the kind schematically represented in Figure 25-1A. Transmembrane volume flow is observed to proceed from warm to cold semicell against a growing pressure head (Figure 25-1B).

Using standard thermodynamic data, it is easy to calculate the difference of chemical potential of water in cold semicell (μ^o_c) and in the warm semicell (μ^o_w). Since $\mu^o_c - \mu^o_w \simeq 3 \times 10^2$ cal•gm/mol^{-1} at the average temperature $T_{av} = 16.5$ C, and with $T_c = 8$ C and $T_w = 25$ C, it is readily seen that the volume flow found in the experiment illustrated in Figure 1B proceeds against the electrochemical potential gradient and against the hydrostatic pressure head. This clear-cut case of "uphill" transport, produced by the thermal radiation force of equation (1), proves point A.

B. From equation (2) it is readily seen that the thermal radiation force is dependent on particle properties such as $\left(H \frac{K}{v} \frac{dT}{dy}\right)_j$ and σ_j. Accordingly, selectivity is to be expected in transport by thermodialysis, notwithstanding the unselective nature of the partition per se. Some experimental verifications of the highly selective nature of this kind of transport, leading to relevant solution fractionation, are illustrated in Figure 25-2.

C. The nature of the liquid medium and its physical state also conditions the intensity of thermal radiation forces, as can be seen from equation (2), where it appears in the term $\left(H \frac{K}{v} \frac{dT}{dy}\right)_{1^*}$. Accordingly, every structural modification of the medium, eg, those induced by interactions with nearby pore walls, is expected to modify the intensity of these forces. Furthermore, according to equation (2) even the sense of the force can be reversed. This is the case when $(K/v)_{1*}$ is smaller than $(K/v)_j$ in the liquid in bulk, but through interactions with pore walls $(K/v)_{1*}$ becomes bigger than $(K/v)_j$. We experimentally proved the occurrence of such cases, thus evidencing one aspect of the influence of solvent-membrane interactions on matter transport (Figure 25-3).

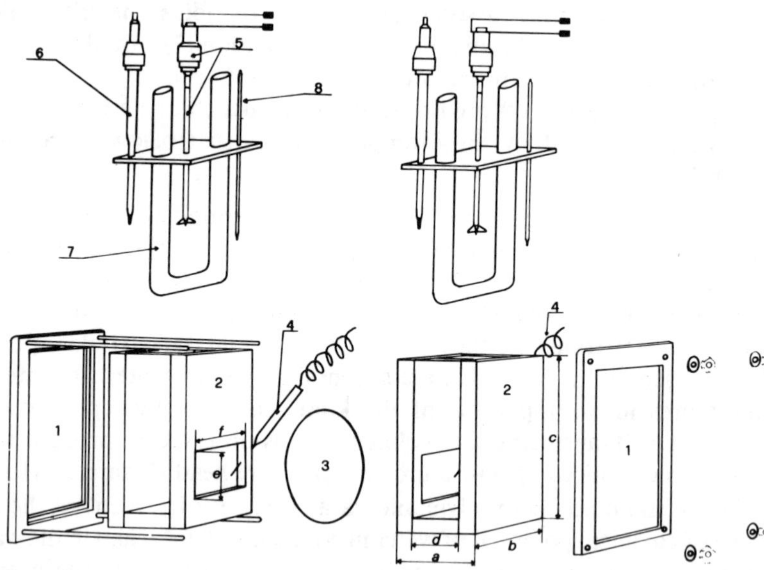

Figure 25-1A

Figure 25-1 **A** Exploded view of the apparatus. Parts indicated by numbers: 1) metal casing; 2) semicells; 3) porous partition; 4) thermocouples; 5) motor and propeller; 6) Vertex; 7) U-shaped glass tube for circulation of thermostating fluid; 8) thermometer a = 35 mm; b = 115 mm; c = 145 mm; d = 15 mm; e = 60 mm; f = 70 mm). **B** Level difference $\Delta h = h_{cold} - h_{warm}$ between semicells, versus time. System: water/AP-20 Millipore porous partition; T_{av} = 24 C; ΔT = 16 C. Water flux is proportional to Δh.

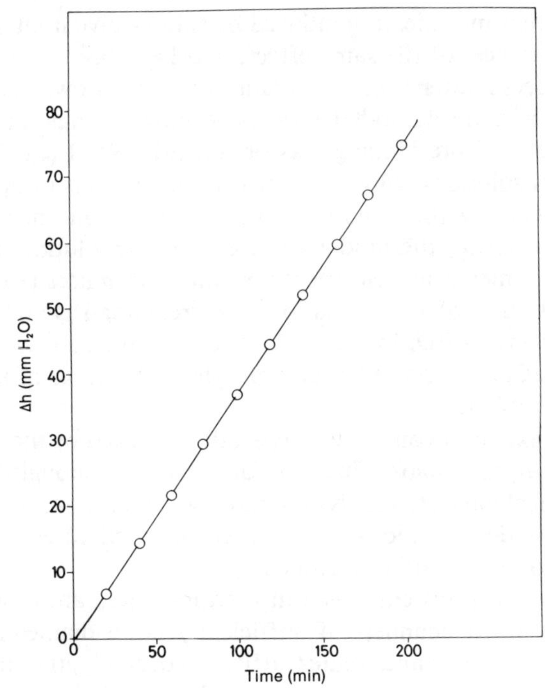

Figure 25-1B

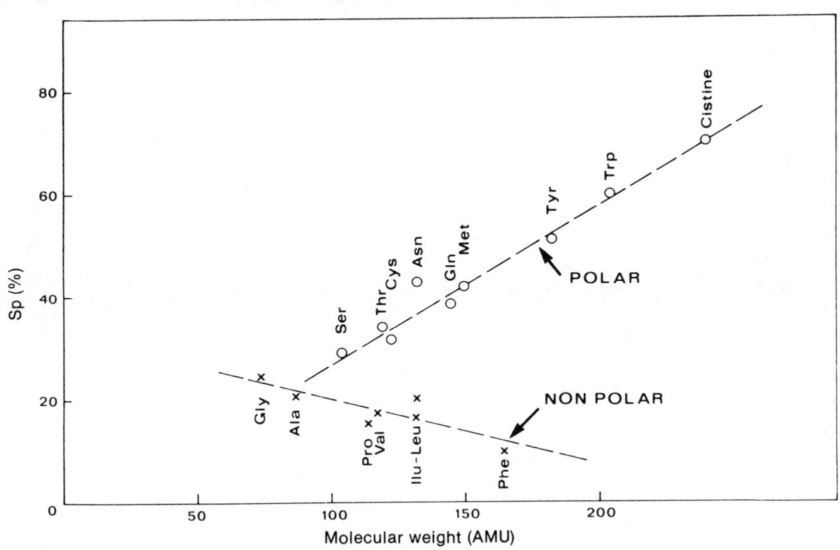

Figure 25-2 Separations obtained with amino acid solutions, represented by the coefficient $S_p = 1 - C_c/C_w$, where C_c and C_w are solute concentrations in the cold and warm semicells, plotted against amino acid molecular weight. System: various amino acids in aqueous puffer-citrate 0.1 M solution/AP-20 Millipore porous partition; duration of runs: 4 hours; T_{av} = 32.5 C; ΔT = 15 C.

D. Solvent modification induced by solute-solvent interactions constitute another facet of the same effect, and have been proven in turn to drastically affect matter transport induced by heat flow, both in thermal diffusion[22,23] and in thermodialysis.[18] A striking demonstration of the influence of water-solute binding was obtained in NaCl and KCl aqueous nonisothermal solutions. Migration of Na^+ and K^+ ions in opposite directions was observed within the range of physiologic concentrations.[22,23]

E. Transport by thermodialysis in electrolytic solutions involves production of transmembrane electrical potential differences (PD). These are produced as a result of a combination of a streaming PD and a concentration (Planck-Nernst) PD. In Figure 25-4 results obtained with an aqueous solution of NaCl are reported as an example of the order of magnitude of such electrical effects.

F. The extent of component separation induced by thermodialysis is often very great, and matter fluxes produced are big enough for the effect to be of practical interest. This is even more so, since it has been possible to fractionate nonideal solutions such as azeotropic mixtures,[14,15] which are difficult to separate by other methods.

G. The partitions employed in thermodialysis are crossed by continuous liquid-filled channels of sufficiently large diameter not to appreciably hinder solvent and solute traffic. Accordingly, the energy barriers of the process are also found to be low. Typical activation energies for the transport of ions and molecules in thermodialysis range from 5 to 7 kcal/mole.

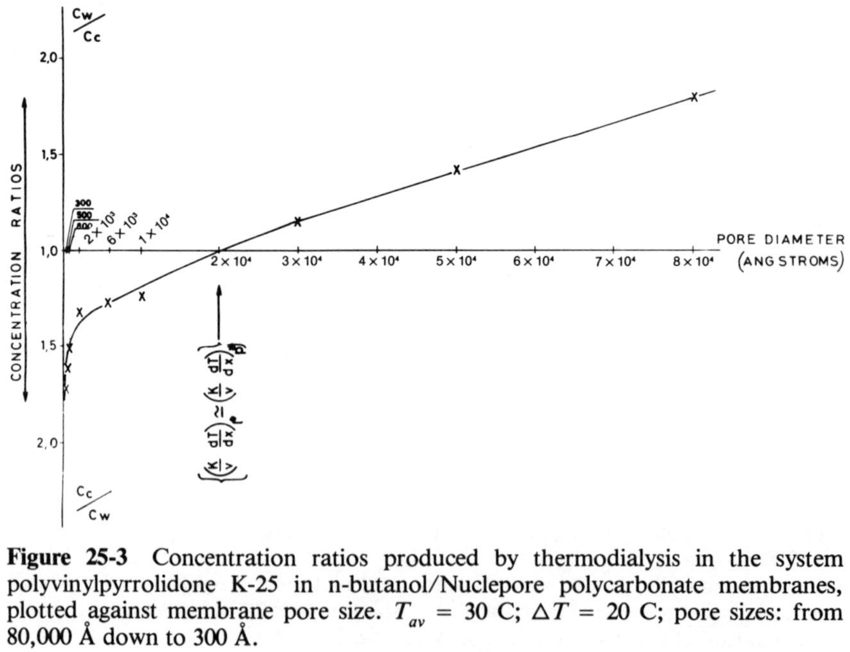

Figure 25-3 Concentration ratios produced by thermodialysis in the system polyvinylpyrrolidone K-25 in n-butanol/Nuclepore polycarbonate membranes, plotted against membrane pore size. T_{av} = 30 C; ΔT = 20 C; pore sizes: from 80,000 Å down to 300 Å.

H. Finally, since the driving force is proportional to gradient temperature, very modest temperature differences can produce transport proceeding uphill against steep gradients of electrochemical potential, when the membrane thickness d is small. The characteristic time scale of the phenomenon decreases proportionally with respect to $1/d^2$.

The possible relevance of the effect of thermodialysis to biologic membrane transport depends on various circumstances: coupling between heat and matter fluxes might significantly contribute to processes of biologic membrane transport, provided the flux of thermal energy generates sufficiently intense forces, and if the structure of the living membrane allows such forms of transport to take place. The greater or lesser significance of thermal transport might also be largely dependent on further circumstances, such as the actual values of membrane permeabilities and of the characteristic time scale of membrane transport induced by heat flow. If thermal transport were too slow, or not selective enough, evolution pressure might have induced the formation of alternative, quicker, and more specific forms of membrane transport, eg, carrier-mediated ones.

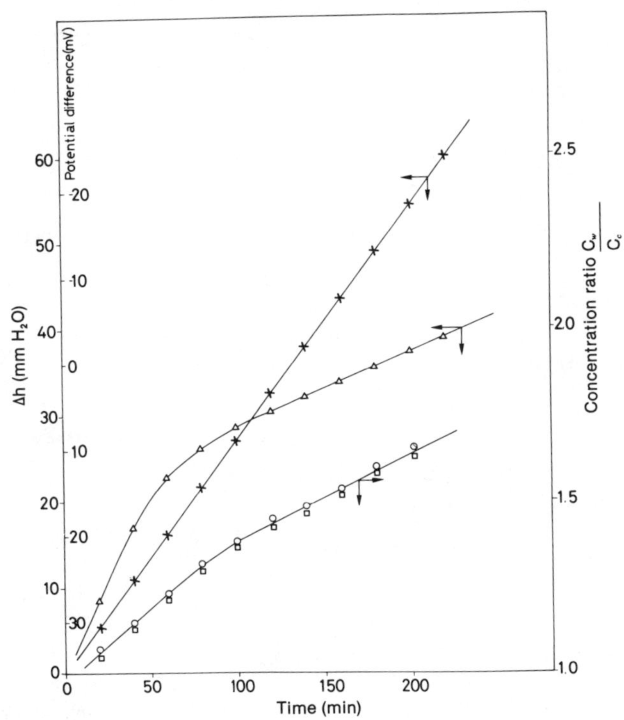

Figure 25-4 Level difference $\Delta h = h_{cold} - h_{warm}$ (x–x), solute concentration ratios of Na^+ (□–□) and Cl (O–O), and transmembrane electric potential difference (△–△), against time, produced by thermodialysis in the system NaCl 0.2 M aqueous solution/AP-20 Millipore porous partition; T_{av} = 30 C; ΔT = 20 C. Arrows indicate ordinate scale for that curve.

Materials and Methods

To test our working hypothesis in vivo we chose the coenocytic alga *Valonia utricularis*. The exceptionally large cellular size, and correlated sluggishness of response, greatly simplified our experimental procedures. The Valoniae are readily available in Naples and have for a long time been a favorite organism for studies of biologic transport.[24-27] On the other hand, possible complications may arise from the fact that *Valonia* is a rather complex system from the point of view of transport studies, consisting of at least two compartments in series—cytoplasm and vacuole—and presenting four successive layers interposed between the vacuole and the outer world. These are: an inner membrane (the tonoplast), a thin cytoplasmic layer, a second membrane (the plasmalemma), and finally an external sturdy cellular wall, made up of cellulose microfibrils (Figure 25-5).

In this study we investigated the kinetics of water exchange between vacuole and external medium by following incorporation and efflux of

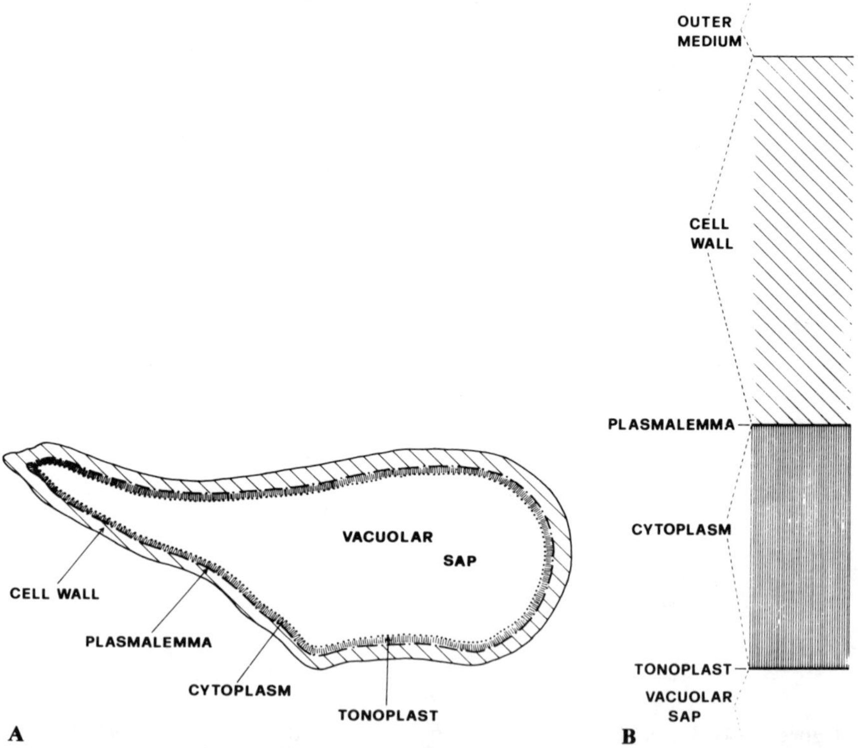

Figure 25-5 **A** Schematic section (not in scale) through one cell, showing the succession of layers interposed between outer medium and vacuolar fluid. **B** The four layers interposed between outer medium and vacuolar fluid in scale: cell wall ≈ 8 μ; plasmalemma ≈ 100 Å, cytoplasm ≈ 5 μ; tonoplast ≈ 100 Å.

tritiated water (^3H_2O). Two different approaches were followed: the philosophy of the first was to investigate normal cell behavior in water transport, and to obtain information on the physicochemical processes involved; the philosophy of the second was instead to determine the effect, if any, of artificially produced heat fluxes on cellular exchange. Isothermal measurements were carried out on the algae at various constant temperatures, in both living intact cells and heat-killed cells. The latter were kept prior to experiment at a temperature of 70 ± 2 C for two hours. The two unit membranes and the cytoplasmic layer are disrupted by this treatment, and what remains of each alga then is just a portion of liquid enclosed by the highly permeable outer cell wall. When the content of such heat-killed cells is extracted, and ionic concentrations and pH are measured, values identical to the ones pertaining to the outer fluid are found, in contrast to the situation in the living Valoniae.

Cells were generally prethermalized overnight, prior to an experiment, at the temperature selected for the run. During the period of stabilization (from 24 hours up to one week) the Valoniae were kept in artificial seawater, prepared according to Harvey's formula[28] in the simplified version of Hyman and Fleming, reported by Pringsheim.[29] Tritium-labeled solution was prepared by adding tritium oxide to obtain a specific activity of the medium, equal to 6.5×10^6 DPM/mole (disintegration per minute). Activity measurements were carried out by an Automatic Counter Tris-Carb 3385 (Packard Ltd). Isothermal incorporation and efflux experiments were conducted both in darkness and at various white-light intensities, up to 7000 lux at the cell's surface.

In the incorporation experiments, cells (prethermalized overnight in thermostated rooms) were transferred to the thermostated bath and kept there for one hour before immersion in the labeled seawater, contained in a separate large beaker in the same bath. After immersion in labeled seawater, aliquots of six algae each were extracted at predetermined time intervals. Some vacuolar fluid was extracted by a syringe from each cell, and introduced into a vial containing the scintillator. The variation in weight of each vial after the introduction of the sap obtained from six algae measured the amount of sap introduced in the scintillator—the value of DPM/gm of sap was thus easily determined.

In the efflux measurements the Valoniae were kept overnight in labeled seawater at the appropriate temperature, after which the activity in the sap became nearly identical to that of the external medium. The efflux experiments were then conducted much in the same way as the incorporation runs, measuring this time the activity decrease over time for the sap, starting from the instant when the cells were immersed in a large vessel of seawater kept in the same thermostated bath. Incorporation and efflux experiments gave essentially similar results, which can be dealt with together, as we shall see shortly.

Coming now to experiments conducted according to our second approach, a method for generating fluxes of thermal energy entering or issuing from the cell had to be devised, compatible with continuation of all normal cell activities. This was made possible by performing both incorporation and efflux runs, while the temperature of the cells and surrounding medium was linearly increased or decreased at a predetermined rate, within a given temperature interval. In every other respect these experiments were conducted as the previously described isothermal runs.

In all experiments the first point was obtained after 30 seconds; successive measurements were effected at one-minute intervals. Every experimental curve is constructed from the data of ten runs (ie, 60 cells per experimental point). In this connection it must be remembered that the reproducibility of the absolute values of the activity is poor, even in very similar conditions. On the other hand, the internal coherence of the data is fortunately much better than that of the absolute values, so that the shape of each curve and their reciprocal positions are fairly well reproducible.

Mathematical Description of the System and Methods of Data Analysis

Equations describing water exchange between vacuole and external medium can be readily obtained, either by assuming that all resistance to water traffic lies in one (or in some) localized barriers, or by assigning to diffusion, through a barrier extended in space, the role of bottleneck. Under the first assumption, when a cell is introduced into a comparatively larger volume of tritium-labeled seawater, the rate of increase of tritium concentration inside the cell is:

$$(3) \qquad \frac{d[Q(t)]_i}{dt} = \left\{ p_{inw}[Q(t)]_o - p_{outw}[Q(t)]_i \right\} \frac{A}{V}$$

where $[Q(t)]_i$ and $[Q(t)]_o$ are intracellular and external tritium activities at time t; P_{inw} and P_{outw} are the inward and outward permeability constants of the cell for water; A and V are respectively the cell surface and volume. Solution of equation (3) is:

$$(4) \qquad [Q(t)]_i = \frac{p_{inw}}{p_{outw}} [Q(t)]_o \cdot \left[1 - \exp\left(- \frac{A}{V} p_{outw} \cdot t \right) \right]$$

having set $[Q(t)]_i = 0$ at $t = 0$. At $t = \infty$, equation (4) gives:

$$(5) \qquad [Q(\infty)]_i = \frac{p_{inw}}{p_{outw}} [Q(\infty)]_o$$

Combining equation (4) and (5) one gets:

$$(6) \qquad \frac{[Q(t)]_i}{[Q(\infty)]_i} = \left[1 - \exp\left(- \frac{A}{V} p_{outw} \cdot t \right) \right]$$

In the case of cells kept overnight in tritiated artificial seawater to reach steady-state internal label concentration $[Q(0)]_i$, and then shifted to a very large container of unlabeled medium, one observes a rate of decrease of internal activity given by:

$$(7) \qquad -\frac{d[Q(t)]_i}{dt} = p_{outw} \cdot \frac{A}{V}[Q(t)]_i$$

from which follows:

$$(8) \qquad \frac{[Q(t)]_i}{[Q(0)]_i} = \exp\left(-\frac{A}{V}p_{outw} \cdot t\right)$$

This equation presupposes again the existence of localized barrier(s) controlling the overall rate of water exchange.

Upon slight remanipulation of equation (8) one has:

$$(9) \qquad 1 - \frac{[Q(t)]_i}{[Q(0)]_i} = 1 - \exp\left(-\frac{A}{V}p_{outw} \cdot t\right)$$

This shows that the fractional uptake in an influx experiment should turn out to be coincident with the complement to unity of the fractional loss in an efflux run. The exponential behavior predicted by equations (6) and (9) will be evidenced by plotting the experimental data on a semi-logarithmic scale. From equation (5) it can also be seen that an influx experiment protracted for a sufficiently long time should reveal a possible vectorial behavior in water exchange.

If, in contrast to the hypothesis of the existence of a space-localized barrier to water exchange, one assumes instead that the rate-limiting step lies in diffusion-controlled transport (eg, through the cytoplasm), then equations (3) through (9) do not apply and a diffusion kinetics should predominate, at least for the period of time required to reach steady-state distribution in the cytoplasmic layer.

From the well-known case of Fickian diffusion it is readily seen that diffusion-controlled transport, through a barrier extended in space over a thickness *h*, would obey the equation:

$$(10) \qquad \frac{\partial[Q(t)]_i}{\partial t} = D\frac{\partial^2[Q(t)]_i}{\partial r^2}$$

where spherical geometry has been assumed. Solutions of equation (10) can be found in textbooks, and consist of exponential functions of time[30] with a characteristic time of the effect given by $\tau = \frac{r_o^2}{\pi^2 D} = \frac{A}{4\pi^3 D}$ being the cell radius, *A* the cell area, and *D* the diffusion constant of the permeant in the cytoplasm. The existence of a diffusional barrier also introduces a time lag in the onset of steady-state fluxes, given by:

$$(11) \qquad L = \frac{h^2}{6D}$$

Plotting $[Q(t)]_i$ on a semi-logarithmic scale against time, the time lag would emerge by extrapolating the rectilinear part of the plot to intersect the abscissa, the interception on this axis giving L.

More complicated patterns than those predicted by equations (6), (9), and (10) are to be expected if the cell consists of more than two compartments (eg, cytoplasm and vacuole in series plus cytoplasmic organelles exchanging water in parallel with the cytoplasm on a comparatively slower time scale).

So much for the evaluation of data relative to the time dependence of intracellular label concentration. Further information can be extracted from these data, when water fluxes are evaluated, and the temperature dependence of the fluxes is obtained. When indeed the results of runs performed at different temperatures are mutually compared, it becomes possible to determine the height of the energy barrier opposing water exchange. The logic underlying the evaluation of the energy barrier is very simple. From each plot of vacuolar activity versus time, at constant temperature, one may calculate the instantaneous values of water flux $\phi(t)$:

$$\Phi(t) = \frac{\pm V}{A\{[Q(t)]_o - [Q(t)]_i\}} \cdot \frac{d[Q(t)]_i}{dt} \tag{12}$$

where the ± sign refers to incorporation and efflux runs, respectively. Remembering equation (3) the expression for $\phi(t)$ can also be written as:

$$\Phi(t) = \frac{p_{inw}[Q(t)]_o - p_{outw}[Q(t)]_i}{[Q(t)]_o - [Q(t)]_i} \tag{13}$$

Since water flux is proportional to the number of molecules effecting the transition through the various energy states in the barrier, one may assume that the number of water molecules effecting the transition (per unit time and surface) will be proportional to the fraction of the population of molecules possessing an energy equal to or higher than the highest energy state E in the barrier. Accordingly:

$$\lim_{t \to 0} \left\{ \frac{1}{C} \ln [\Phi(t)] \right\} = - \frac{E}{RT} \tag{14}$$

where $1/C$ is a numerical constant dependent on the choice of units. The values of the fluexes must be derived from extrapolated initial incorporation rates because of the tendency toward saturation. Plotting the values of $\phi(t)$ corresponding to initial conditions at various temperatures, against $1/t$ (Arrhenius plots), one can thus determine E.

Experimental Results

Water exchange between external medium and vacuole in living cells of

V utricularis is found to be vary active; the equilibrium distribution of tracer is attained in a time interval never exceeding two hours. As can be seen from the ^3H_2O isothermal incorporation curves of Figure 25-6, the approach to steady-state tracer distribution is quicker as the temperature increases; similar features appear from tracer efflux experiments. Incorporation curves appear to level off at about 90% to 95% of external concentration, as can be seen by extrapolating the experimental results of Figure 25-6 (external initial concentration of 6.5×10^6 DPM/mol, equivalent to about 360,000 DPM/gm).

When, on the other hand, Valoniae are kept overnight in labeled medium, their internal activity is found to be slightly less than 95% of external level, in good agreement with the extrapolated values quoted above. When the different composition of external and vacuolar fluid is accounted for, activities calculated per gram of water are found to be approximately coincident.

Isothermal efflux of ^3H_2O is observed to reduce internal activity practically to zero, seemingly in disagreement with incorporation. When one considers the differences in internal and external water concentrations the efflux curves and incorporation curves correspond well. The general trend to higher exchange rates at higher temperatures is also exhibited in efflux.

Summing up, once the concentration correction is introduced, both incorporation and efflux measurements lead, through equations (6) and (9),

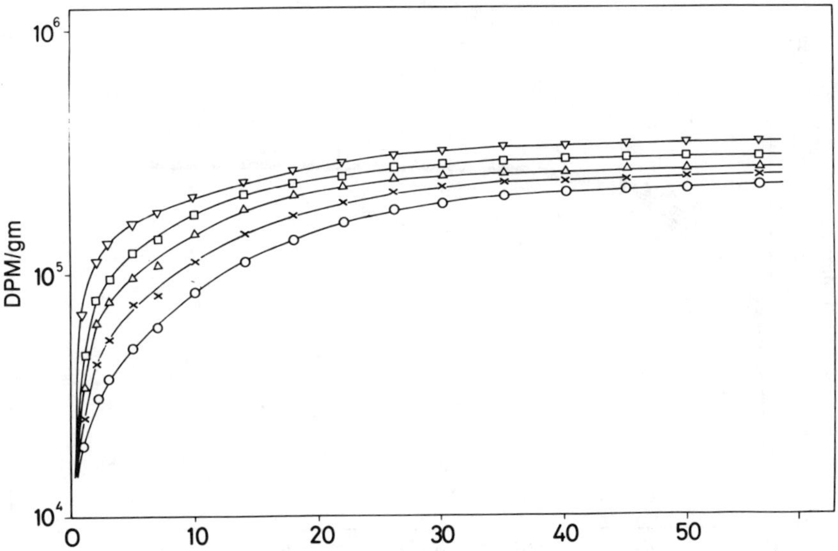

Figure 25-6 Incorporation of ^3H_2O in the vacuolar fluid in living Valoniae at various temperatures: T = 5 C (O—O); T = 15 C (x—x); T = 25C (Δ—Δ); T = 35 C (□—□); T = 45 C (∇—∇).

to approximately equal values of P_{outw}. From equation (5), since $[Q(\infty)]_i \sim [Q(\infty)]_o$, one also finds $P_{inw} \sim P_{outw}$.

Proceeding now to examine tracer uptake in heat-killed cells (Figure 25-7), the higher rates of water exchange and less pronounced temperature dependence of the effect catch the eye. Efflux experiments were found to give results essentially symmetric to uptake in the dead cells. The tendency toward saturation is an obvious consequence of the fact that the tracer eventually becomes uniformly distributed; this also implies that the true rates of water exchange can be deduced only from the initial slopes.

All the experiments reported above were conducted in darkness. Runs performed under various conditions of illumination gave analogous results, the effect of light consisting of a modest increase in the water exchange rate in the living cells. The effects of illumination have been already discussed elsewhere.[31]

The next step is to calculate the water fluxes using equation (12) (or 13). In this connection it is useful to have details of the initial trends of incorporation and efflux measurements, both in living and dead cells. These are given in Figures 25-8 and 25-9. As already mentioned above, initial slopes should be used to get the real values of the fluxes; results of this calculation have been reported in detail elsewhere[31]; values on the order of 5×10^{-4} cm^3/cm•sec are found in the living cells, and values approximately twice as large in the heat-killed algae. These fluxes can then be used to

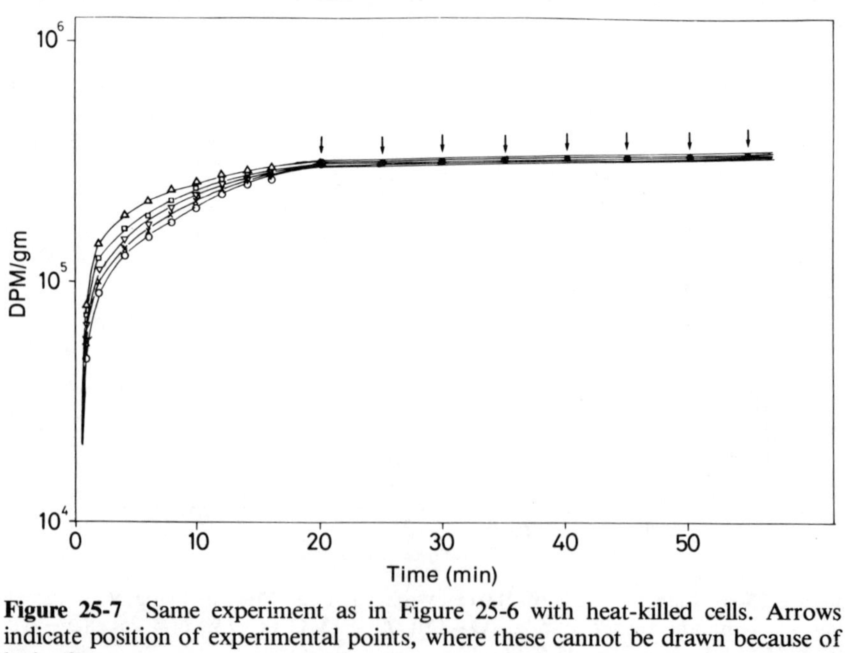

Figure 25-7 Same experiment as in Figure 25-6 with heat-killed cells. Arrows indicate position of experimental points, where these cannot be drawn because of lack of space.

obtain information on the physical nature of the process; this is done by studying the temperature dependence of the effect. When initial fluxes are plotted against $1/T$ (T being the absolute temperature of each run), a linear

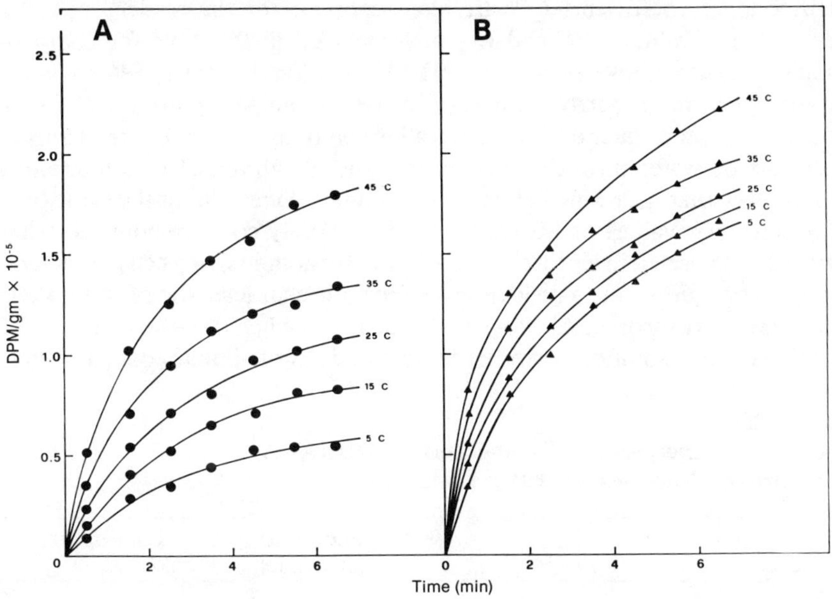

Figure 25-8 Details of the initial part of incorporation experiments in **A** living and **B** heat-killed cells.

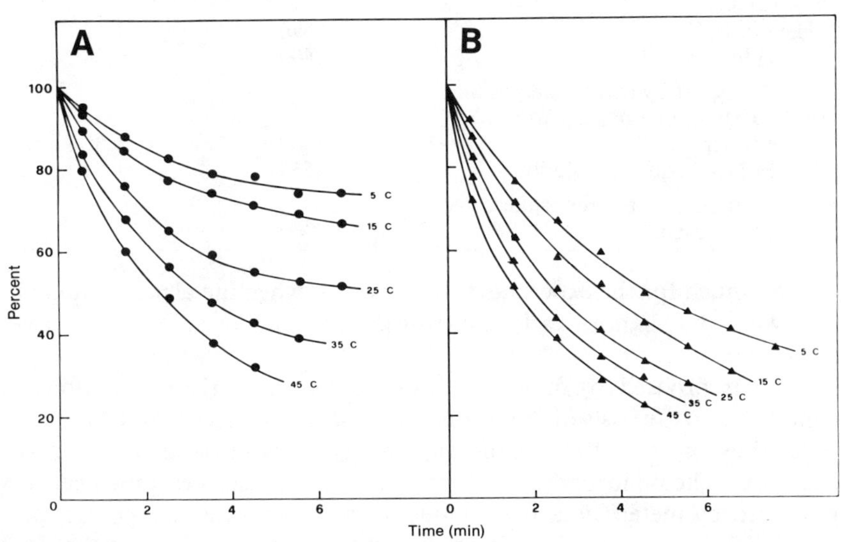

Figure 25-9 Details of the initial part of efflux experiments in **A** living and **B** heat-killed cells.

relationship is found (Figures 25-10 and 25-11), allowing deduction of the apparent Arrhenius activation energy for water transport. The values of these energies are given in Table 25-1, where the corresponding average temperature coefficients Q^{10} are also reported, for the reader's convenience. Other values of E and Q_{10} also included in the table for comparison's sake are those relative to ^3H_2O diffusion in water,[32,33] to water transport by thermodialysis through an AP-20 Millipore porous partition in the case of pure water and of some saline solutions,[15,18] and to the thermoosmosis of water through dense cellophane membranes.[8] It is interesting to observe that heat-killed algae exhibit higher fluxes, but at the same time exhibit lower values of activation energy, closely corresponding to what would be expected in a process of passive diffusion; in living cells the activation energy turns out to be identical to the one characteristic of water thermal transport through AP-20 porous partitions when these separate 0.5 M NaCl aqueous solutions or other ionic solutions of similar concentration.

Table 25-1
Activation Energies and Temperature Coefficients of Various Processes of Water Transport

Process	Activation Energy E (cal/mole)	Temperature Coefficient Q_{10}
Water exchange in *Valonia utricularis*		
Living cells	7.680	1.59
Dead cells	4.250	1.26
Self-diffusion		
H_2O in H_2O	4.890	1.32
^3H_2O in H_2O	4.400	1.28
Water transport by thermodialysis (AP-20 Millipore porous partition)		
Pure water	5.300	1.35
0.5 M NaCl aqueous solution	7.550	1.54
Water transport by thermoosmosis (cellophane membrane)	24.000	4.72

So much for the experiments in darkness; when the algae are illuminated with white light of various intensities, very similar activation energies are found.[31]

Let us proceed to examine the results of another series of experiments, conducted on the basis of the second, more direct approach to the problem of establishing the effect of thermal energy flow on biologic membrane transport. The philosophy of this approach is very simple: if the heat flux produced by metabolic activity induces transmembrane matter transport this should be enhanced or decreased when an issuing (or entering) heat flow is artificially superimposed. To generate such additional transmembrane heat fluxes with the minimum possible disturbance of normal cell

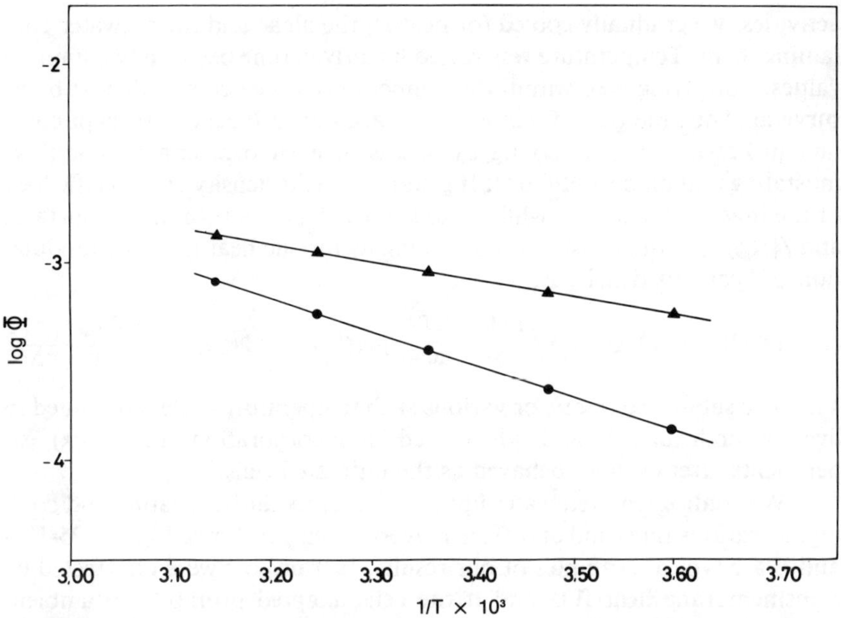

Figure 25-10 Plot of water fluxes against 1/T (T = absolute temperature) calculated by equation (12) from vacuolar activities (data reported in Figure 25-8). Values of initial fluxes have been used (see text). Living cells (• — •); heat-killed cells (▲ — ▲).

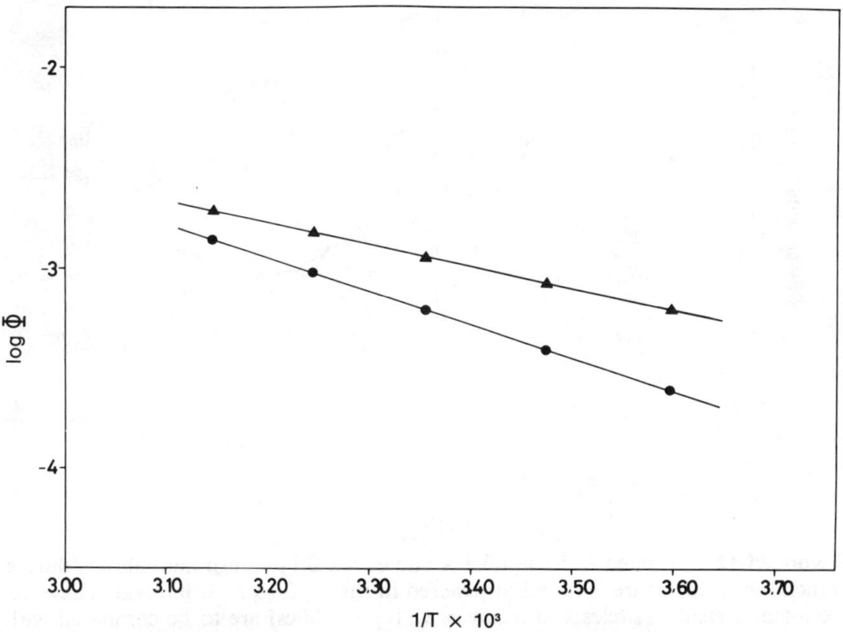

Figure 25-11 Plotted as in Figure 25-10, using data of Figure 25-9.

activities, we gradually cooled (or heated) the algae and the seawater containing them. Temperature was varied linearly in time between two limiting values, both lying well within the temperature range compatible with cell survival. The time rate of temperature variation $\Delta T/\Delta t$ can be kept constant in heating and in cooling by means of a Heto programmable thermostating unit model 02PG623. If ϱ and C_p are the density and specific heat of the inner cell medium, while V and A are the cell's volume and surface, and $[\phi(Q)]_{met}$ is the density of the issuing metabolic heat flux, the modulation of heat flux density is:

$$(15)\quad [\Phi(Q)]_1 = [\Phi(Q)]_{met} + \frac{\rho V C_p}{A} \cdot \frac{\Delta T}{\Delta t} \qquad [\Phi(Q)]_2 = [\Phi(Q)]_{met} - \frac{\rho V C_p}{A} \cdot \frac{\Delta T}{\Delta t}$$

Valoniae submitted to one or various such temperature cycles continued to live on undisturbed, and when used in incorporation (or efflux) experiments after cycling, behaved as the untreated cells.

We studied tritiated water uptake and efflux during heating and cooling, at various rates and at different average temperatures: Figures 25-12A and B are typical examples of our results. If transport were unaffected by transmembrane heat flux and if the cells adapted promptly to ambient temperature, the experiments conducted during heating and cooling should lead to curves lying entirely between the isotherms corresponding to the initial and final temperatures, and should intersect in the middle of the time interval over which the temperature transient is applied.

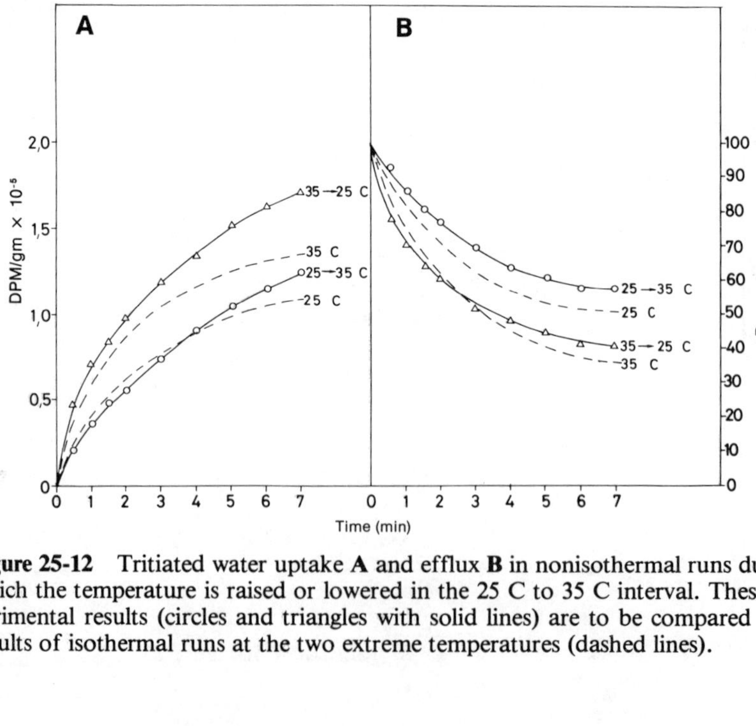

Figure 25-12 Tritiated water uptake **A** and efflux **B** in nonisothermal runs during which the temperature is raised or lowered in the 25 C to 35 C interval. These experimental results (circles and triangles with solid lines) are to be compared with results of isothermal runs at the two extreme temperatures (dashed lines).

The artificially produced heat flux clearly affects the kinetics of water exchange in *Valonia* (Figure 25-12A and B). Incorporation of tritium oxide is greatly speeded up by cooling the cells, ie, an issuing flux of thermal energy enhances water exchange between vacuole and outer medium, as can be seen from the 35 C to 25 C curve of Figure 25-12A. More complicated is the effect of heating: after an initial slowing-down of the process the nonisothermal curve crosses the lower isotherm, the rate of the process becoming intermediate between those at 25 C and 35 C.

Nonisothermal efflux experiments exhibit a behavior antisymmetric to the one found in incorporation: heating the cells leads to a relatively large slowing down of the process (Figure 25-12B; 25 C to 35 C curve), while cooling moderately speeds up water exchange at the beginning, but does not seem to have much effect after the first few minutes (Figure 25-12B; 35 C to 25 C curve). Experiments similar to the ones of Figure 25-12A and B were also conducted at various other temperature intervals, average temperatures, and at various rates of heating and cooling. The general features are the same under all these different conditions. Increasing the rate of heating or cooling seems to give rise to more pronounced effects; on the other hand, the internal coherence and reproducibility of results becomes poorer.

When instead similar nonisothermal runs are performed with heat-killed cells, both incorporation and efflux are found to yield curves intermediate between the two corresponding isotherms, and no significant differences among heating and cooling are observed.

Discussion and Conclusions

We shall now try a critical appraisal of the significance of the results reported above for our working hypothesis. Advantages and difficulties arising from the choice of water as permeant, and of as complicated an organism as *V utricularis,* in this study, must be discussed.

Water readily responds to overall changes in osmolality, and accordingly would be indicative of global trends of membrane transport in the system. This is a desirable aspect in a preliminary investigation of the general physical features of biologic membrane transport, but it must also be remembered that with this choice many interesting details may be obscured and perhaps disappear.

Difficulties of interpretation, dependent on the complexity of cellular structure, are enhanced by the fact that separate measurements of ^3H_2O activity in the cytoplasm of *Valonia* are subject to large random errors, so much so that in this study we limited our measurements to the comparatively much more reproducible assay of label concentration in the vacuolar fluid.

The physical aspects of water exchange that can be determined in this way are extent and temperature dependence of fluxes, from which the kinetics of transport across various barriers in series is determined. The task of sorting out the individual contributions of each barrier is then an additional problem.

Proceeding to the discussion of experimental results, a few prominent features catch the eye. Fluxes and permeabilities are high, in general agreement with previous findings.[25,27] The higher fluxes found in the heat-killed cells reflect the destruction of the two cell membranes and cytoplasmic layer, brought about by thermal treatment. The Arrhenius plot for these dead cells, leading to an estimated value of 4 kcal/mole for the height of the energy barrier, shows indeed that the rate-limiting step in this case is passive diffusion through the sturdy but highly permeable cell wall. This rather banal finding has the merit of reducing a little the complexity of the next step, namely the interpretation of results in living Valoniae, by ruling out the cell wall as a possible bottleneck of the process.

The decreased incorporation and efflux rates of ^3H_2O, as well as the approximately double value of the activation energy for water transport in living cells (about 8 kcal/mole), indicate that the contribution of the outer cell wall to the total hindrance posed to water traffic between vacuole and outer medium is negligible, in comparison with some of the other steps of the process. More interestingly, this numerical value found for the activation energy significantly falls in the range of values characteristic for water transport by thermodialysis. This piece of evidence, taken at its face value, lends support to our working hypothesis.

The problem arising in this connection, however, is the following: in the case of various barriers in series the apparent activation energy found for the process is the one pertaining to the highest of the series. In the present case the process bottleneck may be either at tonoplast, plasmalemma or cytoplasmic layer. Even if it is generally assumed that water transport in the cytoplasm is practically instantaneous, this question should be reexamined here, in light of the mathematical analysis above.

If one hypothetically assumes diffusion in the layer of cytoplasm to be the rate-limiting step in water transport, the following characteristics would be expected in the isothermal curves: vacuolar activity should exhibit an exponential time dependence (in both influx and efflux) with a characteristic time $\tau \leq \frac{r_o^2}{\pi^2 D} = \frac{A^*}{4\pi^3 D}$, which for our algae turns out to be very short indeed ($r_o \approx 10^{-1}$ cm; $D \approx 10^{-5}$ cm^2/sec). This means that the process of water exchange should proceed to tracer equilibrium in about two minutes. The exponential phase of transport should be preceded by a much shorter

*Results reported in this work refer to batches of algae homogeneous in size (0.20 ± 0.03 gm) and as much as possible of similar, elongated shapes (length/radius ≈ 15 ± 1). Diffusive characteristic time in cells of this shape is much shorter than in spherical cells of equal mass. Accordingly the characteristic time τ_{sphere} reported in the text is a large overestimation ($\tau_{elongated} \approx 1/5\ \tau_{sphere}$). This justifies the inequality sign.

transient phase (duration ≈ 10^{-1} sec) during which tracer steady-state distribution is attained in the layer of cytoplasm. The experimental isotherms, on the other hand, exhibit a characteristic time 30 to 50 times longer than expected for the diffusion-controlled process. Thus, we can rule out the possibility that diffusion of water through the cytoplasmic layer constitutes the rate-limiting step.

Accordingly the activation energy of 8 kcal/mol found by us is characteristic of water transport through one of the two biologic membranes (or both). Having identified the plasmalemma and/or tonoplast as the stage(s) that exhibits the energy of activation pertaining to a process of thermodialysis does not unambiguously prove our working hypothesis postulating the existence of heat-flow-driven membrane pumps. It is possible to conceive, for instance, of combinations of various carrier-mediated transport phenomena of different metabolites, whose weight-averaged cumulative influence on the observed kinetics would give the results experimentally found by us.

We tried therefore to gather more direct evidence on the effects of heat flux on biologic membrane transport by performing nonisothermal experiments of the kind described above. We have seen some typical results thus obtained in Figures 25-12A and B. As we already observed, the artificially produced heat flux clearly influences water exchange between vacuole and outer medium. If attention is focused on the results obtained in the first three minutes, it is easy to see that the presence of an artificially generated issuing heat flux (35 C to 25 C curves) speeds up both label entrance in influx, and label outflow in efflux, runs. An entering heat flux (25 C to 35 C curves) in turn exhibits a retarding influence in both cases. Remembering that an issuing heat flux is synergistic with the one produced by metabolism, while an entering flux is antagonistic, the experimental results just discussed clearly lend support to our working hypothesis.

More complicated is the analysis of other prominent experimental features, namely, the behavior after more prolonged periods of heating or cooling; and next, what can be called the asymmetry effect, represented by the fact that the average of the two nonisothermal curves is not coincident with the average of the corresponding isotherms. Only a general comment is possible here concerning these points; a more detailed discussion will have to be postponed until further experimental evidence can be obtained. Since water exchange reflects the combination of all forms of ionic and molecular transport, its time evolution may be complicated by the combination of fluxes of different permeants affecting osmolality, and hence water exchange, each with its own law of time dependence. For instance, the first two or three minutes might reflect exclusively the kinetics of water (active?) transport, and at longer times the curves might be more representative of passive water movement following heat-flux-driven sodium, potassium, and chlorine transport.

As for the asymmetry effect it is even more difficult to analyze; nonisothermal incorporation seems to be, on the average, more rapid than isothermal incorporation, while nonisothermal (average) efflux is slower than isothermal efflux. Does this vectorial effect have to do with the different situations respectively prevailing at plasmalemma and at tonoplast? In this connection it should be remembered that the N localized heat sources q_i, constituted by sites of exothermal chemical reactions, are all situated either in the cytoplasm or at membranes, while practically none of them resides in the vacuole. By applying Gauss' theorem on this system of heat sources one finds that in isothermal runs there is no heat flux at tonoplast, while all the issuing metabolic heat flux $[\Phi(Q)]_{met} = \frac{4\pi}{A} \sum_i^N q_i$ crosses the plasmalemma. In nonisothermal conditions, however, there are heat fluxes across both membranes. In cooling and heating, these fluxes have opposite direction, but equal absolute magnitude at the tonoplast; at the plasmalemma cooling produces a flux that adds with the metabolic heat flux, and the reverse occurs during heating. This circumstance could obviously be the cause of the observed asymmetry. A more conclusive discussion of these points, however, will be possible only after further progress of these studies.

A last point that must be discussed here briefly concerns the nature of membrane pores. As we have seen, the activation energy for water transport in *Valonia* has been found to be comparable with the one characteristic of processes of thermal transport through porous partitions. This means that the "pore" should be an anastomosing liquid-filled channel. The effects of thermodialysis described above suggest that it is still possible to conceive of a liquid phase inside the pores, having well-defined physical properties, such as density and thermal conductivity, even if these may have values different from those pertaining to the liquid in bulk. It would also be possible to use concepts such as hydrated ion and structured or destructured water, referring to the solution within the pores. This is not possible in the case of dense membranes such as those used in thermoosmosis, and this is the essence of the difference between the two processes.

The problem is whether this model of the pore is compatible with the present knowledge of biologic membrane structure. Most current models of aqueous channels in biologic membranes would fit our requirements, although we think that the pore structure is defined by the tertiary conformation of a membrane integral protein, or by the clustering of some membrane proteins to form a funnel-like channel.[31,34,35] We also expect such pores built into the protein component of the membrane to have become intimately associated early in evolution with enzymatic activity, the evolution pressure in this sense obviously being very strong. The enzymatic activity would endow the pore with a gating device, and at the same time provide for intense pulsed local heat sources of very high thermodynamic efficiency in heat-flux-driven matter transport.

The architecture of *Valonia* suggests a good reason for postulating the existence of such active pores, capable of producing localized heat fluxes. Thermal energy generated by metabolic activity occurring in the cytoplasm indeed does not cross the tonoplast; still, active transport is known to take place at this membrane.[24] If thermal energy were produced at sites immediately adjacent the tonoplast, however, there would be local intense heat fluxes, compensated by a generally equivalent but much more dilute generalized flux issuing from the vacuole. This would be compatible with net matter transport at the tonoplast.

As is well known, research carried out for many years has conclusively demonstrated the direct involvement of proteins endowed with enzymatic activity in cellular membrane transport. What these studies have not yet proven is that transport enzymes are in fact mobile "ferry-boat" structures actually conveying molecular or ionic "passengers" across membranes. Increasing difficulties are being encountered by the "carrier" hypothesis, particularly in its literal form of a "ferry-boat" entity.[36,37] Also, every piece of evidence gathered so far in support of the "carrier" hypothesis can be seen to fit nicely the alternative model of stationary enzymatic pore structures, such as those described by us.[34,35] So far our data are concerned with the rate of diffusion in and out of living and dead cells. Further experimental researches are still needed for the understanding of the role of heat in transmembrane transport, since existing data might only indicate its minor importance in biologic processes, in view of the small thermodynamic efficiency of the process.

REFERENCES

1. Prigogine I: Thermodynamics of irreversible processes. New York, International Publishers Co Inc, 1954.

2. De Groot SR, Mazur P: *Non-equilibrium Thermodynamics*. Amsterdam, North Holland Publishing Co, 1962.

3. De Groot SR: *L'effet Soret. Diffusion Thermique dans les Phases Condensées*. Academish Proefschrift (PhD Thesis). Amsterdam, Noord-Hollandsche Uitgevers Maatschappij, 1945.

4. Haase R: Thermoosmose in Flüssigkeiten. I. Grundlagen. *Z Phys Chem* 21:244–269, 1959.

5. Haase R, Greiff HJ: Thermoosmose in Flüssigkeiten. V. Unterschungen am System Cellophan + Methanol. *Z Naturforsch* A26:1773–1774, 1971.

6. Carr C, Sollner K: New experiments on thermoosmosis. *J Electrochem Soc* 109:616–620, 1962.

7. Rastogi RP, Blokhra R, Agarwal RK: Cross-phenomenological coefficients. Part I. Studies on thermo-osmosis. *Trans Faraday Soc* 60:1386–1390, 1964.

8. Rastogi RP, Singh K: Cross-phenomenological coefficients. Part V. Thermo-osmosis of liquids through cellopane membrane. *Trans Faraday Soc* 62:1754–1761, 1966.

9. Rastogi RP, Skukla PC, Yadava B: Membrane permeability of heavy water. *Biochim Biophys Acta* 249:454–461, 1971.

10. Goldstein WE, Verhoff FH: An investigation of anomalous osmosis and thermoosmosis. *Am Inst Chem Eng J* 21:229–238, 1975.

11. Vink H, Chishti S: Thermal osmosis in liquids. *J Membr Sci* 1:149–164, 1976.

12. Gaeta FS, Mita DG, Perna G: Process of thermal diffusion across porous partitions and relative apparatuses. Patents: Italy no. 928656 of 6/25/71. UK no. 23590 of 5/19/72. France no. 72-19189 of 5/29/72. USSR no. 1798775/23-26 of 6/21/72. USA no. 260497 of 5/7/72.

13. Gaeta FS, Mita DG: Non-isothermal mass transport in porous media. *J Membr Sci* 3:191–214, 1978.

14. Bellucci F, Bobik M, Diroli E, et al: Separation by thermodialysis of acetic acid aqueous solutions. *Can J Chem Eng* 56:698–704, 1978.

15. Bellucci F, Drioli E, Gaeta FS, et al: Thermodialysis of non-ideal aqueous solutions: an experimental study. *J Chem Soc [Faraday]* 75:247–260, 1979.

16. Gaeta FS, Mita DG: Thermal diffusion across porous partitions: the process of thermodialysis. *J Phys Chem* 83:2276–2285, 1979.

17. Mita DG, Asprino U, D'Acunto A, et al: Heat-flow-induced mass transport through porous partitions. *Gazzetta Chimica Italiana* 109:475–480, 1979.

18. Bellucci F, Drioli E, Gaeta FS, et al: Temperature gradient affecting mass transport in synthetic membranes. *J Membr Sci* (In press)

19. Gaeta FS: Radiation pressure theory of thermal diffusion in liquids. *Physiol Rev* 182:289–296, 1969.

20. Gaeta FS, Di Chiara A: Dependence of thermal diffusion effects in liquids on the physical properties of the dispersing phase. *J Polymer Sci* 13:163–175, 1975.

21. Gaeta FS: The genesis of radiation forces in liquids and their role in transport processes: some biological implications, in *Proceedings of the IV International Winter School on the Biophysics of Membrane Transport,* part III. Wisa, Poland, Agricultural University of Wrocaw, 1977, pp 67–106.

22. Gaeta FS, Mita DG, Perna G, et al: Thermodiffusive behaviour of NaCl and KCl aqueous solutions: a model for the Na-K pump. Il Nuovo Cimento 30B:153–162, 1975.

23. Gaeta FS, Mita DG, Perna G, et al: Soret effect in homogeneous and heterogeneous (membrane) systems. *Phys chem Hydro* (In press)

24. Gutknecht J: Sodium, potassium and chloride transport and membrane potentials in *Valonia ventricosa. Biol Bull* 130:331–344, 1966.

25. Gutknecht J: Ion fluxes and short-circuit in internally perfused cells of *Valonia ventricosa. J Gen Physiol* 50:1821–1834, 1967.

26. Gutknecht J: Membranes of *Valonia ventricosa:* apparent absence of water-filled pores. *Science* 158:787–788, 1967.

27. Gutknecht J: Permeability of *Valonia* to water and solutes: apparent absence of aqueous membrane pores. *Biochim Biophys Acta* 163:20–29, 1968.

28. Harvey HW: *Recent Advances in the Chemistry and Biology of Seawater.* Cambridge, Cambridge University Press, 1945, p 29.

29. Pringsheim EG: Methods for the cultivation of algae, in Smith GH (ed): *Manual of Phycology.* Waltham, Mass, Chronica Botanica Company, 1951.

30. Jost W: *Diffusion in Solids, Liquids, Gases.* New York, Academic Press, 1960, pp 42–46.

31. Mita DG, Bianco M, Canciglia P, et al: Kinetics of water exchange in *Valonia utricularis. Gazzetta Chimica Italiana* 109:481–489, 1979.

32. Simpson JH, Carr HY: Diffusion and nuclear spin relaxation in water. *Physiol Rev* 111:1201–1202, 1958.

33. Wang JH, Robinson CV, Edelman JS: Self diffusion and structure of liquid water. III. Measurement of the self diffusion of liquid water with H_F, H^3, and O^{18} as traces. *J Am Chem Soc* 75:466–470, 1953.

34. Gaeta FS, Mita DG: The process of thermodialysis and some of its possible consequences in biology. *Chimia* 30:261–272, 1976.

35. Gaeta FS, Mita DG: Non-isothermal membrane transport induced by radiation forces. A model of active transport, in *Proceedings of the IV International Winter School on the Biophysics of Membrane Transport,* part III. Wisa, Poland, Agricultural University of Wrocaw, 1977, pp 107–138.

36. Le Febre PG: The present state of the carrier hypothesis, in Bronner F, Kleinzeller A (eds): *Current Topics in Membranes and Transport,* vol 7. New York, Academic Press, 1975, pp 109–215.

37. Lieb WR, Stein WD: Carrier and non-carrier models for sugar transport in the human red blood cell. *Biochim Biophys Acta* 265:187–207, 1972.

26

Transmitter Release Mechanisms During the Early Stages of Reinnervation of a Fast-Twitch Muscle of the Rat – Effects of Ganglioside Treatments

Giorgio Carmignoto, PhD
Mario Finesso, MS
L. Tedese, MS
Alfredo Gorio, PhD

The maturation of transmitter release is a crucial step during the development and reformation of a synapse. The neuromuscular junction is a particularly well suited preparation for this study, because large amounts of biochemical, ultrastructural, and electrophysiologic data are available.

At the neuromuscular junction sites, motor neurons lose the myelin sheath and give rise to several terminals, which are structures specialized in the release of acetylcholine (ACh). They are characterized by the presence of a large population of synaptic vesicles, which supposedly contain and release the neurotransmitter via an exocytosis mechanism.[1] Vesicles are particularly numerous near synaptic release sites, called active zones.[2,3] Freeze-fracture technique has shown these structures to correspond to ridges bulging from the nerve terminal plasma membrane toward the muscle fiber; these ridges are bordered by double rows of large particles.[4,5]

In several systems, aggregates of particles appear to be associated with sites of membrane fusion, as would occur at the synapse if transmitter release were associated with true exocytosis.[6,7] Freeze-fracture examination of active end-plates reveals protuberances near the rows of particles. Such protuberances are not detected in quiescent junctions, and evidently correspond to transient attachments of synaptic vesicles.

The nerve terminal is characterized by a series of active zone compartments, each designed for rapid release of quantized acetylcholine. The upper part of the post-junctional folds are rich in ACh receptors and face the release sites for a better efficiency of the neuromuscular transmission.[8] ACh is synthesized in the cytoplasm[9] and then packed and stored in synaptic vesicles; however, an appreciable amount is present in a cytoplasmic store, which is also the major source of resting nonquantal release.[10]

The cytoskeleton of the nerve endings is a complex structure, made up of a microfilamentous network, which was termed stereoframework by Gray.[11] In addition, molecules similar to muscle actin and myosin have been discovered in synaptosomal fractions, where they constitute about 10% of the total proteins.[12] Tropomyosin and troponin C, which is the peptide responsible for the calcium sensitivity of the muscle actomyosin complex, have also been identified in brain and cultured neurons.[13] The contractile capabilities of actomyosin complexes and the Ca^{++} sensitivity of tropomyosin have suggested the possibility that contractile mechanisms are responsible for the vesicle movements and quantal release of the transmitter. In fact, there are strong indications that assemblies of microtubules and microfilaments may be involved in the black widow spider venom-induced release.[14–16]

As a result of nerve section or crush, the neuromuscular transmission ceases, and several hours later the spontaneous activity also stops. The changes in synaptic transmission are clearly due to failure of transmitter release from the presynaptic terminals, since the post-synaptic sensitivity after denervation does not decrease at the end-plate area. The first morphologic signs of nerve degeneration are swelling and disintegration of mytochondria, agglutination of synaptic vesicles to form honeycomb-like structures, membranous bodies, and dense masses, and hypertrophy of the Schwann cells. Then, the Schwann cell extends its processes around the axonal fragments and phagocytoes the nerve terminal. Later, the Schwann cells lie in direct apposition to the post-synaptic membrane. After the nerve terminal has disappeared, the end-plate region is still recognizable for several weeks by a cholinesterase reaction (for further details on the effects of denervation on the muscle, see reference 17). A crushed nerve grows back quite rapidly and the regenerating axons return precisely to the old end-plate. The nerve covers about 98% of the original post-synaptic membrane. Up to 50% of these muscle fibers are reinnervated by more than one motor neuron.[18] This phenomenon can be explained as a reaction of the nerve, which, at the damaged site, grows several myelinated branches; hence, the muscle is invaded by a larger number of motor neurons than normal. In addition, the nonmyelinated endings, after innervation of an old end-plate, may sprout to innervate other muscle fibers. This multiple innervation is a transient phenomenon, and several weeks later the situation returns to normal.

Transmitter release mechanisms have been extensively studied on either frog sartorius or cutaneous pectoris muscles.

In the absence of the nerve terminal only rare, spontaneous, miniature end-plate potential (MEPPs) are recorded, and they are due to a quantal release of ACh coming from the Schwann cell MEPPs. The first sign of the end-plate reinnervation is the recovery of the electrical activity indicated by a higher frequency of MEPPs. During this stage end-plate potentials (EPPs) cannot be elicited; therefore it is named "nontransmitting." The amplitude of the MEPPs is drastically reduced compared to normal and the amplitude distribution is skewed to the left, toward the origin. The release mechanism of MEPPs is normal because K^+, lanthanum, and hypertonic solutions induce an increase in the mean frequency, as in normally innervated muscles. Therefore, it appears that at this stage of reinnervation the action potential fails to invade the nerve terminal.[19] Soon after, in some fibers, small subthreshold EPPs can be evoked by nerve stimulation, but the MEPP amplitude is still lower than normal.[20] Similar to what we have now described for the frog are the characteristics of reinnervation of the mouse soleus muscle. During the early stages MEPPs and subthreshold EPPs can be recorded, and only several days later are the EPPs large enough to trigger a muscle action potential.[21]

Materials and Methods

For our studies on the mechanism of transmitter release during reinnervation of mammalian muscles, we have chosen the fast-twitch extensor digitorum longus (EDL).

Extensor muscles were denervated by crushing the sciatic nerve just after the site where the last gluteal nerve buds off. All experiments were performed with male albino rats, Sprague Dowley, weighing 80 to 100 gm at the time of denervation. Animals were decapitated, and the legs were removed and transferred to a dish containing oxygenated Krebs solution. The EDL muscle with a long piece of sciatic nerve was dissected quickly, under a compound microscope. The muscles were mounted in a lucite chamber, stretched over a lens, pinned down, perfused with medium, and equilibrated with 95% O_2 and 5% CO_2. The nerve was drawn into two wells, insulated with petrolatum, and stimulated with platinum wires. After the muscles were equilibrated for about one hour, the electrophysiologic experiment was begun. Suspected neuromuscular junctions were impaled with glass microelectrodes of 10 to 30 MΩ filled with 3M KCL. The responses were displayed on a recording system, composed of a dual-beam storage oscilloscope to monitor the experiment, and a dual-beam oscilloscope, mounted with a C4 Grass movie camera, to record the entire experiment. The standard Krebs solution contained: 113 mM NaCl; 25 mM $NaHCO_3$; 4.7 mM KCl; 1.2 mM KH_2PO_4; 11 mM glucose; 1.2 mM $MgSO_4$; and 2.5 mM $CaCl_2$ The experiments were performed at about 22 C.

The extract from black widow spider venom gland (BWSV) was performed as previously described in detail.[22] The cephalothoraxes of the spiders *(Latrodectus mactans tredecimguttatus)* were kept frozen at −20 C and thawed as needed. At that time eight venom glands were removed and homogenized in 1 ml of isotonic Hepes buffer at pH 7.2. The crude homogenate was stored at 2 C. To apply the venom, 50 μl of the extract was added directly to the recording chamber.

The treatment with gangliosides was performed by injecting intraperitoneally a dose of 50 mg/kg of a ganglioside mixture, with the following composition: 27% GM_1; 39.7% GD_{1a}; 16% GD_{1b}; 17% GT. In two of the trials we employed a 1 mg/kg dose of pure GM_1; however, we did not detect any differences between the two treatments and the data were elaborated together.

Results

Electrophysiologic characteristics of the EDL reinnervation Table 26-1 shows the data concerning the parameters we monitored to study the electrophysiologic characteristics of the reinnervation process. Following nerve degeneration no electrical activity could be recorded at end-plate regions, and the muscle fiber resting potential slowly dropped form the control level of −71 mV to −55 mV. In addition, a profound muscle atrophy was observed and the weight was reduced to about 55% to 60% of the control value.

Table 26-1
Electrophysiologic Characteristics of the EDL Reinnervation

Days from Denervation	Muscle Resting Potential (mV)	30 mM K⁺ MEPP Frequency per Minute	Multiple Innervation (% of Impaled Fibers)	No Quanta Contained in the Terminal
Controls	−71.0 ± 1 (58, 6)	15,000 (15, 6)	10 (50, 7)	60,000
13	−54.7 ± 1 (36, 4)	0	0 (20, 3)	0
14	−56.4 ± 1 (22, 5)	21 ± 8 (43, 5)	0 (58, 6)	4944
16	−60.0 ± 0.65 (74, 7)	222 ± 29 (58, 7)	20 (29, 5)	15,700
18	−63.6 ± 1.7 (57, 4)	620 ± 170 (29, 4)	35 (45, 4)	17,500
21	−68.1 (27, 2)	2700 (19, 2)	60 (21, 2)	40,000
25	−70.5 (33, 2)	4500 (5, 2)	70 (18, 2)	...

Data are shown with SEs. When three muscles were used, no SE was calculated. For multiple innervation the data were very homogeneous, and the SE is not mentioned because it is too small. The numbers in parentheses are, respectively, the numbers of muscle fibers and muscles used.

Denervated frog muscles are characterized by the presence of small spontaneous potentials that can be recorded at denervated end-plates. They originate from the release of small ACh pockets from the Schwann cells. Their frequency is very low, about 1 per minute, and is not altered by various methods of stimulation, which conversely induce a marked increase of the MEPP rate, at normally innervated junctions.[23] However, at suspected end-plates of denervated EDL muscles, we failed to record any Schwann cell MEPP. On the other hand, with our standard recording technique variations of potential, smaller than 0.1 mV, cannot be detected safely, because the baseline noise usually ranges from 0.05 to 0.08 mV. Therefore, Schwann cell MEPPs, if present, must be of the same magnitude as the noise or smaller.

The first sign of the old end-plate reinnervation was the recording of subthreshold EPPs at the site of the nerve entrance, 13 days after denervation (Figure 26-1). Their amplitude ranged from 1 to 3 mV, and usually there was no failure if the nerve was stimulated at a frequency of 1 per second. MEPPs were not recorded even if the same fiber was monitored up to 20 minutes. The EPP grew rapidly and the following day was over threshold, inducing contraction of the innervated muscle fibers. However, the MEPPs were still very rare; in fact, in 11 muscles only 1 out of 70 impaled fibers exhibited MEPPs at rest, with a frequency of 3 per minute (Figure 26-1B). As illustrated in this figure ones were often found. We did

A

B

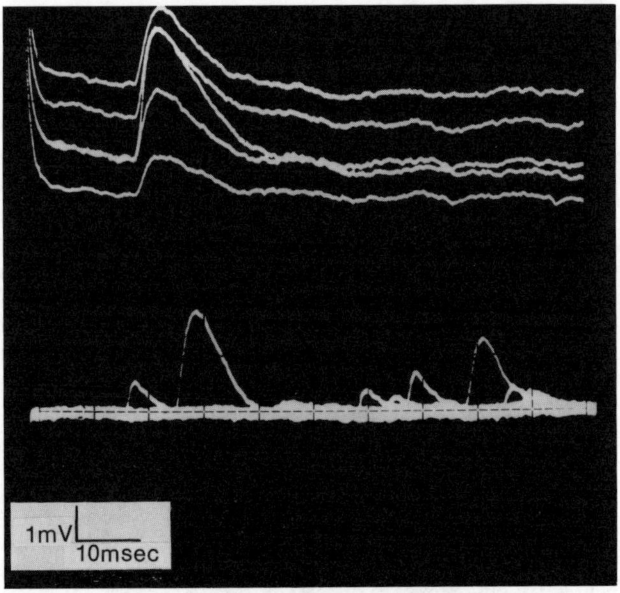

Figure 26-1 Intracellular recordings of EPPs and MEPPs from regenerating nerve terminals. **A** Subthreshold EPP evoked in a regular Krebs solution. The recording was made 13 days after nerve crush. **B** MEPPs recorded 14 days after nerve crush. This was the only fiber with MEPPs at this stage of reinnervation.

not investigate this phenomenon in detail, which has already been described by McArdle and Albuquerque.[24] Only if preparations were stimulated with 30 mM K^+ rich solutions could MEPPs be recorded, but the frequency was still very low, about 21 per minute (Figures 26-2 and 26-3). MEPP frequency at rest returned to normal one month after denervation.

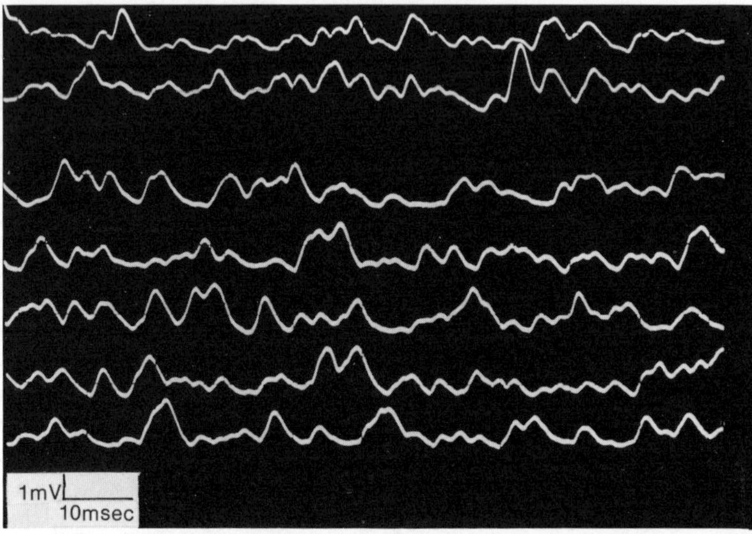

Figure 26-2 MEPPs recorded from a control EDL muscle treated with 30 mM K^+. The frequency is high, about 300 per second.

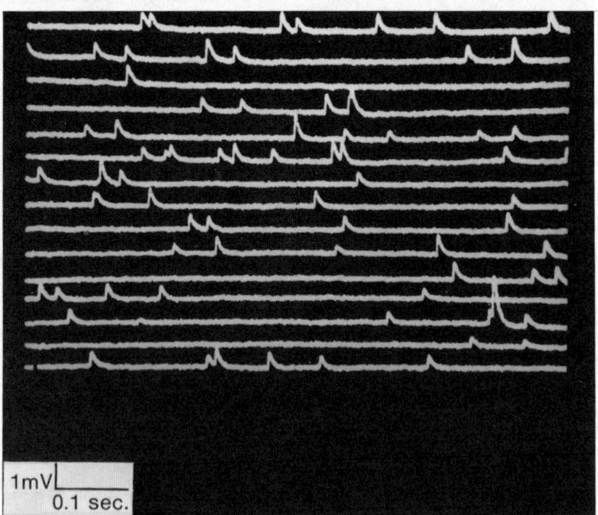

Figure 26-3 MEPPs recorded from an EDL muscle 17 days after denervation and treated with 30 mM K^+. The frequency is still very low (see Figure 26-2).

To summarize, as shown in Table 26-1, on the 14th day, superficial fibers near the nerve entrance were characterized by a resting membrane potential of 56.4 mV. The EPP was large enough to trigger muscle action potential, and when the Mg^{++} was increased to 10 mM, it was still possible to evoke it even with a reduced amplitude of 1 mV. These results suggest a rapid maturation of the transmitter-release mechanisms dependent on the nerve activity. The voltage-dependent Ca^{++} channel must have already been well-developed and similar to the adult state, since transmitter could still be released in spite of the presence of 10 mM Mg^{++}. However, this did not hold true for the spontaneous release of the transmitter; in fact, MEPPs were very rare or not released at all, and even when preparations were stimulated with 30 mM K^+ the frequency was very low compared to control values. These results clearly indicate that both K^+ and the nerve action potential release via different mechanisms.

With time the reinnervation progressed. At day 16 all the superficial fibers and the second layer were innervated, the resting potential was 60 mV, and the K^+ treatment increased the MEPP frequency from 1 to 222 per minute. Only 25 days after innervation the values were close to normal.

We have also followed the maturation of the nerve terminal by measuring the number of quanta available for release. For this purpose we used BWSV.[25] At the end of the experiment a muscle fiber was impaled; if a stable impalement was obtained, 50 μl of BWSV was added. A few minutes after this addition the MEPP frequency dramatically increased; the discharge was sustained for a few minutes at a high rate (Figure 26-4), then subsided and the neuromuscular transmission was blocked. The whole discharge was recorded on film and MEPPs counted. Table 26-1 clearly shows that the maturation of the synapse is parallel to the number of quanta available. Twenty-one days after denervation the content was almost normal.

Another parameter that we followed during the process of the EDL reinnervation was the occurrence of multiple innervation at the same muscle fiber. This phenomenon has been recently evaluated on frog preparations.[18,26] The extensor muscles of the rat also exhibited this phenomenon (Figure 26-5), which progressively increased with time; 25 days after denervation, 70% of the fibers showed polyinnervation (see Table 26-1).

To assess the extent of the polyinnervation, muscles were perfused with Krebs solution, containing 10 mM Mg^{++}, and several muscle fibers were impaled. The recording of multiple EPPs indicated the presence of more than one terminal per fiber. The detection of this phenomenon was facilitated by the slow and various conduction velocity that characterizes the regenerating axons. The polyinnervation was unmasked by the different axon threshold when the conduction velocity was almost identical. The conduction velocity returned to normal at about the 35th day.

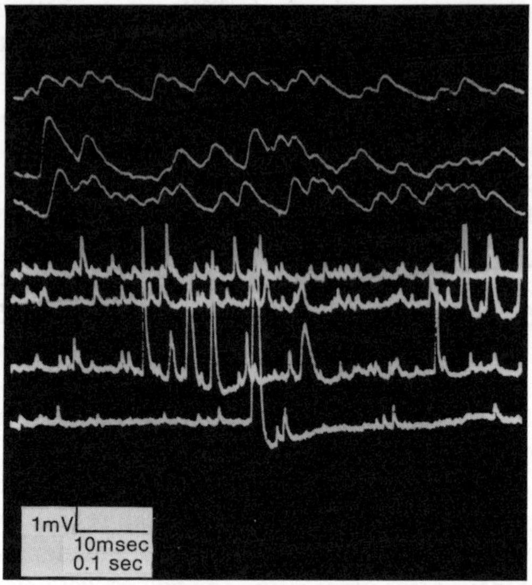

Figure 26-4 Effects of the application of BWSV. An EDL muscle 15 days after denervation was soaked in Krebs containing 10 mM Mg^{++} and treated with 50 μl of BWSV. After a few minutes (lower tracings) the MEPP frequency increased by several orders of magnitude; the discharge was maintained at this rate for one minute and then subsided. In the lower part of the figure several giant and very small MEPPs are noticeable. The 10-msec time scale applies to the upper tracings; the 0.1-sec time scale applies to the lower tracings.

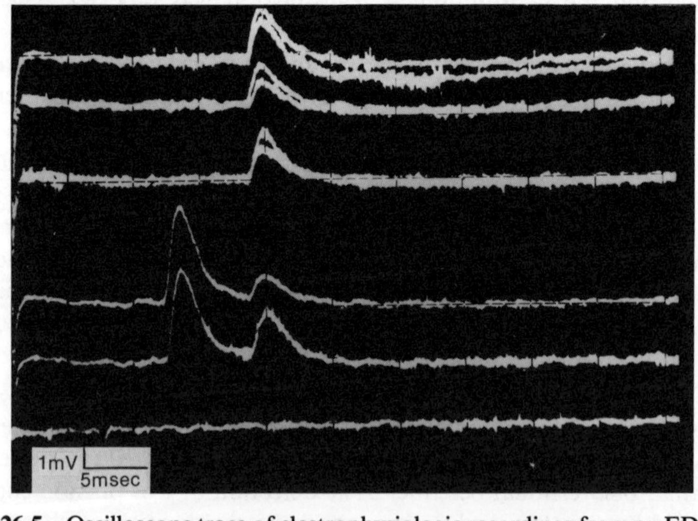

Figure 26-5 Oscilloscope trace of electrophysiologic recordings from an EDL muscle bathed in 10 mM Mg^{++}. The EPPs were evoked by stimulation of the nerve with different and increasing current intensities. It is obvious that at higher currents a second EPP appears. The figure also shows how slow the conduction velocity of the regenerating motor neurons is.

Effects of the ganglioside treatment Our research on the effects of ganglioside treatment during the process of reinnervation initiated from previous results obtained by Ceccarelli et al[27] and Obata et al[28] with two very different experimental models. Ceccarelli et al functionally denervated the cat nictitating membrane by cutting the sympathetic nerve preganglionically or postganglionically, and estimated the recovery of the innervation by measuring the nictitating membrane contracts, induced by electrical stimulation of the nerve. The cut ends of the nerve were joined in an end-to-end anastomosis. Therefore, they observed the reinnervation of a cholinergic synapse at the superior cervical ganglion level, and of an adrenergic one at the nictitating membrane level. Some of the denervated cats were treated with a mixture of bovine brain gangliosides.

In the case of preganglionic anastomosis the treated animals showed complete recovery 45 days after surgery. The pupil, the prolapsed right palpebra, and the nictitating membrane were normal. In addition, the contraction of the nictitating membrane was similar in amplitude and time course to the one of the contralateral unoperated side. This is in complete agreement with the results obtained with regeneration, after the postganglionic denervation. These data are also supported and confirmed by the histochemical analysis of the recovery of the catecholaminergic terminals in the nictitating membrane, by means of the fluorescence technique. In both cases the results strongly suggest that the treatment with gangliosides greatly influences the reinnervation of both cholinergic and adrenergic synapses, enhancing and accelerating it. Obata et al[28] showed that the effect of a particular drug on the formation of synapses can be evaluated directly in vitro, by using cocultures of muscles and spinal cord explants, and recording the spontaneous end-plate potentials. The myotubes, partially covered by the nerve explants, were impaled. Some preparations were incubated for various lengths of times with different concentrations of gangliosides, and the results indicated a marked effect of the treatment. In fact, the function formation was significantly increased. These data strongly suggest the use of gangliosides to enhance and accelerate functional reinnervation. To investigate the mode of action we used the reinnervation of rat EDL muscle as a model, monitoring the same parameters we discussed in the previous section.

The muscle fiber resting potentials, from both treated and untreated animals, did not significantly differ during denervation and reinnervation processes. The lowest value was reached on the 13th day, -55 mV; then the resting potential gradually recovered, and was normal on the 25th day after denervation. However, a marked effect of the ganglioside treatment was observed on the MEPP frequency, both at rest and induced by K^+. In addition, the appearance of the polyinnervation was greatly anticipated in time and increased in value by the gangliosides. Figure 26-6 illustrates the data obtained with 30 mM K^+ stimulation during the EDL reinnervation. It is apparent that EDL from treated animals secretes significantly more quanta than from untreated ones. The difference is maintained up to the 25th day, when data from both treated and untreated preparations were close to the normal value.

During the early stages of reinnervation, from the 14th to the 18th days in resting preparations, the MEPP frequency was two to three times higher in favor of the treated animals (Table 26-2).

Table 26-2
MEPP Frequency in Treated and Untreated Animals

Days from denervation	14	15	16	17	18
MEPPs ratio treated/untreated	MEPPs in treated aniamls only	2.7	2.5	1.9	1

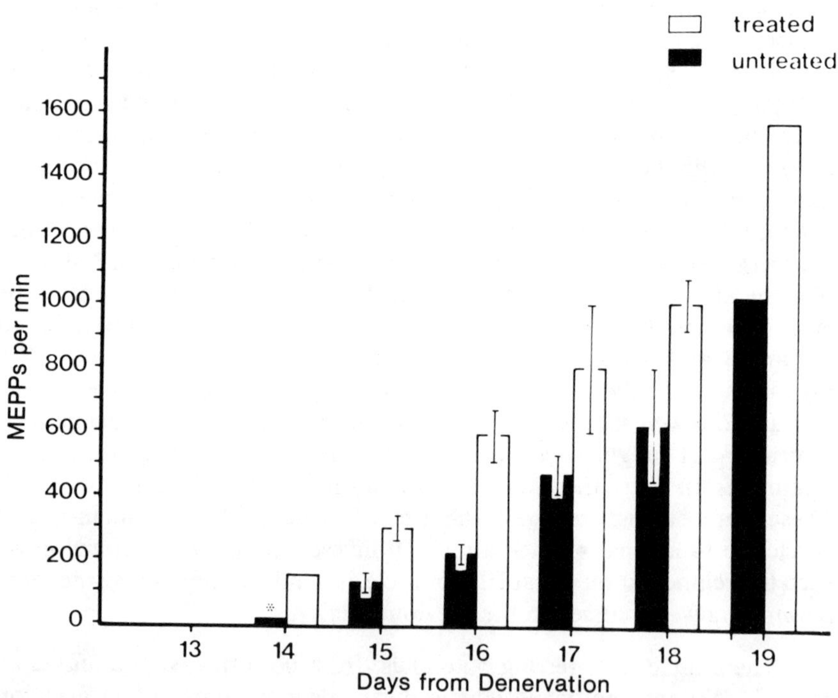

Figure 26-6 Series of histograms showing the gradual increase of MEPP frequency from the day of denervation when muscles were stimulated with 30 mM K^+. The marked effect of ganglioside treatment is obvious. The data shown are $\pm$ SE. On the 14th and 19th days no SE is shown because the data are from 3 muscles only. For the other days a minimum of 5 muscles and 50 fibers were used.
*At this stage of reinnervation the untreated reinervated muscles rarely showed MEPPs, even in 30 mM K^+.

As described earlier, the regenerating axons give rise to several sprouts; therefore, muscle fibers, usually innervated by a single nerve, are polyinnervated. The treatment with the gangliosides induced such a strikingly rapid growth of the terminal axons that even as early as the 14th day, 35% of the muscle fibers exhibited multiple EPPs. In two days, the level of multiple innervation reached 65% to 70%. Only on the 25th day did the untreated muscles match the treated ones (Figure 26-7).

The polyneuronal innervation is probably not permanent; some preliminary data suggested that on the 35th day only a few superficial fibers were polyinnervated, and only deep layers still showed an appreciable amount of double innervation.

Discussion

A large amount of data is available on mammalian muscle reinnervation; however, very little is known about the maturation of neurotransmitter release mechanisms. To study this process and the mode of action of

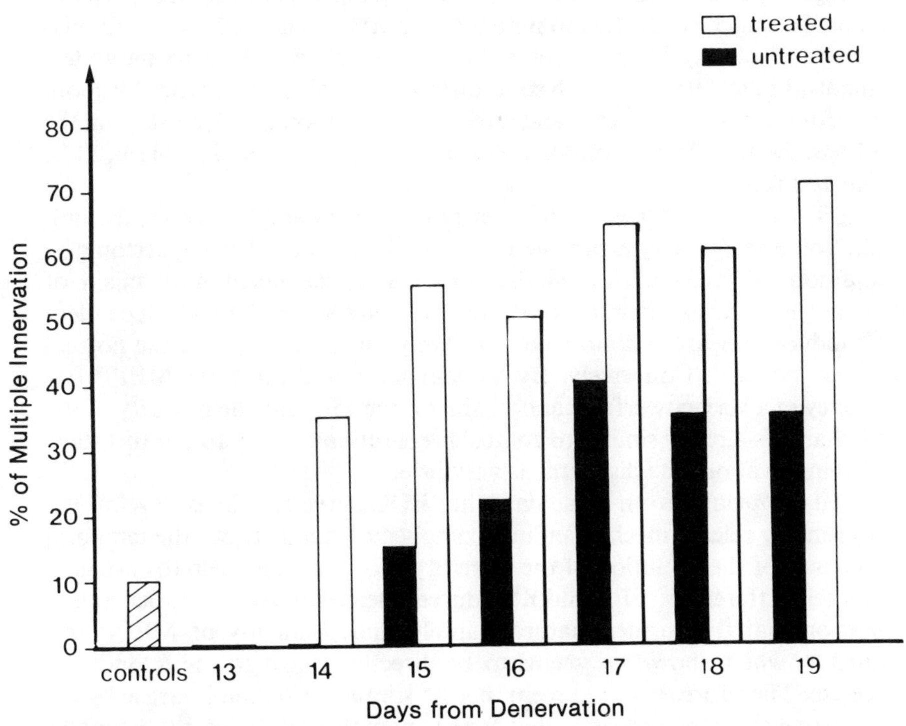

Figure 26-7 This figure demonstrates the percentage of impaled fibers showing multiple innervation. The inductive effect of the ganglioside treatment is very obvious. SEs are not shown because they were very small.

gangliosides on synaptogenesis, we have chosen the model of the rat EDL reinnervation.

Our results show that the terminals of the EDL nerve undergo a peculiar maturation process. The first sign of reinnervation is a subthreshold EPP, which rapidly grows in size and 24 hours later is able to induce a muscle action potential. However, MEPPs are recorded only 24 hours after the EPP is over threshold, ie, on the 15th day, and their frequency is 60 times smaller than controls, mainly about 1 per minute. These results would suggest that EPPs and MEPPs originate from two distinct release mechanisms, and that perhaps the quanta are stored in two different populations, one for spontaneous and the other for evoked release. This suggestion is further confirmed by stimulation with 30 mM K^+-rich solutions. Control preparations release about 250 quanta per second when stimulated with this concentration of K^+; however, reinnervated EDL releases only 21 MEPPs per minute on the 14th day and 130 on the 15th day. This immature state of the MEPP release corresponds to a fully mature mechanism induced by the nerve action potential, since it is possible to record the EPP even in the presence of 10 mM Mg^{++}. Probably the voltage-dependent Ca^{++} channel is already operative, as in control preparations. It is also worthwhile to note that the effectiveness of the K^+ stimulation is absolutely identical for both control and regenerating nerve terminals. In fact 30 mM K^+-rich solutions increase MEPP frequency by about 100-fold in control muscles, and during reinnervation the increase from the 15th to the 18th days, when MEPPs are rare, is respectively 120-, 123-, 115-, and 100-fold.

The results obtained with hypertonic medium and BWSV are dissimilar. Some preliminary experiments with Krebs solution, 50% hypertonic by addition of sucrose, showed that this Ca^{++}-independent mechanism of release is still very immature on the 25th day when the Ca^{++}-dependent K^+-induced release is almost normal. Only on the 35th day is the normal release reached. Conversely, BWSV stimulates and raises the MEPP frequency in a very powerful manner, and on the 15th day the intensity of the discharge is already similar to control preparations; of course, at this stage of reinnervation, the discharge is very brief.

It is obvious from these data that EDL nerve terminals develop the transmitter release mechanism in a complicated manner, but the temporal sequence of the evolution of the secretory processes may help to speculate on their differences. EPP and K^+-induced release are both Ca^{++}-dependent phenomena; the former matures rapidly, independently of MEPP frequency, which, however, seems to be directly related to the K^+-induced release. Therefore, it would seem that K^+ stimulates quantal release by accelerating the same processes that at rest spontaneously release the quanta; perhaps, either by increasing Ca^{++} influx or by acting directly on the mechanisms.

This does not seem to be the case for the Ca^{++}-independent release, induced by hypertonic solutions, which must involve different, slower-to-mature, structures.[15]

However, with both K^+ and hypertonic solutions, BWSV markedly increases the MEPP frequency. A possible explanation is that venom increases nerve terminal permeability[29] and a large influx of divalent cations may induce this powerful release.[30] If the latter explanation is correct the use of Ca^{++} ionophore should also give the same BWSV effect. These experiments are now being performed.

Another interesting parameter is the correlation between recovery of the muscle membrane resting potential and the return toward normality of the spontaneous release, at rest or induced by K^+. The muscle activity first manifests itself on the 14th day, but the resting potential recovers slowly, parallel to the improvement of the MEPP frequency. This may be a coincidence, but the late restoration compared to the early over-threshold EPP and consequent muscle activity suggests that perhaps the spontaneous secretion of either ECh or other factors may also have an influence on the maintenance of the muscle fiber resting potential.

As mentioned above the ganglioside treatment has a very striking role in the reinnervation of the nictitating membrane[27] and in the synaptogenesis in vitro.[28] The results we obtained indicate a marked effect of gangliosides on the maturation of the transmitter release mechanisms that are responsible for the MEPP frequency, both at rest and induced by K^+. These results may explain the findings made by Obata et al[28] and Ceccarelli et al,[27] but the direct effect of the gangliosides on the synaptic functions is as yet unknown. In the central nervous system these molecules comprise up to 10% of the total membrane lipids and may be distributed over the entire neuronal surface.[31] It is well known that when added to the medium, gangliosides are incorporated into the cell membrane and may function as receptors for toxins.[32] In our laboratories it has also been shown that gangliosides are incorporated into neuronal membrane[33] and this process results in a strong stimulation of (Na^+, K^+)-ATPase.[34] However, in our experiments animals were injected with gangliosides intraperitoneally; in this particular case, some molecules were incorporated into the membrane and others were metabolized.[35]

Purpura and collaborators[36-38] have shown that in feline GM_1-gangliosidosis and in human ganglioside storage diseases, there are morphologic alterations of different types of cortical neurons. These cells develop enormous processes, named meganeurites, at the base of the perikaryon and near the site of the axon formation. From the meganeurites there is an extensive outgrowth of neurites. These cortical neurons are characterized by an incomplete gangioside catabolism, and accumulation of either gangliosides or products of their defective metabolism may have induced the exceptional neurite growth.[36-38] Perhaps, the extremely rapid and early achievement of the EDL multiple innervation, during the reinnervation process in animals treated with gangliosides, may reflect an outgrowth of the nerve terminals and enhanced axonal sprouting. This process may be determined either by a direct effect of the gangliosides, incorporated into the neuronal membrane, or by the production of some unknown catabolic products. The axonal sprouting is a rapid and efficient process for restoring innervation and is usually not permanent. At a later stage, some of the redundant innervation is removed and a single terminal will innervate an end-plate.[39,40] This phenomenon is also true for the EDL muscle, as shown by our preliminary data.

All these results viewed together suggest the effect of gangliosides in inducing both a faster maturation of transmitter release mechanisms and a faster growth of the nerve terminals.

Conclusions

The following are the electrophysiologic characteristics of the reinnervation of the rat extensor digitorum longus muscle (EDL).

1. During the absence of the nerve no Schwann cell MEPPs are recorded.

2. The first sign of reinnervation is a subthreshold EPP, which in 24 hours exceeds the threshold level to induce muscle contraction. At this stage of the nerve regeneration MEPPs are almost absent, and stimulation with 30 mM K^+-rich solutions rises the frequency to only a few per minute. Black widow spider venom stimulates release at the normal high frequency.

3. MEPP frequency increases gradually and reaches the normal values about two weeks after the appearance of the EPP. This process is parallel to the recovery of the muscle fiber resting potential, suggesting that, in addition to muscle activity, some factors, spontaneously released from the nerve, play a role in determining some of the electrical characteristics of the post-synaptic membrane.

4. EDL muscles, from animals treated with gangliosides, show a much higher MEPP frequency, both at rest and when stimulated with K^+.

5. Multiple innervation of the EDL muscle is apparent three days after recording of the EPP, after which it gradually reaches the maximum level, in two weeks.

6. Treatment with gangliosides greatly enhances and stimulates the formation of multiple innervations. Two days after reinnervation 60% of the fibers already show multiple EPPs.

7. The redundant innervation is a later stage removed and only a single EPP is recorded.

8. All these results viewed together suggest the effect of gangliosides in inducing both a faster maturation of transmitter release mechanisms and a faster growth of the nerve terminals

The authors wish to thank Miss Roberta Pressato and Mrs Manuela Trainito for their skillful assistance during the preparation of the manuscript.

REFERENCES

1. Ceccarelli B, Hurlbut WP: The vesicle hypothesis of the release of quanta of Ach. *Physiol Rev* (In press)

2. Couteaux R, Pecot-Dechavassine M: Vesicles synaptiques et poche au niveau des "zones actives" de la jonction neuromusculaire. *CR Acad Sci* Sér D 271:2346–2349, 1970.

3. Birks R, Huxley HE, Katz B: The fine structure of the neuromuscular junction of the frog. *J Physiol* 150:134–144, 1960.

4. Peper K, Dreyer F, Sandri C, et al: Structure and ultrastructure of the frog motor end-plate: a freeze-etching study. *Cell Tissue Res* 149:437–455, 1974.

5. Heuser JE, Reese TS, Landis DMD: Functional changes in frog neuromuscular junctions studied with freeze-fracture. *J Neurocytol* 3:109–131, 1974.

6. Poste G, Allison AC: Membrane fusion. Biochim Biophys Acta 300:421–465, 1973.

7. Satir B, Schooley C, Satir P: Membrane fusion in a model system: mucocyst secretion in tetrahymena. *J Cell Biol* 56:153–176, 1973.

8. Fertuck HC, Salpeter MM: Quantitation of junctional and extrajunctional acetylcholine receptors by electron microscope autoradiography after ^{125}I-α-bungarotoxin binding at mouse neuromuscular junctions. *J Cell Biol* 69:144–158, 1976.

9. Hebb C: Biosynthesis of Ach in nervous tissue. *Physiol Rev* 52:918–957, 1972.

10. Gorio A, Hurlbut WP, Ceccarelli B: Acetylcholine compartments in mouse diaphram: a comparison of the effects of black widow spider venom, electric stimulation and high concentration of potassium. *J Cell Biol* 781:716–733, 1978.

11. Gray EG: Synaptic fine structure and nuclear, cytoplasmic and extracellular networks. The stereoframework concept. *J Neurocytol* 4:315–339, 1975.

12. Berl S, Puszkin S, Nicklas WJ: Actomyosin-like proteins in brain. *Science* 179:441–446, 1973.

13. Fine R, Lehman W, Head J, et al: Troponin-C in brain. *Nature* 258:260–262, 1975.

14. Rubin LL, Gorio A, Mauro A: Effect of concanavalin A on black widow spider venom activity at the neuromuscular junction: implications for mechanisms of venom action. *Brain Res* 143:107–124, 1978.

15. Gorio A, Mauro A: Reversibility and mode of action of black widow spider venom on frog neuromuscular junction. *J Gen Physiol* 73:245–263, 1979.

16. Gorio A, Mauro A: Mode of action of black widow spider venom on vertebrate neuromuscular junction, in Ceccarelli B, Clementi F (eds): *Advances in Cytopharmacology,* vol 3. New York, Raven Press, 1979, pp 129–140.

17. Gorio A: Muscle innervation and reinnervation as a model for development and specificity of neuronal connections, in Di Benedetta C (ed): *A Multidisciplinary Approach to Brain Development.* Amsterdam, Elsevier North-Holland, (In press)

18. Letinsky MK, Fischbeck GD, McMahan UJ: Precision of reinnervation of original post-synaptic sites in muscle after a nerve crush. *J Neurocytol* 5:691–718, 1976.

19. Dennis MJ, Miledi R: Non-transmitting neuromuscular junctions during an early stage of end plate reinnervation. *J Physiol* 239:553–570, 1974.

20. Dennis MJ, Miledi R: Characteristics of transmitter release at regenerating frog neuromuscular junctions. *J Physiol* 239:571–594, 1974.

21. Tonge DA: Physiological characteristics of re-innervation of skeletal muscle in the mouse. *J Physiol* 241:141–153, 1974.

22. Frontali N, Ceccarelli B, Gorio A, et al: Purification from black widow spider venom of a protein factor causing the depletion of synaptic vesicles at neuromuscular junctions. *J Cell Biol* 68:462–479, 1976.

23. Bevan S, Grampp W, Miledi R: Properties of spontaneous potentials at denervated motor end-plates of the frog. *Proc R Soc Lond* 194:195–210, 1976.

24. McArdle JJ, Albuquerque EX: A study of the reinnervation of fast and slow mammalian muscles. *J Gen Physiol* 61:1–23, 1973.

25. Ceccarelli B, Hurlbut WP, Mauro A: Turnover of transmitter and synaptic vesicles at the frog neuromuscular junction. *J Cell Biol* 57:499–524, 1973.

26. Rotshenker S, McMahan UJ: Altered pattern of innervation in frog muscle after denervation. *J Neurocytol* 5:719–730, 1976.

27. Ceccarelli B, Aporti F, Finesso M: Effects of brain gangliosides on functional recovery in experimental regeneration and reinnervation, in Porcellati G, Ceccarelli B, Tettamanti G (eds): *Ganglioside Function: Biochemical and Pharmacological Implications*. New York, Plenum Publishing Corporation, 1976.

28. Obata K, Oide M, Handa S: Effects of glycolipids on in vitro development of neuromuscular junction. *Nature* 266:369–371, 1977.

29. Gorio A, Rubin L, Mauro A: Double mode of action of black widow spider venom on frog neuromuscular junction. *J Neurocytol* 7:193–205, 1978.

30. Misler S, Hurlbut WP: Action of black widow spider venom on quantized release of acetylcholine at the neuromuscular junction: dependence upon external Mg^{++}. *Proc Natl Acad Sci USA* 76:991–995, 1979.

31. Ledeen RW: Ganglioside structures and distribution: are they located at the nerve ending? *J Supramol Struct* 8:1–17, 1978.

32. Cuatrecasas P: Gangliosides and membrane receptors for cholera toxin. *Biochemistry* 12:3558–3566, 1973.

33. Toffano G, Benvegnù D, Bonetti AC, et al: Interaction of monomeric GM_1 ganglioside with crude rat brain neuronal membranes. *J Neurochem* (In press)

34. Bonetti AC, Benvegnù D, Leon A, et al: Modifications of (Na^+, K^+)ATPase activity by GM_1 incorporation into neuronal membranes, in Schauer R, Boer P, Buddecke MF, et al (eds): *Proceedings of the Fifth International Symposium on Glycoconjugates*. Stuttgart, Georg Thieme Verlag KG, 1979, pp 576–577.

35. Orlando P, Cocciante G, Ippolito G, et al: The fate of tritium labelled GM_1 ganglioside injected in mice. *Pharmacol Res Commun* 11(9):759–773, 1979.

36. Purpura DP, Pappas GD, Baker HJ: Fine structure of meganeurites and secondary growth processes in feline GM_1-gangliosidoses. *Brain Res* 143:1–12, 1978.

37. Purpura DP: Ectopic dendritic growth in mature pyramidal neurones in human ganglioside storage disease. *Nature* 276:520–521, 1978.

38. Purpura DP, Baker HJ: Meganeurites and other aberrant processes of neurons in feline GM_1-gangliosidosis: a Golgi study. *Brain Res* 143:13–26, 1978.

39. Brown MC, Ironton R: Sprouting and regression of neuromuscular synapses in partially denervated mammalian muscles. *J Physiol* 278:325–348, 1978.

40. Thompson W: Reinnervation of partially denervated rat soleus muscle. *Acta Physiol Scand* 103:81–91, 1978.

INDEX